DICTIONARY OF
MILITARY TERMS
Second Edition

Richard Bowyer

PETER COLLIN PUBLISHING

First published in Great Britain 1999
Second edition published 2002

Peter Collin Publishing is an imprint of
Bloomsbury Publishing Plc
38 Soho Square, London, W1D 3HB

British Library Cataloguing-in-Publication Data

A catalogue record for this book is available from the British Library

ISBN 1-903856-20-5

Text computer typeset by Bloomsbury Publishing
Printed and bound in Italy by LegoPrint
Cover artwork by Gary Weston

Preface to the First Edition

English is the language most frequently used in international military operations. This dictionary aims at providing the user with a basic vocabulary of British and American military terms. These cover the three services, and deal with subjects such as rank, organization, training, operations in the field, logistics, radio communications, as well as some of the more common weapons and equipment currently in use. In addition, there are selected items of general vocabulary relating to geography, terrain, weather and other relevant matters.

Definitions are written in simple English, making them easily accessible to anyone with a basic knowledge of the language, and phonetic symbols are used to show the correct pronunciation. Encyclopaedic comments are provided where necessary and most entries include example sentences, showing how the words and expressions are used in practice.

At the back of the book is a supplement of further useful information.

Preface to the Second Edition

This new edition has been completely revised to reflect the changes in military strategy, weapons, tactics and equipment that have been developed and used over the past three years. The dictionary also includes the new terms used in the media when reporting on or describing recent conflicts (for example, *mission creep*) and provides a fully updated reference book for anyone involved in teaching English to the military or armed police forces, dealing with this level of specialist English vocabulary.

Phonetics

The following symbols have been used to show the pronunciation of the main words in the dictionary.

The main stress of an entry has been indicated by a main stress mark (ˈ) in front of the syllable; the second-most stressed syllable is indicated by (ˌ). Note that these are only guides as the stress of the word may change according to its position in the sentence.

Vowels		*Consonants*	
æ	back	b	buck
ɑː	harm	d	dead
ɒ	stop	ð	other
aɪ	type	dʒ	jump
aʊ	how	f	fare
aɪə	hire	g	gold
aʊə	hour	h	head
ɔː	course	j	yellow
ɔɪ	loyalty	k	cab
e	head	l	leave
eə	fair	m	mix
eɪ	make	n	nil
ɜː	word	s	save
iː	keep	ʃ	shop
ɪ	fit	t	take
ɪə	near	tʃ	change
u	supreme	χ	loch
uː	pool	θ	theft
ʊ	book	v	value
ʌ	shut	w	work
		z	zone
		ʒ	measure

ALPHA - Aa

A-10 [ˌeɪ 'ten] *noun* American ground-attack aircraft (NOTE: also known as the **Thunderbolt,** or unofficially as the **Warthog;** plural is **A-10s** [ˌeɪ 'tenz])

A-4 [ˌeɪ 'fɔː] *noun* American-designed multirole attack aircraft, designed to operate from an aircraft carrier (NOTE: also known as the **Skyhawk;** plural is **A-4s** [ˌeɪ 'fɔːz])

A-40 [ˌeɪ 'fɔːti] *noun* Soviet-designed airborne early warning and control (AEW & C) aircraft with a large disk-like antenna (radome) mounted on the fuselage (NOTE: known to NATO as **Mainstay**)

A-6 [ˌeɪ 'sɪks] *noun* American-designed ground-attack aircraft, designed to operate from an aircraft carrier (NOTE: also known as the **Intruder;** plural is **A-6s** [ˌeɪ 'sɪksɪz])

A-7 [ˌeɪ 'sevən] *noun* American-designed ground-attack aircraft (NOTE: also known as the **Corsair;** plural is **A-7s** [ˌeɪ 'sevənz])

AA [ˌeɪ 'eɪ] = ANTI-AIRCRAFT

AAA [ˌeɪ eɪ 'eɪ] = ANTI-AIRCRAFT ARTILLERY

AAAV [ˌeɪ eɪ eɪ 'viː] *noun* = ADVANCED AMPHIBIOUS ASSAULT VEHICLE American-designed amphibious infantry fighting vehicle (IFV), which is designed to disembark from a landing ship at a considerable distance from the shore and is capable of travelling on water at high speed

AAC = ARMY AIR CORPS

AAM [ˌeɪ eɪ 'em] = AIR-TO-AIR MISSILE

AAR [ˌeɪ eɪ 'ɑː(r)] *noun US* = AFTER ACTION REVIEW debriefing held at the conclusion of an exercise, in which the participants discuss their performance with the umpires

Aardvark ['ɑːdvɑːk] *noun* unofficial name for the American-designed F-111 fighter bomber

AAV-7A1 [ˌeɪ eɪ viː ˌsevən eɪ 'wʌn] *noun* = AMPHIBIOUS ASSAULT VEHICLE SEVEN American-designed amphibious tractor (NOTE: also known as **Landing Vehicle Tracked Personnel (LVTP-7A1)**)

AAW [ˌeɪ eɪ 'dʌb(ə)ljuː] = ANTI-AIR WARFARE

AAWS [ˌeɪ eɪ ˌdʌb(ə)ljuː 'es] ANTI-AIR WARFARE SYSTEM

AB [ˌeɪ 'biː] = ABLE RATING

abandon [ə'bændən] *verb* **(a)** to leave a vehicle or ship (usually for reasons of safety); ***the captain gave the order to abandon ship*** **(b)** to leave behind; ***we had to abandon large quantities of ammunition during the retreat*** **(c)** to terminate an enterprise before it has been completed; ***poor visibility forced us to abandon the remainder of patrol***; *see also* ABORT

abandoned [ə'bændənd] *adjective* deserted; ***we found the enemy trenches abandoned***

abatis ['æbətɪs *or* ə'bætɪs] *noun* obstacle constructed by felling trees to block a likely approach; ***the road was blocked by an abatis***

Abbot ['æbət] *noun* British 105mm self-propelled howitzer (SPH)

ABCCC *noun US* = AIRBORNE COMMAND, CONTROL AND COMMUNICATIONS command team operating from a C-130 aircraft, in order to coordinate close air support

abeam [ə'biːm] *adverb* at right angles to the length of a ship *or* aircraft; ***the torpedo***

missed because we weren't properly abeam of the target

able rating (AB) [ˌeɪbl 'reɪtɪŋ] *noun GB* lowest non-commissioned rank in the navy (equivalent of an experienced or well qualified private soldier in the army) (NOTE: also called **able seaman;** they are referred to as **Able Seaman Smith,** etc.; Note also that the ranks of **ordinary rating** and **junior rating** were abolished in April 1999)

ablutions [ə'bluːʃ(ə)nz] *noun* room *or* building containing washing facilities and toilets; ***the ablutions were in a filthy state***

aboard [ə'bɔːd] *adverb* = ON BOARD on *or* onto a boat *or* ship *or* aircraft *or* vehicle; ***he is already aboard***; ***come aboard***

abode [ə'bəʊd] *noun* home; **no fixed abode** = not having a permanent home *or* address

abort [ə'bɔːt] *verb* to stop an enterprise before it has been completed; ***we were forced to abort the mission***; *see also* ABANDON

abortive [ə'bɔːtɪv] *adjective* unsuccessful; ***the enemy launched several abortive attacks***

about [ə'baʊt] *adverb* the opposite direction to that which you are now facing; **about turn !** = *(drill command)* turn around and face the opposite direction

Abrams (M-1) ['eɪbrəmz] *noun* American 1980s-era main battle tank

abreast [ə'brest] *adverb* side by side and facing the same direction; ***the infantry advanced in line abreast; the tanks halted abreast of each other***

abseil ['æbseɪl] *verb* to descend, using a rope; ***we had to abseil down the cliff***; *see also* RAPPEL

absent ['æbsənt] *adjective* away from a military unit (usually without permission); ***he's been absent for three days***; *see also* AWOL

AB triple C [ˌeɪ biː ˌtrɪp(ə)l 'siː] *see* ABCCC

AC [ˌeɪ 'siː] **1** = AIRCRAFTMAN **2** = HYDROGEN CYANIDE type of blood agent (NOTE: also known as **HCN**)

A/C = AIRCRAFT

AC-130 [ˌeɪ siː wʌn 'θɜːtɪ] *noun* ground-attack variant of the Hercules C-130 transport aircraft

> COMMENT: despite its age and old-fashioned appearance, the AC-130 possesses an enormous amount of firepower and earned itself a fearsome reputation during the Gulf War of 1991. Its slow speed makes it vulnerable to surface-to-air missiles, so it is most effective at night. AC-130 aircraft are often referred to as **gunships**

ACC [ˌeɪ siː 'siː] *noun US* = AIR COMBAT COMMAND department of the US forces with overall responsibility for the use of all combat aircraft of the US Air Force

> COMMENT: Air Combat Command is an amalgamation of the former **Strategic Air Command (SAC)** which was responsible for heavy long-range bombers and ICBMs and the former **Tactical Air Command (TAC)** which was responsible for fighters and attack aircraft. Inter-continental ballistic missiles are now the responsibility of **Strategic Command (STRATCOM)**

access ['ækses] **1** *noun* **(a)** way of approaching or entering; ***access to the building is at the rear*** **(b)** opportunity to look at or use; ***I need access to the company records*** **2** *verb* to activate a programme or open a file on a computer; ***I cannot access the database*** (NOTE: as a noun, **access** is often used without the definite or indefinite article)

accessible [ək'sesəbl] *adjective* **(a)** approachable; ***the castle is only accessible in good weather*** **(b)** readily available; ***that information is not accessible at the moment***

accident ['æksɪdənt] *noun* **(a)** physical harm or damage as a result of an error in judgement, defective equipment or bad luck; ***there has been an accident on the grenade range*** **(b)** event which occurs by chance or without apparent cause; ***he was there purely by accident***

accidental [ˌæksɪ'dentl] *adjective* happening by chance or as a result of an error in judgement or defective equipment; ***there will be an official enquiry over the accidental shooting of Sgt Jones***

accidentally [ˌæksɪ'dentəli] *adverb* by accident; ***he shot the woman accidentally***

acclimatization [əklaɪmətaɪ'zeɪʃn] *noun* process of becoming acclimatized

acclimatize [ə'klaɪmətaɪz] *verb* **(a)** to allow someone to become accustomed to new conditions (especially climate); ***we will need at least ten days to acclimatize the troops to the heat*** **(b)** to become accustomed to new conditions; ***the men will need at least ten days to acclimatize***

accommodate [ə'kɒmədeɪt] *verb* to provide shelter; ***we were accommodated in a factory***

accommodation [əˌkɒmə'deɪʃn] *noun* **(a)** place to live; ***the officer inspected the soldiers' accommodation*** **(b)** act of providing shelter; ***you are responsible for the accommodation of refugees***

accomplice [ə'kʌmplɪs] *noun* someone who helps another person to carry out an illegal act; ***the gunman had an accomplice***

accomplish [ə'kʌmplɪʃ] *verb* to complete a task successfully; **mission accomplished** = mission completed

accusation [ˌækjuː'zeɪʃn] *noun* act of saying that a person has carried out an illegal act; ***he has made an accusation of sabotage against you***

accuse [ə'kjuːz] *verb* to say that a person has carried out an illegal act; ***he was accused of war crimes***

ACE [ˌeɪ siː 'iː] = ALLIED COMMAND EUROPE, ARMORED COMBAT EARTHMOVER

acetate ['æsɪˌteɪt] *noun* clear plastic sheeting, used for map overlays

achieve ['æktɪv] *verb* to complete a task successfully; ***we have achieved our mission***; **achieve a solution** = to have a target in the sights of a guided weapon, with the guidance system activated; ***he achieved a solution on the leading plane***; *see also* LOCK-ON

achievement [ə'tʃiːvmənt] *noun* successful completion of a task; ***it was a splendid achievement***

acknowledge [ək'nɒlɪdʒ] *verb* **(a)** to confirm that a piece of information has been received and understood **(b)** to recognize that something is true

acknowledgement [ək'nɒlɪdʒmənt] *noun* **(a)** confirmation that a piece of information has been received and understood **(b)** recognition that something is true

ACLANT = ALLIED COMMAND ATLANTIC

ACMI [ˌeɪ siː em 'aɪ *or* 'ækmɪ] *noun* = AIR COMBAT MANOEUVRING AND INSTRUMENTATION computerized data system, which is used during air-to-air combat exercises to provide exercise controllers with information concerning an individual aircraft's speed, altitude, heading, etc; ***each aircraft was fitted with an ACMI pod***

ACP [ˌeɪ siː 'piː] = AMMUNITION CONTROL POINT

acquire [ə'kwaɪə] *verb* **(a)** to obtain something; ***where did you acquire that food?*** **(b)** to have a target in the sights of a weapon; ***he acquired the tank with his thermal imaging sight*** **(c)** to select and lock onto a target using a weapon-guidance system; ***this missile can acquire a target automatically***

acquired immunodeficiency syndrome [əˌkwaɪəd ɪmjʊnəʊdɪ'fɪʃənsi ˌsɪndrəʊm] *see* AIDS

acquisition [ækwɪ'zɪʃn] *noun* act of acquiring; ***he is responsible for the acquisition of food from local sources***; **acquisition logistics** = the logistics involved in acquiring supplies, buildings and other materiel for armed forces; **target acquisition** = act of selecting and locking onto a target with a weapon guidance system

ACR [ˌeɪ siː 'ɑː(r)] = ARMORED CAVALRY REGIMENT

ACR [ˌeɪ siː 'ɑː(r)] *noun GB* = AIRCRAFT CONTROL ROOM admin centre on an aircraft carrier, which deals with administration concerning the actual aircraft

act [ækt] **1** *noun* something which is done; **act of aggression** = unprovoked attack; **in the act** = in the process of doing something; ***we caught them in the act of planting the bomb*** **2** *verb* **(a)** to do something; ***we were forced to act when the man was shot*** **(b)** to do someone else's job on a temporary basis; ***he acted as platoon sergeant during the final exercise***

acting ['æktɪŋ] *adjective* doing someone else's job on a temporary basis; ***he is the acting platoon sergeant at the moment***; *compare* SUBSTANTIVE

action ['ækʃn] **1** *noun* **(a)** something which is done; ***you are responsible for your own actions***; **to take action** = to do something **(b)** engagement with the enemy; ***there were several small actions during the afternoon***; **action stations** = ready for battle; **killed in action** = killed during an engagement **2** *verb* to cause something to be done (by another person); ***I have actioned your request for a transfer***

activate ['æktɪveɪt] *verb* to make a device operate; ***the device is activated by pressing this button***

active ['æktɪv] *adjective* **(a)** operating; ***enemy special forces are active in your area*** *GB* **active list** = list of officers available for active service; **active service** = service in a war zone; **active service unit (ASU)** = small group used by the Irish Republican Army (IRA) to carry out a terrorist attack (NOTE: in the British armed forces, the term **active service** is only used when the nation is officially at war; For counter-insurgency and peacekeeping operations, the term **operational service** is used) **(b)** working or ready to work; ***the device is now active*** **(c)** relating to night-viewing devices which require an external source of infrared (IR) light in order to operate; *compare* PASSIVE

activist ['æktɪvɪst] *noun* person, normally holding extreme views, who believes in the use of action rather than debate in order to achieve their political aims; ***the rioting was started by left-wing activists***

activity [æk'tɪvəti] *noun* several different actions or an action which is carried out repeatedly or over a period of time; ***there has been very little enemy activity today***

Adamsite ['ædəmzaɪt] *see* DM

adapt [ə'dæpt] *verb* **(a)** to alter or to modify; ***the APC has been adapted to carry surveillance equipment*** **(b)** to modify your behaviour in order to meet a change in circumstances; ***we must adapt to these new tactics***

ADC [ˌeɪ diː 'siː] = AIDE-DE-CAMP

add [æd] *verb* **(a)** to join one thing to another in order to increase the quantity **(b)** to correct artillery or mortar fire so that the rounds land further away from the observer; **Add 200!** = add 200 metres; *compare* DROP

ad hoc [ˌæd 'hɒk] *Latin phrase meaning* 'formed for a specific purpose'; ***they were organized into an ad hoc unit***

Adj = ADJUTANT

adjust [ə'dʒʌst] *verb* **(a)** to change the position of something; ***he adjusted the straps on his rucksack*** **(b)** to direct artillery or mortar fire onto a target by observing the fall of shot and sending corrections back to the gun line; ***the enemy artillery is adjusting onto B Company's position***; **adjusting fire** = firing of a single round by one gun or mortar so that a forward observer can observe the fall of shot

adjustment [ə'dʒʌstmənt] *noun* **(a)** act of changing the position of something; ***he made several minor adjustments to the weapon sight*** **(b)** process of adjusting artillery or mortar fire; ***we will need at least fifteen minutes for adjustment*** **(c)** corrections calculated by a forward observer after observing the fall of shot of an artillery or mortar round; ***the guns were on target after my first adjustment***

adjutant (Adj) ['ædʒətənt] *noun* battalion officer (usually a captain) who acts as the commanding officer's assistant and is also responsible for discipline; **Adjutant-General** = top administrative post in the army (NOTE: in certain other armies, **adjutant** is synonymous with **aide-de-camp (ADC)**)

Adm = ADMIRAL

admin ['ædmɪn] *noun (informal)* administration; ***I've got a lot of admin to do***

administer [əd'mɪnɪstə] *verb* to manage a unit or organization

administrate [əd'mɪnɪstreɪt] *verb* to administer on a national or regional level; ***he was sent out to administrate the region***

administration [ədmɪnɪ'streɪʃn] *noun* **(a)** day-to-day management of a unit or organization; ***I am responsible for administration*** **(b)** government; ***the local administration is extremely unhelpful***

administrative [əd'mɪnɪstrətɪv] *adjective* relating to administration; **administrative tasks** = tasks other than actual fighting (such as the resupply of ammunition, food or fuel, personnel matters, etc.)

admiral (Adm) ['ædmrəl] *noun GB & US* senior officer in the navy (usually in command of a fleet); *GB* **Admiral of the Fleet** *US* **Fleet Admiral** = highest rank in the Navy; *see also* REAR-ADMIRAL, VICE-ADMIRAL

Admiralty ['ædmrəlti] *noun GB* department which administers the Royal Navy

advance [əd'vɑːns] **1** *adjective* **(a)** early; ***we will need advance warning for any ammunition requests*** **(b)** forward; ***advance units of the enemy have been seen***; **advance guard** = small military force which advances between the reconnaissance units and the main body of an advancing force, in order to engage the enemy and occupy his attention while the main body deploys into attack formation (NOTE: do not confuse with **vanguard**) **2** *noun* **(a)** movement towards the enemy; ***the advance will begin at first light*** **(b)** progress; ***in recent years there have been great advances in the development of armour*** **(c) in advance** = early; ***ammunition bids must be submitted well in advance*** **3** *verb* **(a)** to move forward **(b)** to move towards the enemy; **advance to contact** = method of locating the enemy by advancing into his territory until contact is made, whereupon the leading units or sub-units engage the enemy, while the main force deploys to mount an attack

adverse ['ædvɜːs] *adjective* causing difficulty; ***we were unable to fly because of adverse weather conditions***

advice [əd'vaɪs] *noun* suggestion as to what should be done; ***he refused to take my advice***

advise [əd'vaɪz] *verb* **(a)** to suggest what should be done; ***he advised the colonel to stop the attack*** **(b)** to inform someone; ***he was advised that the situation would not improve***; ***you should be advised that the bridge is not suitable for tanks***

adviser [əd'vaɪzə] *noun* person appointed to give advice

A Echelon ['eɪ ˌeʃəlɒn] *noun* logistical elements of a tactical grouping

Aegis ['iːdʒɪs] *noun* American-designed integrated naval air defence system (AAWS), consisting of computerized radar and other surveillance systems, fire control systems and surface-to-air missiles (SAM)

AEO [ˌeɪ iː 'əʊ] *noun GB* = AIR ENGINEERING OFFICER officer responsible for ensuring that the aircraft of a squadron are fit to fly

aerial ['eəriəl] **1** *adjective* relating to the air; **aerial bombardment** = bombing attack by aircraft; **aerial photography** = photography involving the use of aircraft; **aerial reconnaissance** = reconnaissance using aircraft **2** *noun* metal wire, rod, mast or structure used in the receiving and transmission of radio signals; ***enemy command tanks usually have two aerials***; *see also* ANTENNA

aerodrome ['eərədrəʊm] *noun* small airbase, airfield or airport (NOTE: this term is now obsolete)

aerodynamics [ˌeərəʊdaɪ'næmɪks] *noun* study of the effects caused by a solid object moving through air

aeroplane ['eərəpleɪn] *noun GB* fixed-wing aircraft (NOTE: American English is **airplane**)

aerosol ['eərəsɒl] *noun* tiny particles of solid *or* liquid matter, which are suspended in the atmosphere (eg. mist, smoke, vapour); ***the agent is delivered as an aerosol***

AEW & C [eɪ iː ˌdʌb(ə)ljuː ənd 'siː] *noun* = AIRBORNE EARLY WARNING AND CONTROL technology which detects enemy aircraft and missiles and then controls interception by friendly fighters (NOTE: also known as **Airborne Warning and Control System (AWACS)**)

affirmative [ə'fɜːmətɪv] **1** *adjective* true or accurate; ***that is affirmative*** **2** *adverb (radio terminology)* that is correct; ***'Hullo 22, this is 2, are you in position yet, over?'***

- *'22, affirmative, over'*; *compare* NEGATIVE; *see also* COPY, ROGER

aft [ɑːft] *adverb* towards the rear of an aircraft or ship; ***he went aft***

AFV [ˌeɪ ef 'viː] = ARMOURED FIGHTING VEHICLE

AFV-432 *or* **FV-432** [ˌeɪ ef ˌviː fɔː θriː 'tuː] *noun* British 1960s-era armoured personnel carrier (APC) (NOTE: normally referred to simply as a **432**)

agent ['eɪdʒənt] *noun* **(a)** someone who poses as a civilian in order to gather information, carry out assassinations or acts of sabotage; ***we arrested two enemy agents near the missile base***; *see also* SPY **(b)** chemical used as a weapon; **Agent Orange** = American defoliating agent; **blister agent** = chemical designed to cause severe blisters; **blood agent** = chemical designed to deprive the body of oxygen; **defoliating agent** = chemical designed to kill vegetation; **choking agent** = chemical designed to damage the lungs; **G-agent** = non-persistent nerve agent; **incapacitating agent** = chemical designed to cause mental confusion; **nerve agent** = chemical which attacks a person's central nervous system; **tear agent** = chemical agent designed to irritate the eyes and throat, normally used for crowd control; **V-agent** = persistent nerve agent; **vomiting agent** = chemical agent designed to make a person feel ill and vomit

agent provocateur [ˌæʒɒn prɒvɒkə'tɜː] *French words meaning* 'agent who provokes': person who provokes others to start civil disorder or to commit a crime (often by taking part himself) in order to start a revolution, or to find out who is not reliable, or in order to encourage people to commit crimes for which they will be arrested

aggression [ə'greʃn] *noun* hostile behaviour; **act of aggression** = unprovoked attack

aggressive [ə'gresɪv] *adjective* offensive (rather than defensive); ***the divisional commander has called for an increase in aggressive patrolling***; **aggressive delay** = tactic involving the aggressive use of small units to slow down an advancing enemy force so that a main line of defence can be prepared or strengthened

aggressor [ə'gresə] *noun* nation which attacks another nation without provocation; ***Ruritania is seen as the aggressor in this conflict***

aggro ['ægrəʊ] *noun (informal)* disorderly behaviour leading to violence; ***we're expecting aggro tonight***

AGM [ˌeɪ dʒiː 'em] = AIR-TO-GROUND MISSILE

agreement [ə'griːmənt] *noun* contract made between the authorities of different nations, or between NATO commanders and national authorities; *see also* TECHNICAL AGREEMENT

AH-1 [ˌeɪ eɪtʃ 'wʌn] *see* HUEY COBRA

AH-64 [ˌeɪ eɪtʃ sɪksti 'fɔː] *see* APACHE

AI [eɪ 'aɪ] = AIRBORNE INTERCEPTION

aid [eɪd] **1** *noun* help; **aid to the civil authorities** = military assistance in maintaining public services (such as fire-fighting, refuse collection, etc.); **aid to the civil powers** = military assistance to the police in maintaining law and order; **aid worker** = person involved in foreign aid; **first aid** = emergency measures to treat an injured person until proper medical treatment is available; **foreign aid** = assistance in the form of money, food or other necessities, provided by one nation to another in times of trouble **2** *verb* to help; *(legal term)* **to aid and abet** = to help and encourage someone to commit a crime

aide-de-camp (ADC) [ˌeɪd də 'kɒn] *French noun* officer (usually a captain) who acts as a personal assistant to a general (NOTE: **plural form is aides-de-camp)**

aide-mémoire [ˌeɪd me'mwɑː(r)] *French noun* book *or* card containing useful *or* specialist information in an easy-to-read format; ***he left his aide-mémoire in the briefing room***

AIDS *or* **Aids** [eɪdz] *noun* = ACQUIRED IMMUNODEFICIENCY SYNDROME infection caused by the HIV virus which attacks the body's immune system

AIFV [ˌeɪ aɪ ef 'viː] *noun* American-designed infantry fighting vehicle

aileron ['eɪləˌrɒn] *noun* moving part of an aircraft wing, which is used to control lateral balance

aim [eɪm] **1** *noun* **(a)** the act of directing a weapon; ***his aim was unsteady*** **(b)** intention; ***their aim was to disrupt our communications*** **2** *verb* **(a)** to direct a weapon at something; ***he aimed at the tank*** **(b)** to intend; ***we aim to capture the bridge intact***

AIM [eɪm *or* ˌeɪ aɪ 'em] *noun* = AIR INTERCEPT MISSILE another name for an air-to-air missile (AAM)

air [eə] **1** *adjective* **(a)** relating to the atmosphere; **air-cooled** = cooled by a current of air or simply by exposure to the atmosphere; ***this machine-gun is air-cooled***; **air waves** = entire range of radio frequency; ***the enemy will be scanning the air waves for our radio transmissions*** **(b)** relating to the use of aircraft; *US* **air cavalry** = air-assault infantry, whose regimental origins stem from a former cavalry regiment; *GB* **air chief marshal** = senior officer in the air force (equivalent of a general in the army); *GB* **air commodore** = senior officer in the air force, below an air vice marshal; **air controller** = *see* AIR TRAFFIC CONTROLLER; **air tasking order (ATO)** = daily programme of all air tasks, including routes, targets, frequencies, call-signs, logistical details, etc.; **forward air controller (FAC)** = air-force or artillery officer or NCO operating from an aircraft or attached to ground troops in order to direct close air support; **air defence** *US* **air defense** = defence against enemy aircraft; ***the divisional artillery includes an air defence battery***; *GB* **air marshal** = senior officer in the air force, below an air chief marshal; *GB* **air officer commanding (AOC)** = commander of a large air-force grouping; **air operation** = military operation involving aircraft; **air photograph** = photograph of an area of ground taken from an aircraft; **air raid** = attack by aircraft against a target on the ground (usually with bombs); **air-sea rescue** = the use of aircraft and helicopters to rescue someone from the sea; **air sentry** = sentry assigned to look out for enemy aircraft; **air support** = **(i)** attack by aircraft in support of ground troops; **(ii)** any assistance given by aircraft to ground troops; **air-to-air refuelling** = process by which an aircraft s refuelled in mid-air from a tanker aircraft; **close air support** = attack by aircraft on a target which is close to friendly ground forces; **air-to-air missile (AAM)** = anti-aircraft missile designed to be fired from an aircraft; **air-to-ground missile (AGM)** = missile designed to be fired by an aircraft at a target on the ground; **air-to-surface missile (ASM)** = missile designed to be fired from an aircraft at a target on the ground or on the surface of the sea; *see also* SURFACE-TO-AIR; **air traffic controller** = civilian or military official who controls the passage of aircraft through a defined air-space; *GB* **air vice marshal** = senior officer in the air force, below an air marshal **2** *noun* **(a)** the earth's atmosphere; ***the air is contaminated with radioactive dust*** **(b)** place where aircraft or birds can fly; ***he was ordered to watch the air***; **in the air** = flying, in flight; ***the plane is already in the air*** **(c)** using aircraft; ***the battalion deployed by air*** **3** *adverb* **in the air** = (relating to the flank of an army *or* formation) exposed *or* unprotected; ***the enemy's right flank is in the air***

air-assault [ˌeər ə'sɔːlt] *adjective US (of infantry)* equipped with their own transport helicopters and supporting attack helicopters; ***this is an air-assault battalion***; *see also* AIR CAVALRY; *compare* AIRBORNE, AIR-PORTABLE

airbase ['eəbeɪs] *noun* base for the operation of military aircraft

airborne ['eəbɔːn] *adjective* **(a)** carried by aircraft; **AWACS is an airborne warning and control system** **(b)** deployed by parachute; **airborne troops** = paratroopers; **airborne warning and control system (AWACS)** = electronic equipment, carried in specially designed aircraft, which detects enemy aircraft or missiles at long ranges and then coordinates their interception by friendly aircraft or missiles (NOTE: also known as **airborne early warning and control (AEW & C)**); *compare* AIR-ASSAULT, AIR-PORTABLE

airburst ['eəbɜːst] *noun* explosion of an artillery round or missile in the air (above its target)

aircraft ['eəkrɑːft] *noun* machine capable of flight; **aircraft carrier** = large ship designed to carry aircraft and equipped with maintenance facilities and a runway for take-off and landing; **attack aircraft= aircraft which is designed to drop bombs on or fire missiles at targets on the ground, and is also capable of defending itself against enemy fighter aircraft; strike aircraft** = fighter aircraft used to attack targets on the ground (NOTE: the word **aircraft** is used for both singular and plural); *see also* FIGHTER-BOMBER, STRIKE AIRCRAFT

> COMMENT: aircraft are classified as **fixed-wing aircraft** (aircraft with wings) and **rotary-wing aircraft** (helicopters)

aircraftman (AC) ['eəkrɑːftmən] *noun* *GB* lowest non-commissioned rank in the air force (equivalent to a private soldier in the army); **leading aircraftman** = junior non-commissioned rank in the air force (equivalent to an experienced or well qualified private soldier in the army)

aircrew ['eəkruː] *noun* personnel who man an aircraft

airdrop ['eədrɒp] *noun* dropping cargo or personnel from an aircraft which is flying

airfield ['eəfiːld] *noun* area of ground (often unprepared) where aircraft can take off and land and be maintained

air force ['eə ˌfɔːs] *noun* branch of a state's armed forces which operates in the air; **Royal Air Force (RAF)** = the British air force; **United States Air Force (USAF)** = the American air force

air-force ['eə ˌfɔːs] *adjective* relating to an air force; ***all air-force personnel on the base should report to the duty officer***

airframe ['eəfreɪm] *noun* body of an aircraft

Air-Land Battle [ˌeə ˌlænd 'bætl] *noun* current military doctrine in which aircraft and long-range missiles are used to attack the enemy's reserves and logistical support at the same time as his forward elements are being engaged by ground forces using all the principles of manoeuvre warfare

airlift ['eəlɪft] **1** *noun* movement of men or equipment or supplies using aircraft; ***the Government authorized a massive airlift of grain to the disaster area*** **2** *verb* to move men or equipment or supplies using aircraft; ***it was decided to airlift the guns to their new positions***

airman ['eəmən] *noun* **(a)** member of an aircrew **(b)** member of the air force **(c)** *US* lowest non-commissioned rank in the air force; **airman first class** = junior non-commissioned rank in the air force

airmobile [eə'məʊbaɪl] *adjective* deployed to an area of operations by transport aircraft

airplane ['eəpleɪn] *noun* *US* fixed-wing aircraft (NOTE: British English is **aeroplane**)

airport ['eəpɔːt] *noun* complex of runways for the take-off and landing of civil aircraft, with facilities for aircraft maintenance and the care of passengers

air-portable ['eə ˌpɔːtəbl] *adjective* capable of deploying to an area of operations by transport aircraft; **air-portable battalion** = infantry battalion which is not equipped with armoured fighting vehicles and can therefore be transported by air to an area of operations; *compare* AIRBORNE, AIR-ASSAULT

> COMMENT: the term **air-portable** is usually applied to non-armoured or non-mechanized infantry

airspace ['eəspeɪs] *noun* aerial territory controlled by an air force or subject to the jurisdiction of a state; ***we are now entering enemy airspace***

airstrike ['eəstraɪk] *noun* attack by aircraft against a target on the ground (usually with air-to-ground missiles); ***the UN Council has authorized airstrikes on the gun positions in the Demilitarized Zone***

> COMMENT: during the UN peacekeeping operation in Bosnia (1992-95), **air strikes** meant retaliatory attacks on multiple targets within a specified area, while **close air support,** meant limited attacks on

individual positions *or* vehicles which were actually firing at UN peacekeepers. The difference between these two definitions was crucial during negotiations between the UN and the warring factions throughout this conflict

airstrip ['eəstrɪp] *noun* area of ground cleared of vegetation and levelled in order to allow the take-off and landing of small aircraft; ***the engineers prepared an airstrip close to the field hospital***

airtight ['eətaɪt] *adjective* which does not allow air to enter; ***this equipment must be stored in an airtight container***

airwoman ['eəwʊmən] *noun GB & US* lowest female non-commissioned rank in the air force

airworthy ['eəwɜːði] *adjective* fit to fly; ***this helicopter is not airworthy***

AK-47 [ˌeɪ keɪ fɔːti'sevən] *noun* Soviet-designed 7.62mm assault weapon; *see also* KALASHNIKOV

a.k.a. [ˌeɪ keɪ 'eɪ] *adverb* also known as; ***we are looking for Sidney Logan, a.k.a. Michael Higgins***; *see also* ALIAS

Alamo ['æləməʊ] *noun* Soviet-designed medium-range air-to-air missile (AAM)

alarm [ə'lɑːm] **1** *noun* **(a)** warning of threat or danger; ***we had a gas alarm last night***; *see also* ALERT **(b)** signal to be given when there is an alarm; ***the gas alarm is given by banging two mess tins together*** **(c)** electronic or mechanical device designed to detect a specific danger; **intruder alarm** = device designed to detect movement **(d)** state of fear and disorder caused by imminent danger; ***there is no cause for alarm*** **2** *verb* to give someone a reason to be frightened; ***we must not alarm the civilians***

ALARM [ə'lɑːm] *noun* = AIR-LAUNCHED ANTI-RADIATION MISSILE British-designed air-to-ground anti-radar missile (ARM)

alarmed [ə'lɑːmd] *adjective* frightened; ***do not be alarmed***

alert [ə'lɜːt] **1** *adjective* watchful and ready to take action; ***we must be alert tonight*** **2** *noun* warning of a threat or danger; ***there is an NBC alert*** **3** *verb* to warn someone of a danger or change in situation, or to inform someone about an incident; ***I alerted the duty officer***

Alfa ['ælfə] *US see* ALPHA

alias ['eɪliəs] **1** *adverb* also known as; ***we are looking for Sidney Logan, alias Michael Higgins***; *see also* a.k.a. **2** *noun* false name; ***he uses Kurt Baumann as an alias***

alien ['eɪliən] **1** *adjective* **(a)** belonging to or coming from another country; ***alien ships are being impounded***; *see also* FOREIGN **(b)** from another planet; ***he claims to have seen an alien spacecraft*** **2** *noun* **(a)** person who is a citizen of another country; ***all aliens must register at their local police station***; *see also* FOREIGNER, FOREIGN NATIONAL **(b)** life-form from another planet; ***some people believe in the existence of aliens***

alienate ['eɪliəneɪt] *verb* to cause someone to become unfriendly; ***the soldiers managed to alienate the villagers***

align [ə'laɪn] *verb* to bring something into line with something else; ***he aligned his sights on the church***; ***the mortars are not properly aligned***

alignment [ə'laɪnmənt] *noun* act of aligning; ***he went to check the alignment of the guns***

allegiance [ə'liːdʒəns] *noun* loyalty to a person or cause; ***the rebels owe their allegiance to the former president***

alliance [ə'laɪəns] *noun* cooperation between two or more nations or ethnic or political groups, usually as a result of a formal agreement known as a treaty

allied ['ælaɪd] *adjective* related by an alliance; ***allied troops entered the capital yesterday***; **Allied Publication (AP)** = standardized document accepted by various NATO countries; **Allied Rapid Reaction Corps (ARRC)** = British-led NATO force designed to react at short notice to any crisis involving NATO countries

all-round defence [ɔːl ˌraʊnd dɪ'fens] *noun (of positions)* having all the approaches to your position covered by fire, including those from the flanks and rear; ***this position offers excellent all-round defence***

all-terrain [ˌɔːl təˈreɪn] *adjective* capable of operating in all types of terrain; ***we will need all-terrain vehicles for this operation***

ally [ˈælaɪ] **1** *noun* member of an alliance; **the Allies** = states which form an alliance (used during the Second World War to refer to states which opposed Germany, Italy and Japan) **2** *verb* to make an alliance; ***Austria has allied itself to Germany***

ALO [ˌeɪ el ˈəʊ] *noun* = AIR LIAISON OFFICER air-force officer attached to an army headquarters to coordinate close air support (CAS)

Alpha *US* **Alfa** [ˈælfə] first letter of the phonetic alphabet (Aa)

Alphajet [ˈælfəˌdʒet] *noun* French/German-designed light fighter aircraft

alter [ˈɒltə] *verb* to change something; ***it's too late to alter the fireplan now***

alternate 1 [ɒlˈtɜːnət] *adjective* one after the other (often repeatedly); **alternate bounds** = movement in bounds, with one person or vehicle or sub-unit stationary and giving or prepared to give covering fire, while the other moves past to occupy a fire position beyond **2** [ˈɒltəneɪt] *verb* to change from one thing or activity to another (often repeatedly); ***he had to alternate between manning the gun and operating the radio***

alternately [ɒlˈtɜːnətli] *adverb* one after the other (often repeatedly); ***the guns were firing alternately***

alternative [ɒlˈtɜːnətɪv] **1** *adjective* different or additional; ***that gully would make a good alternative position for the mortars*** **2** *noun* one of two or more possible courses of action; ***we have no alternative: we must attack now***

altitude [ˈæltɪtjuːd] *noun* vertical distance from the ground or sea level (NOTE: altitude is usually measured in **feet**)

ambassador [æmˈbæsədə] *noun* diplomat sent by a state to act as its senior representative in a foreign country

amber [ˈæmbə] *adjective* colour similar to orange; *see also* AWLS

ambulance [ˈæmbjʊləns] *noun* **(a)** vehicle used to transport injured persons to a hospital **(b)** medical unit; **field ambulance** = battalion-sized medical unit (usually attached to a brigade)

ambush [ˈæmbʊʃ] **1** *noun* **(a)** surprise attack by troops who wait in a concealed position for the enemy to come to them; ***the patrol was caught in an ambush*** **(b)** troops who carry out an ambush; ***the ambush hasn't returned yet***; **ambush patrol** = large well-armed patrol sent out to lay an ambush; *see also* ANTI-AMBUSH DRILL **2** *verb* to carry out an ambush; ***we were ambushed on our return journey***

American [əˈmerɪkən] *adjective* relating to the United States of America (USA)

amidships *US* **amidship** [əˈmɪdʃɪps] *adverb* in the middle of a ship; ***the torpedo struck the ship amidships***

AMM [ˌeɪ em ˈem] = ANTI-MISSILE-MISSILE

ammo [ˈæməʊ] *noun (informal)* ammunition

ammunition [æmjuːˈnɪʃn] *noun* quantity of munitions (especially projectiles such as bullets, shells, missiles); ***they have enough ammunition left for six days***; ***we are going to run out of ammunition soon***; **ammunition dump** = temporary store of ammunition (usually in the field); **ammunition state** = quantity of ammunition held by a unit or sub-unit; **out of ammunition** = to have fired all your ammunition

amphibious [ˈæmfɪbɪəs] *adjective* suitable for use both on water and on land; **amphibious engineers** = engineer troops who specialize in the construction of bridges and river crossing; **amphibious operation** = operation involving ground forces landed from the sea; **amphibious tractor (Amtrac)** = amphibious armoured personnel carrier (APC) which is capable of travelling on water from a landing ship to the shore; **amphibious vehicle** = vehicle designed for use both on water and on land

amputate [ˈæmpjʊteɪt] *verb* to remove a person's limb (normally by surgical operation); ***the doctor decided to amputate his leg***

amputation [æmpjʊˈteɪʃn] *noun* act of amputating a person's limb

amputee [æmpjʊ'tiː] *noun* person who has had a limb amputated

AMRAAM ['æmræm] *noun* = ADVANCED MEDIUM-RANGE AIR-TO-AIR MISSILE American-designed radar-guided air-to-air missile (AAM)

Amtrac ['æmtræk] *noun* = AMPHIBIOUS TRACTOR amphibious armoured personnel carrier (APC) which is capable of travelling on water from a landing ship to the shore

AMX [ˌeɪ em 'eks] *noun* French series of armoured fighting vehicles; **AMX-30** = 1960s-era main battle tank (MBT); **AMX-10** = infantry fighting vehicle; **AMX-13** = light tank; **AMX-13 DCA** = self-propelled anti-aircraft gun (SPAAG); **AMX-40** = 1980s-era main battle tank (MBT); **AMX-105** = self-propelled gun

anchor ['æŋkə] **1** *noun* heavy metal weight, which is lowered to the bottom of the sea, in order to stop a stationary ship from drifting; **at anchor** = moored with an anchor **2** *verb* to moor a ship with an anchor; ***we anchored in the estuary***

ANGLICO ['æŋglɪˌkəʊ] *noun US* = AIR NAVAL GUNFIRE LIAISON COMPANY small Marine Corps team, which is trained to direct close air support, naval gunfire support and artillery fire

annotate ['ænəʊteɪt] *verb* to add explanatory notes to a document; ***he annotated the fire plan***

Antarctic [æn'tɑːktɪk] **1** *noun* **the Antarctic** = the continent at the South Pole, the region south of the Antarctic Circle largely covered in snow and ice; *compare* ARCTIC **2** *adjective* referring to the Antarctic; **the Antarctic Circle** = the parallel running round the earth at latitude 66° 32 S, to the south of which lies the Antarctic region

antenna [æn'tenə] *noun* metal rod, mast or structure used in the transmission of radio signals; *see also* AERIAL (NOTE: the plural of **antenna,** in this context, is **antennas,** rather than **antennae**)

ante-room ['æntɪˌrʊm] *noun* drawing-room in an officers' mess; ***the CO is in the ante-room***

anthrax ['ænθræks] *noun* a disease of cattle and sheep which is transmissible to humans

> COMMENT: caused by a bacillus, *Bacillus anthracis,* anthrax can be transmitted by touching infected skin, meat or other parts of an animal. It causes pustules on the skin or in the lungs. Certain nations are known to have developed anthrax for use as a biological weapon

anti- ['ænti] *prefix* designed to counter

anti-aircraft (AA) [ˌænti 'eəkrɑːft] *adjective* designed to damage or destroy an aircraft in flight; ***anti-aircraft guns fired at the incoming bombers***

anti-aircraft artillery (AAA) [ˌæntɪ ˌeəkrɑːft ɑː'tɪlərɪ] *noun* cannon *or* heavy machine-gun (often self-propelled), which is designed to shoot down aircraft; ***the radar site is surrounded by anti-aircraft artillery***; *see also* TRIPLE-A, SELF-PROPELLED ANTI-AIRCRAFT GUN (SPAAG)

anti-air warfare (AAW) [ˌænti eə 'wɔːfeə(r)] *noun* naval term for air defence

> COMMENT: the air defence of a naval force is organized in depth, with three distinct defence zones. The first line of defence is the **Aircraft Defence Zone** which is patrolled by friendly fighter aircraft (usually operating from aircraft carriers). If the enemy aircraft manage to evade the fighters, they then enter the **Area Defence Zone** which is covered by the warships' long-range surface-to-air missiles (SAM) for the mutual defence of the entire force. If the enemy get through this, they enter the **Point Defence Zone** in which individual warships use their short-range SAMs and other weapons systems (eg. CIWS) for self-defence.

anti-ambush drill [ˌænti 'æmbʊʃ ˌdrɪl] *noun* standard countermeasure for troops who find themselves caught in an ambush

anti-dim *or* **anti-dimmer** [ˌænti 'dɪmə(r)] *noun* grease designed to stop condensation forming on the eyepieces of a respirator

anti-missile-missile (AMM) [ˌænti 'mɪsaɪl ˌmɪsaɪl] *noun* missile designed to shoot down an enemy ballistic missile

anti-personnel [ˌænti pɜːsə'nel] *adjective* designed to injure or kill a person; **anti-personnel mine** = mine designed to injure or kill a person

anti-radar missile (ARM) [ˌænti 'reɪdɑː ˌmɪsaɪl] *noun* missile designed to home in on an enemy's radar transmissions

anti-ship missile (ASM) [ˌænti 'ʃɪp ˌmɪsaɪl] *noun* missile designed to damage *or* destroy a ship (usually launched from an aircraft *or* other ship)

anti-tank [ˌænti 'tæŋk] *adjective* designed to damage or destroy or obstruct an armoured vehicle; **anti-tank ditch** = ditch dug as an obstacle to tanks and other armoured vehicles; **anti-tank mine** = mine designed to damage or destroy an armoured vehicle; **anti-tank platoon** = specialist platoon of an infantry battalion, whose specific role is the destruction of enemy armour; *see also* LAW

AOC [ˌeɪ əʊ 'siː] = AIR OFFICER COMMANDING

AOCC [ˌeɪ əʊ siː 'siː] = AIR OPERATIONS COORDINATION CELL

AOR [ˌeɪ əʊ 'ɑː(r)] = AREA OF RESPONSIBILITY

AP = 1 ARMOUR-PIERCING 2 ANTI-PERSONNEL 3 ALLIED PUBLICATION

Apache (AH-64) [ə'pætʃi] *noun* American attack helicopter

APC [ˌeɪ piː 'siː] = ARMOURED PERSONNEL CARRIER; ***I can see three APCs on the edge of the wood***

APDS [ˌeɪ piː diː 'es] = ARMOUR-PIERCING DISCARDING-SABOT

APFSDS [ˌeɪ piː ˌef es diː 'es] = ARMOUR-PIERCING FIN-STABILIZED DISCARDING-SABOT

Aphid ['eɪfɪd] *noun* Soviet-designed short-range air-to-air missile (AAM)

appoint [ə'pɔɪnt] *verb* **(a)** to assign a person to a job; ***he was appointed as a platoon sergeant*** **(b)** to arrange a time and place for something; ***the O Group was appointed for 1600 hours***

appointment [ə'pɔɪntmənt] *noun* **(a)** act of assigning a person to a job; ***he is responsible for all appointments within the unit*** **(b)** arrangement to meet at a specific time and place; ***I have an appointment with the doctor*** **(c)** job; ***he is not suitable for this appointment***

appreciation [əˌpriːʃi'eɪʃn] *noun* systematic decision-making process involving a careful examination of all the factors involved, the identification of all the available options, and finally, the selection of the most suitable option as the basis for a plan; ***the company commander is making his appreciation at the moment***

approach [ə'prəʊtʃ] **1** *noun* **(a)** act of coming near; ***we could hear the approach of enemy tanks***; **approach lights** = series of lights on the ground which show a pilot the route to an airport or landing strip **(b)** route towards a specific location; ***we must cover all the likely approaches*** **2** *verb* to come near; ***they were approaching the enemy position***

arable ['ærəbl] *adjective* relating to the cultivation of crops; **arable land** = farmland devoted to the cultivation of crops

arc [ɑːk] *noun* part of the circumference of a circle; **arc of fire** = designated area of ground covered by an individual weapon; ***each soldier was shown his arcs of fire***

> COMMENT: a defensive position is normally sited so that the arcs of each weapon or position or sub-unit overlap with those of its neighbours; these are called **interlocking arcs of fire**

Archer ['ɑːtʃə] *noun* Soviet-designed short-range air-to-air missile (AAM)

Arctic ['ɑːktɪk] **1**; **the Arctic** = *noun* area of ice and snow around the North Pole, the region north of the Arctic Circle; *compare* ANTARCTIC **2** *adjective* **(a)** referring to the Arctic; **the Arctic Circle** = the parallel running round the earth at latitude 66° 33 N, to the north of which lies the Arctic region **(b)** relating to conditions of extreme cold; **arctic clothing** = clothing designed for use in extremely cold climates; **arctic warfare** = military operations carried out near both Arctic and Antarctic regions

area ['eəriə] *noun* piece of ground; **area defence** = naval anti-air warfare (AAW) term for warships' use of their long-range surface-to-air missiles (SAM) for the mutual defence; **area of influence** = ground occupied by the enemy which will probably affect a unit's current operations; **area of interest** = ground occupied by the enemy which could affect a unit's future operations; **area weapon** = weapon which can deliver a quantity of projectiles over a wide area and thus effectively engage several targets simultaneously (eg. machine gun, artillery, mortar, cluster bomb)

ARM [ˌeɪ ɑː 'rem] = ANTI-RADAR MISSILE

arm [ɑːm] **1** *noun* **(a)** weapon; ***the right to bear arms is protected by the constitution***; **side arm** = pistol; **skill at arms** = weapons training; **small arms** = arms which can be carried, such as rifles, machine-guns and sub-machine-guns **(b)** *(military instructions)* **to order arms** = to hold a rifle with the butt resting on the ground beside the right foot; **to port arms** = to hold a rifle diagonally across the chest; **to present arms** = to salute someone by holding a rifle in front of the body in a vertical position; **to reverse arms** = to hold a rifle with the butt facing upwards and the muzzle pointing at the ground (used at funerals); **to shoulder arms** = to carry a rifle by resting it on the shoulder **(c) arms** = military service in general; **under arms** = serving in the armed forces; ***they have an army of 100,000 permanently under arms*** **(d)** branch of the armed forces (for example, armour, artillery, infantry); **combined arms** = two or more arms working together; ***this will be a combined arms operation***; **supporting arms** = arms which support the teeth arms (for example, engineers, signals, transport); **teeth arms** = arms that actually do the fighting (for example, armour, artillery, infantry) **2** *verb* **(a)** to equip with weapons; ***the government is starting to arm the police*** **(b)** to prepare a shell, bomb, etc., by removing any safety mechanism; ***to arm the shell you must remove the safety pin***

Armalite (M-16) ['ɑːməlaɪt] *noun* American 5.56mm assault weapon

armament ['ɑːməmənt] *noun* **(a)** general term for weapons; **armaments factory** = factory making guns, tanks, etc. **(b)** process of equipping with weapons

armd = ARMOURED

armed [ɑːmd] *adjective* **(a)** equipped with a weapon; ***the man is armed and dangerous***; **armed forces** = general title for all military forces (army, navy, air force, etc.); **armed insurrection** = resistance to established authority, involving the use of weapons **(b)** ready to fire or explode; ***the shell is armed when the safety pin is removed***

armistice ['ɑːmɪstɪs] *noun* agreement by both sides to stop fighting; ***an armistice was signed to end the war***; *see also* CEASEFIRE, TRUCE

armor *US see* ARMOUR

armored *US see* ARMOURED

armored cavalry regiment (ACR) [ˌɑːməd 'kævəlri ˌredʒɪmənt] *noun US* tactical organization of three combined-arms groupings, each of battalion strength, known as 'cavalry squadrons', plus one air cavalry squadron of helicopters; a highly mobile force specializing in the roles of reconnaissance, advance guard and covering force

armored combat earthmover (ACE) [ˌɑːməd ˌkɒmbæt 'ɜːθmuːvə] *noun* American-designed armoured bulldozer

armorer *US see* ARMOURER

armor-piercing *US see* ARMOUR-PIERCING

armory *US see* ARMOURY

armour *US* **armor** ['ɑːmə] *noun* **(a)** defensive covering designed to protect a vehicle from bullets, shrapnel and other projectiles; ***the frontal armour on this tank is 150mm thick***; **body armour** = vest fitted with panels of synthetic material (such as Kevlar) designed to protect a soldier from shrapnel and low-velocity bullets; **combination armour** = armour composed of layers of steel and other substances (such as ceramics, plastics, other types of metal, etc.); **composite armour** = another name for combination armour; **compound armour** = another name for combination armour; *see also* EXPLOSIVE REACTIVE ARMOUR, ROLLED HOMOGENEOUS ARMOUR; **secondary armour** = additional armour fitted onto an armoured vehicle in order to increase its protection **(b)** collective word for armoured fighting vehicles (especially tanks); ***enemy armour is concentrating to the south of Mistelbach***

armoured *US* **armored** ['ɑːməd] *adjective* **(a)** protected by armour; **armoured bridgelayer** = armoured vehicle fitted with a folding bridge; **armoured car** = light wheeled armoured fighting vehicle, normally used for reconnaissance; **armoured fighting vehicle (AFV)** = armoured vehicle equipped with some form of weapon (eg. anti-tank gun, heavy machine-gun); **armoured personnel carrier (APC)** = armoured vehicle used to transport troops or police (NOTE: normally referred to as an **APC: I can see three APCs on the edge of the wood**); **armoured recovery vehicle (ARV)** = armoured vehicle designed to tow a disabled or broken-down armoured vehicle away from the battlefield; **armoured repair and recovery vehicle (ARRV)** = updated version of the armoured recovery vehicle (ARV) fitted with additional lifting equipment to assist in the repair of armoured vehicles in the field; **Armoured Vehicle Launched Bridge (AVLB)** = British armoured vehicle based on a Chieftain tank and fitted with a folding bridge; **Armoured Vehicle Royal Engineers (AVRE)** = British armoured vehicle based on the Centurion tank and fitted with one or more specialist pieces of engineer equipment (such as a demolition gun, mine plough, fascines) **(b)** equipped with armoured fighting vehicles; **armoured battalion** = tank battalion; **armoured infantry battalion** = infantry battalion equipped with infantry fighting vehicles (IFVs); **armoured regiment** = **(i)** tactical grouping of two or more armoured battalions, possibly including armoured infantry; **(ii)** *GB* tank battalion

COMMENT: In the British Army, the **brigade** is used instead of the regiment as a tactical grouping of two or more battalions. Battalion-sized units of tanks or artillery are known as **regiments** for traditional reasons, while an infantry regiment is purely an historical and administrative grouping for two or more battalions which normally serve in different brigades. Armoured regiments and armoured infantry battalions are usually organized into armoured brigades at a ratio of 2:1, depending upon the tactical requirement

armourer *US* **armorer** ['ɑːmərə] *noun* technician who services and repairs weapons

armour-piercing *US* **armor-piercing** ['ɑːmə ˌpiːəsɪŋ] *adjective* capable of penetrating armour; ***the tank was hit by a 120mm armour-piercing round***; **armour-piercing discarding-sabot (APDS)** = anti-armour projectile consisting of a long-rod penetrator, fitted with a stabilizing metal collar (sabot) which falls away once the projectile is in flight; **armour-piercing fin-stabilized discarding-sabot (APFSDS)** = armour-piercing discarding-sabot in which the long-rod penetrator is fitted with metal fins for extra stability; *see also* LONG-ROD PENETRATOR

armoury *US* **armory** ['ɑːməri] *noun* secure location where weapons are stored

army ['ɑːmi] *noun* **(a)** branch of a state's armed forces which operates on land (for example infantry, armour and artillery); ***the latest Defence Review will have serious implications for the Army***; **Army Air Corps (AAC)** *GB* = air force (mainly helicopters) which is part of the army rather than the RAF; *see also* GROUND FORCES **(b)** tactical grouping of two or more corps; ***the US Third Army was commanded by General Patton***

arrangement [ə'reɪnʒmənt] *see* TECHNICAL ARRANGEMENT

ARRC [ˌeɪ ɑː ɑː 'siː] = ALLIED RAPID REACTION CORPS

arrest [ə'rest] **1** *noun* act of arresting someone; ***a police spokesman admitted that the arrest had been a mistake***; **close arrest** = state in which a person is detained in a secure location (such as a prison, police station or guardroom); **house arrest** = state in which a person is detained in his own home; **open arrest** = state in which a person is considered to be in custody and his movements are restricted, but he is allowed to go about his normal daily business; **under arrest** = detained in custody by the authorities; ***you are under arrest!*** **2** *verb* to seize a person and take him into custody; ***the patrol arrested two suspected terrorists***; **arrest warrant** = document issued by a judge, magistrate or other official which authorizes the security forces to arrest a specified person; **powers of arrest** = conditions under which a member of the security forces may legally arrest a person; ***there is no power of arrest for this offence***

arrestable [ə'restəbl] *adjective* for which you can be arrested; **arrestable offence** = illegal act for which someone may be arrested without an arrest warrant

ARRV [ˌeɪ ɑːr ɑː 'viː] = ARMOURED REPAIR AND RECOVERY VEHICLE

arrowhead ['ærəʊˌhed] *noun* tactical formation of men *or* vehicles, in the form of an inverted letter V (L); ***the platoon moved across the open ground in arrowhead formation***

arsenal ['ɑːsnəl] *noun* **(a)** room or building where weapons and ammunition are stored; ***the platoon will parade at the arsenal at 0800hrs*** **(b)** government establishment for the manufacture of weapons; ***the arsenal in Birmingham has been closed down*** **(c)** figurative term for the weaponry available to a government or paramilitary organization; ***this is the most powerful weapon in the terrorists' arsenal***

Arsine ['ɑːsiːn] *noun* = ARSENIC TRIHYDRIDE type of blood agent

arson ['ɑːsən] *noun* criminal offence of setting fire to something; ***there have been several arson attacks***; ***he was arrested for arson***

artificer [ɑː'tɪfɪsə] *noun* mechanic or technician

artificial [ɑːtɪ'fɪʃl] *adjective* man-made

artillery [ɑː'tɪləri] *noun* **(a)** general title for large calibre guns, missiles and air-defence weapons; **artillery piece** = large calibre gun used as an indirect-fire weapon; **artillery raid** = tactic using artillery, where the guns move into enemy territory to attack a specific target and then withdraw before the enemy can retaliate **(b)** branch of the army which uses these weapons; ***he served in the Royal Artillery***; ***"Great battles are won with artillery." Napoleon (an ex-gunner)***

artilleryman [ɑː'tɪlərimən] *noun* soldier serving in the artillery

arty = ARTILLERY

ARV [ˌeɪ ɑː 'viː] = ARMOURED RECOVERY VEHICLE

ASAP ['eɪsæp *or* ˌeɪ es eɪ 'piː] = AS SOON AS POSSIBLE

ASM [ˌeɪ es 'em] = ANTI-SHIP MISSILE, AIR-TO-SURFACE MISSILE

asphalt ['æsfɒlt] *noun US* road surface made of a mixture of tar and gravel (NOTE: British English is **tarmac**)

ASRAAM ['æzræm] *noun* = ADVANCED SHORT-RANGE AIR-TO-AIR MISSILE British-designed radar-guided air-to-air missile (AAM)

assault [ə'sɒlt] **1** *adjective* designed for use in combat; **assault boat** = light, man-portable boat designed to carry a section of infantry; **assault course** = series of obstacles used by infantry training establishments to practise obstacle-crossing; **assault craft** = small boat designed for amphibious operations; **assault force** = group of troops, tanks, etc., which attacks a position; **assault river crossing** = act of crossing a river while in contact with the enemy; **assault weapon** = semi-automatic rifle, equipped with a magazine holding 20-30 rounds and fitted with a bayonet **2** *noun* final stage of an attack onto an enemy position; ***the assault on the farm was a complete success*** **3** *verb* to use force in order to occupy an enemy position; ***B Company will assault the village***

assemble [ə'sembl] *verb* **(a)** to come together; ***the battalion assembled in the gymnasium*** **(b)** to bring together; ***Sgt Jones assembled the platoon in the briefing room*** **(c)** to put together; ***they were killed as they were assembling the mortar***

assembly [ə'sembli] *noun* act of coming together; **assembly area** = specified location where sub-units of a tactical grouping assemble in order to prepare themselves for the next phase of an operation

assign [ə'saɪn] *verb* to appoint to a position or task; ***he was assigned to the mortar platoon***

assignment [ə'saɪnmənt] *noun* task or job; ***my first assignment was to update the brigade security orders***

assist [ə'sɪst] *verb* to help

assistance [ə'sɪstəns] *noun* help; ***we had to provide assistance to the civil authorities***

assy area; = ASSEMBLY AREA

astern [ə'stɜːn] *adverb* to the rear of a boat or ship; ***he went astern***

ASU = ACTIVE SERVICE UNIT

ASW [ˌeɪ ˌes 'dʌb(ə)ljuː] = ANTI-SUBMARINE WARFARE

AT- *prefix* anti-tank

ATAF ['eɪtæf] *noun* = ALLIED TACTICAL AIR FORCE large NATO air-force grouping

ATGM [ˌeɪ tiː dʒiː 'em] = ANTI-TANK GUIDED MISSILE

ATGW [ˌeɪ tiː dʒiː 'dʌb(ə)ljuː] = ANTI-TANK GUIDED WEAPON

A/Tk = ANTI-TANK

ATO ['eɪtəʊ] *noun GB* = AMMUNITION TECHNICAL OFFICER officer or non-commissioned officer (NCO) who is trained to make bombs, booby traps and unexploded munitions safe

ATO = AIR TASKING ORDER

ATOC ['eɪtɒk] *noun* = AIR TASK OPERATIONS CENTRE department of an air-force headquarters which is responsible for allocating tasks to squadrons

atoll ['ætɒl] *noun* ring-shaped tropical island

atom ['ætəm] *noun* smallest unit of a chemical element, which can be used as a source of nuclear energy

atomic [ə'tɒmɪk] *adjective* **(a)** relating to the structure of atoms **(b)** relating to the use of nuclear energy; **atomic energy** = nuclear energy; **atomic submarine** = submarine driven by nuclear power; **atomic warfare** = warfare involving the use of atomic weapons; **atomic weapon** = bomb, missile or other device which utilizes the release of nuclear energy

> COMMENT: although their meanings are not identical, the word **atomic** has now been superseded by **nuclear** for most general contexts

atrocity [ə'trɒsəti] *noun* act considered by normal people to be extremely wicked (such as murder of civilians, rape, etc.)

atropine ['ætrəpiːn] *noun* substance injected as first aid for someone who has been exposed to a nerve agent

attach [ə'tætʃ] *verb* **(a)** to fasten one object to another; ***our sleeping bags were attached to the side of the vehicle*** **(b)** to assign a soldier or sub-unit to another unit for a specific role or task; ***we have a troop of tanks attached to us for this attack***; *compare* DETACH

attaché [ə'tæʃeɪ] *French noun* specialist member of an ambassador's staff; **military attaché** = officer attached to an ambassador's staff in order to deal with military matters

attachment [ə'tætʃmənt] *noun* **(a)** something which is attached to another object for a special purpose **(b)** the act of sending a person or sub-unit to work with another unit for a specific task; ***he was sent on attachment to the navy***

attack [ə'tæk] **1** *adjective* designed for offensive action; **attack aircraft** = aircraft which is designed to drop bombs on or fire missiles at targets on the ground, and is also capable of defending itself against enemy fighter aircraft; **attack helicopter** = helicopter equipped with weapons to attack other helicopters or targets on the ground **2** *noun* offensive use of force in order to achieve an objective (for example the capture of ground); ***the attack was a complete success***; **attack in echelon** *or* **echelon attack** = attack made by several units deployed side by side, where one unit sets off first, followed after an interval by the second, followed after another interval by the third, and so on; **deliberate attack** = attack which is mounted once full reconnaissance, planning and preparation have been carried out; **fighter ground-attack (FGA)** = attack by fighter aircraft on a target on the ground; **flanking attack** = attack on the enemy's flank; **frontal attack** = attack on the front of an enemy position (as opposed to the flank); **ground attack** = attack by aircraft on a target on the ground; **hasty** *or* **quick attack** = attack which is mounted without the opportunity to first carry out full reconnaissance, planning and preparation; **holding attack** = attack mounted to halt the advance of an enemy and keep him occupied, while other friendly forces conduct operations elsewhere; **spoiling attack** = attack mounted on an advancing enemy force in order to disrupt its activities and prevent it carrying out its intentions **3** *verb* to act offensively against an enemy, a position, etc.; ***C Company will attack the village at first light***; *see also* FIGHTER-BOMBER, STRIKE AIRCRAFT

attempt [ə'tempt] **1** *noun* **(a)** act of trying (usually unsuccessful); ***there were no more attempts to escape*** **(b)** attack (usually unsuccessful); ***the enemy made several attempts on the bridge*** **2** *verb* to try; ***he attempted to climb the fence***

attend [ə'tend] *verb* to be present at; ***he attended the conference***

attention [ə'tenʃn] *noun* **(a)** the act of applying your mind to something; ***may I have your attention, please?***; **pay attention** = to concentrate on something **(b)** *(military command)* **attention!** = stand to attention!; ***Parade, attention!***; **position of attention** = act of standing erect, with the feet together and the arms held in to the sides; **to stand at attention** = to stand in the position of attention; **to stand to attention** = to change position to stand at attention

attic ['ætɪk] *noun* space below the roof of a house, normally used for storage; ***there was a sniper in the attic***

attrition [ə'trɪʃn] *noun* **(a)** damage caused to an object as a result of repeated contact with another object; ***this grease will reduce the rate of attrition*** **(b)** gradual destruction of an enemy force by repeated attacks or by stubborn defence

attritional warfare [əˌtrɪʃnl 'wɔːfeə] *noun* outdated military doctrine which seeks to destroy an enemy's will to fight simply through the use of attrition; *compare* MANOEUVRE WARFARE

Auftragstaktik ['aʊftrɑːgzˌtæktɪk] *German noun* (literally mission tactics) German expression for DIRECTIVE COMMAND (NOTE: German nouns are always spelt with a capital letter)

> COMMENT: many English-speakers prefer to use this word, since *directive command* was very much a German invention. In fact, it was developed by the Prussian general staff and used to great effect during the war with Austria in 1866. Surprisingly, the British Army clung to the doctrine of *restrictive control* until the Falklands conflict in 1982. Now the British are also firm exponents of Auftragstaktik

augmentation forces [ˌɔːgmən'teɪʃn ˌfɔːsɪz] *noun* forces brought from Europe or from North America to provide reinforcements to NATO forces already in a certain area

Aussie ['ɒzi] *noun (informal)* Australian soldier

authenticate [ɔː'θentɪkeɪt] *verb* to carry out authentication

authentication [ɔːθentɪ'keɪʃn] *noun* radio procedure (usually involving a code), designed to establish whether a radio message or radio user is genuine

authority [ɔː'θɒrəti] *noun* **(a)** official power to do something; ***I do not have the authority to make that decision*** **(b)** organization which exercises power; **the authorities** = police or other law enforcement organization **(c)** strength of character which makes other people obey your orders; ***he lacks authority***

authorize ['ɔːθəraɪz] *verb* to give someone official permission to do something; ***I am not authorized to do that***

autojet ['ɔːtəʊˌdʒet] *noun* syrette, fitted with a mechanism which injects the dose automatically (usually by striking it against the flesh); ***each man was issued with an autojet of morphine***

automatic [ɔːtə'mætɪk] **1** *adjective* capable of performing a function by itself; **automatic weapon** = firearm which will continue to reload itself and fire for as long as pressure is applied to the trigger (for example a machine-gun); **semi-automatic weapon** = firearm which reloads itself after each shot (for example a self-loading rifle) **2** *noun* firearm which will continue to reload itself and fire for as long as pressure is applied to the trigger (for example a machine-gun); ***he was armed with an automatic***; ***there was a burst of automatic fire from the woods***

> COMMENT: many contemporary assault weapons have both an automatic and a semi-automatic capability

automatically [ɔːtə'mætɪkli] *adverb* **(a)** by itself; ***the door locks automatically*** **(b)** immediately and without having to think; ***you should be able to do that automatically***

AV-8 [ˌeɪ viː 'eɪt] *see* HARRIER

avenue ['ævənjuː] *noun* **(a)** road with a line of trees on both sides **(b)** approach route; ***we must cover all the likely avenues of approach***

AVGAS ['ævgæs] *noun* aircraft fuel

aviation [eɪvi'eɪʃn] *noun* use of aircraft

aviator ['eɪvieɪtə] *noun* pilot or other member of an aircrew

avionics [eɪvɪ'ɒnɪks] *noun* general term for all electronic systems on an aircraft

AVLB [ˌeɪ viː el 'biː] = ARMOURED VEHICLE LAUNCHED BRIDGE

AWACS ['eɪwæks] *noun* = AIRBORNE WARNING AND CONTROL SYSTEM electronic equipment, carried in specially designed aircraft, which detects enemy aircraft or missiles at long ranges and then coordinates their interception by friendly aircraft or missiles

award [ə'wɔːd] **1** *noun* official recognition of an achievement (for example a medal, commendation); ***he has been recommended for a gallantry award*** **2** *verb* to give a prize or punishment to someone; ***he was awarded a medal for bravery***; ***he was awarded ten days' restriction of privileges***

AWI [ˌeɪ ˌdʌb(ə)ljuː 'aɪ] = AIR WARFARE INSTRUCTOR

AWLS [ˌeɪ ˌdʌb(ə)ljuː ˌel 'es] *noun* = AMBER WARNING LIGHT SYSTEM amber coloured warning light required by law to be fitted to all armoured vehicles in Germany

AWO ['eɪwəʊ] *noun GB* officer on a warship who coordinates the air battle; *compare* PWO

AWOL ['eɪwɒl] *adverb* = ABSENT WITHOUT LEAVE away from a military unit without permission; ***he's been AWOL for three days***

axis ['æksɪs] *noun* real or imaginary line on the ground used to indicate the primary direction for a unit or sub-unit which is deployed in a tactical formation; ***our axis is the main road***

aye aye [ˌaɪ 'aɪ] *adverb* traditional sailors' expression, meaning 'Yes, I will carry out your instruction'

azimuth ['æzɪməθ] *noun US* direction in mils or degrees of an object on the ground; *see also* BEARING

BRAVO - Bb

B-1 [ˌbiː ˈwʌn] *noun* American-designed long-range strategic bomber aircraft (NOTE: also known as the **Lancer**)

B-2 [ˌbiː ˈtuː] *noun* American-designed stealth bomber aircraft (NOTE: also known as the **Spirit**)

B-52 [ˌbiː fɪfti ˈtuː] *noun* American-designed bomber aircraft (NOTE: plural is **B-52s** [ˌbiː fɪfti ˈtuːz])

BAA [ˌbiː eɪ ˈeɪ] = BRIGADE ADMINISTRATION AREA

back-bearing [ˈbæk ˌbeərɪŋ] *noun* bearing from a reference point to your own location

> COMMENT: a back-bearing is calculated by taking a bearing from your location to the reference point and then adding that bearing to 180 degrees *or* 3,200 mils if the bearing is less than that amount, *or* alternatively, by subtracting 180 degrees *or* 3,200 mils from the bearing if the bearing is greater. Once you have calculated two *or* more back-bearings from known *or* probable reference points, your exact location should be where they all intersect on the map. Remember to subtract the magnetic variation.; *see also* TRIANGULATE

backblast [ˈbækblɑːst] *noun* gasses and heat released to the rear when a rocket launcher is fired (which can injure a person standing in their way)

Backfire [ˈbækfaɪə] *noun* NATO name for a strategic variant of the Soviet-designed TU-22 medium bomber aircraft (Blinder)

backup [ˈbækʌp] *noun* additional assistance or resources available in the event of difficulty or failure; ***B Company can provide backup if necessary***

bacteriological warfare [ˌbæktɪərɪəˌlɒdʒɪkəl ˈwɔːfeə(r)] *noun* another name for biological warfare

badge [bædʒ] *noun* insignia worn on a uniform or displayed on a vehicle; **badge of rank** = insignia showing the wearer's rank (for example bars, chevrons, stars, etc.)

Badger [ˈbædʒə(r)] *noun* NATO name for the Soviet-designed TU-16 medium bomber aircraft

bag [bæg] *noun* soft container made of paper, fabric or other material; **bag charge** = fabric bag containing propellant for an artillery or tank round; **bivvy bag** = waterproof sleeping-bag cover; *see also* KITBAG, SLEEPING-BAG

baggage [ˈbægɪdʒ] *noun* spare clothing and other personal effects packed up for transportation; ***each company was allocated a lorry for baggage***

bagpipes [ˈbægpaɪps] *plural noun* musical instrument, traditionally used by Irish and Scottish regiments, and also by some Indian and Arab regiments (they are played by blowing air into a bag and then pumping it through a set of pipes)

bail *or* **bale out** [ˌbeɪl ˈaʊt] *verb* **(a)** to escape from a damaged vehicle or aircraft; ***the pilot baled out*** **(b)** to clear water from a leaking boat; ***they used their helmets to bale out***

balaclava *or* **Balaclava helmet** [ˌbæləˈklɑːvə(r)] *noun* warm woollen garment which covers the head and neck, but leaves the face *or* parts of the face free, and is therefore sometimes used to conceal a person's identity; ***the gunman was wearing a balaclava***; *see also* SKI-MASK

ball [bɔːl] *noun* **(a)** spherical object (normally used in sport); **ball-bearing =**

small solid metal ball used to reduce friction in machinery **(b)** standard bullets for a rifle, machine-gun or pistol; ***we need five thousand rounds of 5.56mm ball***; *see also* CANNONBALL (NOTE: no plural in this meaning)

ballistic [bəˈlɪstɪk] *adjective* **(a)** relating to projectiles; ***we have received the ballistic report on the shooting of Corporal Jones*** **(b)** moving by the force of gravity; **ballistic bomb** = bomb which is simply dropped onto a target by an aircraft; *see also* GP BOMB, IRON BOMB; **ballistic missile** = guided missile which ends its flight in a ballistic descent; *see also* INTERCONTINENTAL

ballistics [bəˈlɪstɪks] *noun* science of projectiles and firearms

balloon [bəˈluːn] *noun* large bag filled with gas to make it rise in the air; *see also* BARRAGE

ban [bæn] **1** *noun* law which makes an activity or object illegal; ***we want an international ban on biological weapons*** **2** *verb* to make an activity or object illegal; ***many nations wish to ban the use of anti-personnel mines***

band [bænd] *noun* **(a)** group of musicians; ***the band of the Coldstream Guards played at the reception*** **(b)** group of people who have organized themselves for a specific purpose (usually criminal or paramilitary); ***there are several bands of rebels operating in the area*** **(c)** strip of plastic, metal or other material put around an object to keep it together; ***he removed the bands from the packing case*** **(d)** range of radio frequencies; ***which bands are you monitoring?***; **citizens' band (CB)** = range of frequencies allocated to the general public for the use of two-way radios

bandage [ˈbændɪdʒ] **1** *noun* strip of fabric used to bind a wound or other injury; ***the nurse put a bandage round his knee*** **2** *verb* to apply a bandage; ***she bandaged the wound***

bandit [ˈbændɪt] *noun* **(a)** robber (usually a member of a gang) who operates in rural areas **(b)** *(air-force slang)* enemy aircraft

bandoleer *or* **bandolier** [bændəˈlɪə] *noun* belt which goes over one shoulder, designed to carry ammunition

bandsman [ˈbændzmən] *noun* member of a musical band

> COMMENT: military bandsmen are usually employed as stretcher-bearers on the battlefield

bang [bæŋ] *noun* noise made by an explosion; ***we heard a loud bang***

Bangalore torpedo [ˌbæŋgəlɔː tɔːˈpiːdəʊ] *noun* device for clearing wire entanglements, consisting of piping filled with explosive, which is pushed into the obstacle and then detonated

bank [bæŋk] *noun* **(a)** artificial mound of earth used to enclose a field; ***we took cover behind a bank*** **(b)** margin of a river or lake; ***the far bank of the river has been mined*** **(c)** place where people can deposit or store money; ***the bank has been robbed***; **blood bank** = building or vehicle where blood for transfusion is stored

banner [ˈbænə] *noun* **(a)** ceremonial flag **(b)** piece of fabric attached to two poles and bearing a written message; ***the soldiers unfurled a banner showing instructions in Arabic for the crowd to disperse***

BAOR [ˌbiː eɪ əʊ ˈɑː] *noun* = BRITISH ARMY OF THE RHINE

bar [bɑː] **1** *noun* **(a)** rod of metal or wood used as an obstruction; ***the window was protected with metal bars*** **(b)** something which is long, thin and rigid (for example a bar of chocolate, bar of gold); **bar mine** = type of anti-tank mine **(c)** sandbank in a river or estuary; ***the landing craft had to navigate between sand bars*** **(d)** place where alcohol may be bought and consumed **(e)** badge of rank for junior officers in the US Army (a single bar denotes lieutenant, while a double bar denotes captain) **2** *verb* **(a)** to obstruct; ***the road was barred by fallen trees*** **(b)** to forbid an activity; ***soldiers were barred from all the pubs in the town***

barbed wire [ˌbɑːbd ˈwaɪə] *noun* wire with sharp spikes attached to it, used as an obstacle; **barbed-wire entanglement** = obstacle to infantry made out of barbed wire

barge [bɑːdʒ] *noun* long flat-bottomed boat used for carrying freight

barn [bɑːn] *noun* large farm building (normally used for storage)

barracks ['bærəks] *noun* non-operational military base; **confined to barracks (CB)** = punishment by which a soldier is not allowed to leave the barracks; ***he was awarded 10 days CB***; **barrack dress** = everyday uniform consisting of a sweater and service-dress trousers

barrage ['bærɑːʒ] *noun* **(a)** concentrated artillery attack (usually lasting for some time); ***a barrage of mortar fire was directed at the enemy positions***; **creeping barrage** = artillery bombardment which is constantly adjusted, so that the shells continue to land in front of friendly troops as they advance **(b)** man-made barrier in a river or estuary; **barrage balloon** = balloon which is secured to the ground by a wire cable, and used as an obstacle to low-flying aircraft

> COMMENT: in the literal sense of the word, the purpose of an artillery barrage is to prevent, or at least hinder the movements of the enemy, rather than to destroy his men, equipment and positions. If the latter effect is desired, then the word bombardment would be more appropriate instead.

barrel ['bærəl] *noun* **(a)** tube part of a gun, down which the bullet or shell slides when it is fired; ***he spent hours cleaning the barrel of his rifle*** **(b)** large cylindrical container; ***the bomb was attached to a barrel of oil***

Barrett (M-82) ['bærət] *noun* American .50 calibre sniper rifle

barricade [bærɪ'keɪd] **1** *noun* improvised obstacle or fortification; ***the street was blocked by a barricade*** **2** *verb* to make an obstruction (with whatever materials happen to be available); ***we barricaded the door***

barrier ['bæriə] *noun* obstacle which prevents forward movement; ***there was a barrier across the road***; ***the mountains form a natural barrier between France and Spain***

barrow ['bærəʊ] *noun* huge man-made mound of earth, marking the site of an ancient grave

base [beɪs] **1** *noun* **(a)** secure location from which military operations can be conducted; ***we have several bases in that region*** **(b)** part on which an object rests; ***he examined the base of the container*** **2** *verb* **(a)** to station a soldier at a base; ***I was based in Germany*** **(b)** to use as a starting point for a calculation or development process; ***the plan was based on the belief that the enemy would not fight***; ***this engineer vehicle is based on the Chieftain tank***

base bleed (BB) [beɪs 'bliːd] *noun* system which increases the range of an artillery shell by means of a small gas generator fitted to the base of the shell; the generator expels gas at low pressure to reduce drag caused by the vacuum which forms at the base of the shell while it is in flight

baseline ['beɪslaɪn] *noun* offensive manoeuvre carried out under fire, in which men *or* vehicles move forward into extended line in order to engage the enemy; *compare* FOOTHOLD LINE

basement ['beɪsmənt] *noun* part of a building which lies below ground level

baseplate ['beɪspleɪt] *noun* firing platform of a mortar

basha ['bæʃə] *noun GB* improvised shelter made from a poncho

basic training [ˌbeɪsɪk 'treɪnɪŋ] *noun* period of training for new recruits; ***he has just completed his basic training***

basket ['bɑːskɪt] *noun* device used in air-to-air refuelling; the basket is a receptacle fitted to the end of a fuel pipe,into which an aircraft must insert its refuelling probe in order to receive fuel

batman ['bætmən] *noun* soldier who cleans an officer's kit; *see also* ORDERLY

baton ['bætɒn *or* 'bætɪn] *noun* **(a)** stick carried as a mark of rank; ***a painting of the Field-Marshal with his marshal's baton hangs in the mess*** **(b)** stick made of wood or other material for use as a weapon; ***they carried riot-shields and batons***; **baton round** = large projectile made of plastic or rubber which is fired from a special gun and is designed to knock a person over but not to cause a serious injury (NOTE: also known as **plastic bullet** *or* **rubber bullet**)

> every soldier carries a marshal's baton in his rucksack
>
> *Napoleon*

battalion (Bn) [bə'tæljən] *noun* tactical and administrative army grouping of three or more companies or equivalent-sized groupings; **air-portable battalion** = infantry battalion which is not equipped with armoured fighting vehicles and can therefore be transported by air to an area of operations; **armoured battalion** = tank battalion; **armoured infantry battalion** = infantry battalion equipped with infantry fighting vehicles (IFVs); **battalion landing team (BLT) = US combined arms grouping based on a marine infantry battalion, including artillery, armoured reconnaissance, tanks and engineers; air-assault battalion** = US infantry battalion equipped with its own transport helicopters and supporting attack helicopters; **mechanized battalion** = infantry battalion equipped with armoured personnel carriers (APCs) or infantry fighting vehicles (IFVs)

COMMENT: British tank and artillery battalions are known as **regiments,** as are battalion-sized units of certain supporting arms (such as engineers). American armored cavalry battalions are known as **squadrons, although normal armored units use the term battalion.** In some contexts, British infantry battalions traditionally use the word **regimental** as an adjective relating to the battalion: eg. Regimental Sergeant Major (RSM), regimental aid post (RAP). A British armoured brigade might consist of two armoured or mechanized infantry battalions and one armoured regiment or, alternatively, two armoured regiments and one infantry battalion, with artillery and supporting arms. On operations, these units are broken down and combined into **battle groups**. As an example, an armoured infantry battle group might consist of two infantry companies and one squadron of tanks, which are organized into two **company and squadron groups** and a **squadron and company group** under the command of the infantry battalion HQ. The exact composition will vary according to the tactical requirement at the time. In the US Army, a battle group is known as a task force, while company and squadron groups and squadron and company groups are known as company teams

batter ['bætə] *verb* to cause damage or injury by hitting repeatedly; ***our trenches were battered by the enemy artillery***

battery ['bætri] *noun* **(a)** company-sized artillery grouping with six or more guns; ***we have been allocated two batteries to support the attack*** (NOTE: shortened to **Bty** in this meaning) **(b)** power source for portable electrical equipment; ***this radio needs a new battery***

battle ['bætl] *noun* prolonged engagement between large numbers of opposing troops; ***during the tank battle, several enemy tanks were put out of action***; ***he served in the British Fleet at the Battle of Jutland***; **battle casualty replacement (BCR)** = soldier who remains on stand-by in order to take the place of a soldier who is killed or wounded; **battle fatigue** = mental and physical exhaustion resulting from a long period in battle (NOTE: also called **shell shock** or **post-traumatic stress disorder)**; **battle handover point (BHP)** = point, during the passage of lines, where the passing unit takes over or, in the case of a rearward passage of lines, hands over responsibility for the battle; **battle honour** = official recognition of a unit's achievements or conduct during a battle, which gives that unit the right to carry the name of the battle on its colours; **battle inoculation** = preparing soldiers for battle by the use of live rounds and simulated battle effects; **battle stations** = state of readiness for battle; ***the brigade remained at battle stations for most of the night***; ***Next to a battle lost, the greatest misery is a battle gained - Wellington***

battlefield ['bætlfi:ld] *noun* ground on which a battle is fought; ***the dead and wounded were removed from the battlefield***

battle group (BG) ['bætl ˌgru:p] *noun* **(a)** *GB* combined arms grouping based on an armoured regiment or infantry battalion **(b)** tactical grouping of warships

> COMMENT: as an example, an armoured infantry battle group might consist of two infantry companies and one squadron of tanks, which are organized into two **company and squadron groups** and a **squadron and company group** under the command of the infantry battalion HQ. The exact composition will vary according to the tactical requirement at the time Note: US Army equivalent is task force; US Marine Corps equivalent is battalion landing team (BLT)

battleship ['bætlʃɪp] *noun* large armoured warship, equipped with heavy guns, which is used to destroy enemy warships and provide naval gunfire support (NGS) to land forces

bay [beɪ] *noun* **(a)** part of a coastline where the sea curves inland; ***they selected a lonely bay for the landing*** **(b)** space set aside for a specific purpose; **unloading bay** = place where weapons may be loaded and unloaded safely

bayonet ['beɪənət] **1** *noun* stabbing blade attached to the muzzle of a rifle or assault weapon; **bayonet charge** = charge with the intention of using the bayonet; **to charge bayonets** = to level the bayonet at an enemy prior to charging at him **2** *verb* to stab someone with a bayonet; ***he was bayoneted to death*** (NOTE: **bayoneting - bayoneted**)

bazooka [bə'zuːkə] *noun* hand-held anti-tank rocket launcher

BB 1 BATTLESHIP **2** BASE BLEED

BCR [ˌbiː siː 'ɑː] = BATTLE CASUALTY REPLACEMENT

BDA [ˌbiː diː'eɪ] = BATTLE-DAMAGE ASSESSMENT

Bde = BRIGADE

Bdr = BOMBARDIER

BDU [ˌbiː diː 'juː] *noun US* = BATTLEDRESS UTILITIES camouflage combat uniform; ***he was wearing BDUs*** (NOTE: British English is **DPM**)

beach [biːtʃ] *noun* strip of sand or gravel at the edge of the sea, lake or river; **beach landing** = act of disembarking troops and vehicles onto a beach; **beach-master** = officer who controls the movement of troops and vehicles during a beach landing

beachhead ['biːtʃˌhed] *noun* defensive position established around the site of a beach landing, which is used as a secure base for subsequent operations

beacon ['biːkən] *noun* **(a)** bonfire or light used as a signal or warning **(b)** lamp designed for use as a beacon **(c)** radio transmitter which acts as a guide to shipping or aircraft; **hazard beacon** = warning beacon indicating that there is some danger to aircraft **(d)** *GB* hill traditionally used for beacon fires

Bear [beə] *noun* NATO name for the Soviet-designed TU-95 strategic bomber aircraft

bearing ['beərɪŋ] *noun* direction, in mils or degrees, of a feature on the ground in relation to north; ***the church is on a bearing of 1825 mils***; **grid bearing** = bearing obtained from a map using a protractor; **magnetic bearing** = bearing obtained using a compass; *see also* AZIMUTH; BACK-BEARING

> COMMENT: a grid bearing may be converted into a magnetic bearing by adding the **magnetic variation**

bearskin *or* **bearskin cap** ['beəskɪn] *noun* tall ceremonial headdress traditionally worn by guards infantry soldiers

> COMMENT: the bearskin should never be confused with the *busby*, which is a similar but much shorter headdress traditionally worn by cavalry soldiers

beat [biːt] *verb* **(a)** to hit something repeatedly; **beaten zone** = area of ground which is hit by the bullets from an automatic weapon; **to beat someone up** = to injure a person by repeated punching and kicking; ***he was badly beaten up*** **(b)** to win a victory over someone else; ***we've been beaten*** (NOTE: **beating - have beaten**)

B Echelon ['biː ˌeʃəlɒn] *noun* administrative elements of a tactical grouping

belt [belt] *noun* **(a)** strip of leather, webbing or other material, worn around the waist and used to support a person's trousers or to carry equipment-pouches; *see also* SAM BROWNE **(b)** ammunition which is linked together by metal clips or fastened by loops to a strip of canvas, in order to be fired by a machine-gun; **belt-fed** = designed to fire belts of ammunition

beret ['bereɪ *US* bə'reɪ] *noun* soft peakless hat; **Blue Berets** = soldiers of a United Nations force; **Green Berets** = American special forces unit or British marines; **Red Berets** = British paratroopers

bergen ['bɜːgən] *noun* large fabric container suspended from a metal frame, which is designed to be carried on a person's back; *see also* PACK, RUCKSACK

berm [bɜːm] *noun* artificial bank of earth or sand used as a barrier or fortification

> COMMENT: berms were extensively used by both the Iraqis and coalition forces during the Gulf War of 1991

besiege [bɪ'siːdʒ] *verb* to surround an enemy town or fortress with troops in order to prevent anyone entering or leaving, with the ultimate intention of capturing the place; *see also* INVEST

> COMMENT: "besiege" is not normally used in modern military English; it has now been largely replaced by the verb "invest"

Betalight™ ['biːtəˌlaɪt] *noun* tiny hand-held apparatus, containing a luminous substance which gives off a very weak light and is therefore suitable for map-reading *or* signalling when you are close to the enemy

betray [bɪ'treɪ] *verb* **(a)** to reveal a secret; ***we were betrayed by the villagers*** **(b)** to abuse someone's trust; ***the general betrayed his men by agreeing to surrender***

betrayal [bɪ'treɪəl] *noun* act of betraying; *see also* TREACHERY

beyond [bɪ'jɒnd] *adverb* on the far side of something; ***the enemy position is 100 metres beyond that line of trees***

BFA [ˌbiː ef 'eɪ] = BLANK-FIRING ATTACHMENT

BFV [ˌbiː ef 'viː] *noun* = BRADLEY FIGHTING VEHICLE M2 Bradley infantry fighting vehicle; *compare* CFV

BG = BATTLE GROUP

BHP [ˌbiː eɪtʃ 'piː] = BATTLE HANDOVER POINT

bid [bɪd] **1** *noun* formal request for something; ***you must submit your ammunition bids at least 24 hours in advance*** **2** *verb* to make a formal request for something; ***I will bid for two places on the next anti-tank course*** (NOTE: **bidding - bid - have bid**)

billet ['bɪlɪt] **1** *noun* place (usually a civilian home) where a soldier is accommodated; ***he went back to his billet*** **2** *verb* to arrange accommodation for a soldier; ***we were billeted on the local priest***

bind [baɪnd] *verb* to fasten around something; ***the containers were bound with metal strips*** (NOTE: **binding - bound**)

binoculars [bɪ'nɒkjʊləz] *plural noun* optical instrument with a lens for each eye, designed for looking at distant objects; *see also* FIELD-GLASSES, TELESCOPE

binos ['baɪnəʊz] *plural noun* binoculars; ***I lost my binos during the attack***

biological [baɪə'lɒdʒɪkl] *adjective* relating to biology or living organisms; **biological warfare** = the use of disease as a weapon; **biological weapon** = disease such as anthrax, developed for use as a weapon

> COMMENT: biological weapons are unstable, difficult to deliver with any precision and impossible to control once they are delivered. Furthermore, anyone contemplating the use of such weapons can expect retaliation in its rest form

bio-weapon ['baɪəʊˌwepən] *noun* biological weapon

bipod ['baɪpɒd] *noun* two-legged stand designed to support a weapon or other piece of equipment

bird-strike ['bɜːdstraɪk] *noun* collision between a bird and an aircraft; ***the crash was caused by bird-strike***

Birthday Parade ['bɜːθdeɪ pəˌreɪd] *noun GB* ceremonial parade held by the Household Troops in London on the official birthday of the Monarch, during which a battalion from the Brigade of Guards troops its colour

COMMENT: this ceremony is more popularly known as "Trooping the Colour"

bivouac ['bɪvʊæk] **1** *noun* **(a)** improvised shelter **(b)** campsite of improvised shelters **2** *verb* to sleep outside without proper tents; ***they bivouacked in the corner of a field*** (NOTE: **bivouacking - bivouacked**)

bivvy ['bɪvi] *noun (informal)* bivouac; **bivvy-bag** = waterproof sleeping-bag cover

BK [bi: 'keɪ] *noun GB* = BATTERY KAPITAN second in command of a battery; ***the BK has been killed***

BL-755 [ˌbi: el ˌsev(ə)n faɪv 'faɪv] *noun* British-designed cluster bomb

black [blæk] **go black** = *US* to exhaust your ammunition; ***our recon platoon has gone black***

Blackbird ['blækbɜ:d] *see* SR-71

Blackhawk ['blækhɔ:k] *noun* American-designed UH-60 utility/transport helicopter

Blackjack ['blækdʒæk] *noun* NATO name for the Soviet-designed TU-160 strategic bomber aircraft

black market [ˌblæk 'mɑ:kɪt] *noun* illicit trade in articles which are illegal *or* rationed *or* difficult to obtain, usually at a considerable profit; ***respirators and NBC suits are fetching very high prices on the black market***

blackout ['blækaʊt] *noun* measures designed to ensure that no lights are showing after dark; ***all units must observe the blackout***; **information blackout** = withholding all information from the media and general public (usually for reasons of security)

bladder ['blædə(r)] *noun* huge inflatable rubber container, which is used to store fuel *or* water at a POL *or* water point

blank [blæŋk] *noun* **blank** *or* **blank round** = training ammunition, consisting of the propellant but no projectile, which is designed to simulate the firing of a weapon; ***we will need 5000 rounds of 7.62mm blank***; ***they were firing blanks***; **blank-firing attachment (BFA)** = device fitted to an automatic or semi-automatic weapon to enable it to operate with blank rounds; *compare* LIVE, LIVE ROUND

blast [blɑ:st] **1** *noun* **(a)** wave of heat and gasses released by an explosion, and the debris carried by it; ***the blast broke all the windows in the vicinity*** **(b)** explosion; ***several people were killed in the blast*** **2** *verb* to use explosives; ***we will have to blast a way through***

bleed [bli:d] *verb* to lose blood; ***the wound is bleeding badly*** (NOTE: bleeding - bled)

blend [blend] *verb* to mix together; **to blend in** = to look the same as everyone or everything else; ***camouflage enables the soldiers to blend in with the woodland***

blind [blaɪnd] **1** *adjective* unable to see **2** *noun* **(a)** *US* camouflaged screen designed to conceal a soldier or piece of equipment; ***they erected a blind in front of the tank*** **(b)** missile, shell or other projectile which has been fired but has failed to explode; ***the last shell was a blind*** **3** *verb* to make someone blind, either temporarily or permanently; ***he was blinded by a piece of shrapnel***; **blinding agent** = chemical agent designed to make people blind

Blinder ['blaɪndə(r)] *noun* NATO name for the TU-22 medium bomber aircraft

blindfold ['blaɪndfəʊld] **1** *noun* piece of fabric tied over a person's eyes *or* head so that he cannot see; ***they used a sandbag as a blindfold*** **2** *verb* to tie a blindfold on someone; ***he was blindfolded***

blind spot ['blaɪnd ˌspɒt] *noun* **(a)** location which cannot be observed **(b)** location in which it is impossible to send or receive radio transmissions

blister ['blɪstə] *noun* liquid-filled swelling on the skin caused by a burn, friction or chemical agent; **blister agent** = chemical agent designed to cause serious blisters

blitzkrieg ['blɪtskri:g] *noun Germany* offensive operation making maximum use of firepower, manoeuvre warfare and all-arms cooperation; ***the enemy favour blitzkrieg tactics***

blivet ['blɪvɪt] *noun* inflatable rubber container, which is used to store fuel

blizzard ['blɪzəd] *noun* combination of heavy snow and strong wind

blob [blɒb] *noun* tactical infantry formation, in the form of a rough circle; ***they moved through the scrub in blob formation***

bloc [blɒk] *noun* group of nations which share a common purpose; **Eastern Bloc** = term sometimes given to the Warsaw Pact; **Western Bloc** = term sometimes applied to NATO

block [blɒk] **1** *noun* **(a)** obstruction; *see also* ROAD BLOCK **(b)** solid piece of hard material; ***a block of wood*** **2** *verb* to obstruct; ***the road is blocked by fallen trees***

blockade [blə'keɪd] **1** *noun* obstruction of another country's coastline or borders in order to prevent the movement of goods and supplies; ***only two ships managed to get through the blockade***; **blockade runner** = ship, vehicle or person who tries to enter or exit a blockaded country **2** *verb* to carry out a blockade; ***the enemy is blockading our entire coast***

blockhouse ['blɒkhaʊs] *noun* fortified structure; ***lines of concrete blockhouses were built along the Atlantic coast***; *see also* BUNKER

blood [blʌd] *noun* red liquid in the body; **blood agent** = chemical agent designed to deprive the body of oxygen; **blood bank** = place or vehicle used for the storage of blood for blood transfusions; **blood donor** = someone who gives blood for blood transfusions; **blood group** = type of blood (for example A, B, O, AB); **blood transfusion** = injection of blood, taken from a blood donor, into the vein of another person; **blood vessel** = vein, artery or capillary carrying blood around the body

bloodbath ['blʌdbɑːθ] *noun* massacre, the killing of large numbers of people

bloodless coup [ˌblʌdləs 'kuː] *noun* seizure of power achieved without bloodshed; ***the army took over after a bloodless coup in 1994***

bloodshed ['blʌdʃed] *noun* action which results in physical injury or death; ***the mission was achieved without bloodshed***

bloodthirsty ['blʌdθɜːsti] *adjective* eager to kill; ***the Gurkhas have the reputation of being bloodthirsty fearless soldiers***

blow [bləʊ] *verb* to destroy with explosives; ***the bridge has been blown*** (NOTE: **blowing - blew - have blown**)

Blowpipe ['bləʊpaɪp] *noun* British-designed hand-held optically-tracked surface-to-air missile (SAM)

blow up [ˌbləʊ 'ʌp] *verb* **(a)** to destroy with explosives; ***they blew up the fuel dump***; ***the railway track has been blown up in several places*** **(b)** to explode; ***the tank blew up***

BLT [ˌbiː el 'tiː] *noun* *US* = BATTALION LANDING TEAM combined arms grouping based on a marine infantry battalion, including artillery, armoured reconnaissance, tanks and engineers (NOTE: US Army equivalent is **task force (TF)**)

blue [bluː] *noun* **Blue Berets** = soldiers of a United Nations force; **blue forces** = friendly forces; **blue on blue** = incident where friendly forces fire on their own troops or vehicles by mistake; *see also* FRATRICIDE, FRIENDLY FIRE

> COMMENT: the positions of friendly forces are usually marked on a map in blue, while those of the enemy are marked in red

bluey ['bluːɪ] *noun* *GB* air-mail letter; ***he was writing a bluey***

bluff [blʌf] **1** *noun* **(a)** *US* steep, almost vertical slope (usually above a stream *or* river); ***the enemy is dug in on the bluffs above the town*** **(b)** attempt to deceive; ***the enemy withdrawal is just a bluff*** **2** *verb* to attempt to deceive; ***the enemy is trying to bluff us into thinking that he is going to withdraw***

BMD [ˌbiː em 'diː] *noun* Soviet air-portable infantry fighting vehicle (IFV)

BMNT [ˌbiː em en 'tiː] *noun* *US* = BEGINNING OF MORNING NAUTICAL TWILIGHT first light

BMP [ˌbiː em 'piː] *noun* Soviet series of infantry fighting vehicles (IFVs); **BMP-1** = 1960s-era IFV; **BMP-2** = 1980s-era IFV; **BMP-3** = 1990s-era IFV

Bn = BATTALION

board [bɔːd] *verb* to attack and climb onto a ship; ***the enemy boarded our ship during cover of darkness***; **boarding party** = group of marines, sailors, etc., who attack and board a ship; **on board** = on *or* onto a boat *or* ship *or* aircraft; ***the Admiral is spending the night on board HMS Ardent***; *see also* ABOARD

boat-people ['bəʊt ˌpiːpəl] *plural noun* political refugees who try to escape from an oppressive regime by sea

boatswain *or* **bosun** ['bəʊs(ə)n] *noun* officer *or* petty officer in charge of equipment and the crew; **boatswain's chair** = seat suspended by ropes for work on the side of a ship; **boatswain's pipe** = metal whistle traditionally used by the boatswain for signalling and salutes

body armour ['bɒdi ˌɑːmə] *noun* vest fitted with panels of synthetic material (eg. Kevlar) designed to protect a soldier from shrapnel and low-velocity bullets; *see also* BULLETPROOF VEST, FLAK JACKET

body bag ['bɒdi ˌbæg] *noun* strong waterproof bag designed for transporting a dead body

body-count ['bɒdi ˌkaʊnt] *noun* **(a)** check to ensure that all the members of a sub-unit are present; ***he took a quick body-count before moving on*** **(b)** *US* number of enemy killed; ***this company has the highest body-count in the battalion***

bodyguard ['bɒdigɑːd] *noun* person or group assigned to guard a dignitary or other important person; ***the general's bodyguard was killed in the attack***

Bofors ['bəʊfəz] *noun* Swedish-designed light anti-aircraft gun

bog [bɒg] *noun* area of permanently wet ground

bogey ['bəʊgɪ] *noun (slang)* enemy fighter aircraft; ***Watch out ! There's a bogey on your tail !***; *see also* BANDIT

bolt [bəʊlt] *noun* part of the firing mechanism of a firearm, consisting of a movable metal block which houses the firing-pin and which is used to push a round into the breech and then seal in the gases which are released when the round is fired; **bolt-action rifle** = rifle where the bolt must be operated by hand for each round (as opposed to a semi-automatic rifle); ***most of the rebels are armed with bolt-action rifles***

bomb [bɒm] **1** *noun* explosive device used as a weapon, consisting of a strong metal container containing explosive material together with a priming device; ***they dropped two tons of bombs on the castle***; ***terrorists placed bombs in the city centre***; **bomb-aimer** = member of an aircrew responsible for the aiming and release of bombs; **bomb-bay** = compartment in an aircraft used to hold bombs; **bomb disposal** = disarming and safe destruction of unexploded bombs; **bomb-disposal unit** = small group of soldiers trained to make unexploded bombs safe; **bomb-sight** = optical instrument in an aircraft for the aiming of bombs; **bomb-site** = area where buildings have been destroyed by bombs; **bomb squad** = bomb-disposal unit; **ballistic bomb** = bomb which is simply dropped onto a target by an aircraft; **car bomb** = terrorist bomb concealed in a vehicle; **cluster bomb** = aircraft-dropped device containing a quantity of small bombs or bomblets which are released in mid-air over a target area; **fire-bomb** = bomb designed to set buildings alight; **general purpose (GP) bomb** = BALLISTIC BOMB; **glide bomb** = aerodynamic bomb which is released by an aircraft several kilometres from its target and which then makes a ballistic descent to the target controlled by a guidance system; **incendiary bomb** = bomb designed to set buildings alight; **iron bomb** = BALLISTIC BOMB; **letter-bomb** = explosive device concealed in a letter or package and designed to explode when the letter is opened; **pipe-bomb** = home-made grenade consisting of a piece of metal pipe filled with explosive; **proxy bomb** = terrorist bombing tactic, where an innocent civilian is forced by the terrorists to carry an explosive device or drive a car containing an explosive device up to a target (eg. security force base); ***the device is then initiated by a timer or by remote control***; **suicide bomb** = terrorist bombing tactic, where a terrorist carries an explosive device or drives a vehicle containing an explosive device up to a target (eg. security force base) and initiates it, deliberately killing himself in the process; **time-bomb** = bomb detonated by a time mechanism **2** *verb* to attack with bombs; ***the base has been bombed twice in the past two days***; ***enemy aircraft bombed our positions***; **to**

bomb up = to resupply a fighting vehicle or aircraft with ammunition

bombard ['bɒmbɑːd] *verb* to attack with artillery; ***the enemy started to bombard our positions***; *see also* SHELL

bombardier [bɒmbə'diːə] *noun* **(a)** *GB* corporal in the artillery (NOTE: shortened to **Bdr** in this meaning); **lance-bombardier (L/Bdr)** = lance corporal in the artillery **(b)** *US* bomb-aimer in an aircraft

bombardment [bɒm'bɑːdmənt] **1** *noun* artillery attack (usually lasting some time); **creeping bombardment** = artillery bombardment which is constantly adjusted, so that the shells continue to land in front of friendly troops as they advance **2** *adjective* *US* referring to bomber aircraft; ***34th Bombardment Squadron***

bomber ['bɒmə] *noun* **(a)** large aircraft designed to drop bombs; **fighter-bomber** = aircraft which is designed to drop bombs on or fire missiles at targets on the ground, and is also capable of defending itself against enemy fighter aircraft **(b)** person who takes part in a bomb attack; ***two of the bombers have been arrested***; *see also* ATTACK, AIRCRAFT, STRIKE AIRCRAFT

bombing ['bɒmɪŋ] *noun* action of dropping bombs on a target; ***the bomber squadron undertook several bombing raids on enemy positions;*** **carpet bombing** = dropping bombs *or* bomblets evenly over a wide area of ground; **dive-bombing** = attack where the aircraft makes a steep descent to drop a bomb directly onto a target; **laydown bombing** = low altitude attack in which the aircraft passes very low over its target and releases bombs fitted with parachutes or other devices to slow down the descent, so that the aircraft can get clear before the bombs explode; **tactical bombing** = bombing carried out in direct support of ground forces; **toss-bombing** = attack where bombs are released as the aircraft is making a shallow climb at high speed; the bombs' trajectories then carry them forward a considerable distance before they hit the ground, making it unnecessary for the aircraft to pass directly over its target

bomblet ['bɒmlət] *noun* small bomb released in mid-air by a cluster bomb or missile

bonnet ['bɒnət] *noun* Scottish military head-dress; ***some Scottish regiments wear tartan flashes on their bonnets***

booby trap ['buːbi ˌtræp] *noun* hidden or harmless-looking device (often explosive) designed to kill or injure anyone who touches it

booby-trap ['buːbi ˌtræp] *verb* to set a booby trap (in a house, under a car, etc.); ***most of the houses had been booby-trapped***

boom [buːm] *noun* **(a)** floating barrier; ***there was a boom across the entrance to the harbour*** **(b)** refuelling-probe on an aircraft

boot [buːt] *noun* strong footwear reaching above the ankle; *US (informal)* **boot camp** = army training establishment for new recruits (usually with a particularly harsh regime)

Bora ['bɔːrə] *noun* strong cold wind which blows in the Balkans

border ['bɔːdə] *noun* frontier between two countries; ***two tank divisions crossed the border***; **border patrol** = patrol sent out to prevent or provide warning of border incursions

bore [bɔː] *noun* measurement across the inside of a tube, such as the barrel of a gun; *see also* SMALL-BORE

boresight ['bɔːsaɪt] **1** *noun* device which is inserted into the barrel of a weapon and then aligned on an aiming mark, so that the weapon's sighting systems can also be aligned on the same mark **2** to adjust the sights of a weapon using a boresight

> COMMENT: boresighting is only a very rudimentary method of aligning the sights of a weapon. To ensure accuracy, you need to **zero** the weapon

bosun ['bəʊs(ə)n] *see* BOATSWAIN

botulism ['bɒtjuːˌlɪz(ə)m] *noun* fatal disease, which is normally associated with food poisoning

> COMMENT: caused by a bacillus, *Clostridium botulinum*, symptoms include paralysis of the muscles, vomiting, hallucinations and death. Certain nations are known to have developed botulism as a biological weapon

bound [baʊnd] *noun* **(a)** single movement made by a person, sub-unit or vehicle, usually from fire position to fire position or from cover to cover; ***the troop moved in bounds, with one tank covering while the other two were moving***; **alternate bounds =** movement in bounds, with one person, vehicle or sub-unit stationary, and giving or prepared to give covering fire, while the other moves past to occupy a fire position beyond; **tactical bound =** distance which ensures that one group is close enough to support another group without the risk of both coming under effective fire from the same enemy; ***Platoon HQ was moving a tactical bound behind the point section*** **(b)** limit; **in bounds =** where one is allowed to go; ***that pub is in-bounds to troops***; **out of bounds (OOB) =** where one is not allowed to go; ***that pub is out of bounds to troops***

boundary ['baʊndrɪ] *noun* real *or* imaginary line which marks the limits of a grouping's area of responsibility; ***that road is the brigade boundary***; ***our mission is to destroy the enemy within boundaries***

bow *or* **bows** [baʊ] *noun* front end of a ship; **shot across the bows =** shot fired in front of a ship as a warning; *compare* STERN

bowser ['baʊzə] *noun* cylindrical container mounted on a trailer, designed to carry fuel or water

box [bɒks] *noun* **(a)** square *or* rectangular container; ***the used ammunition boxes to strengthen the position*** **(b)** tactical vehicle formation, in the form of a square *or* rectangle; ***we usually assault in box formation***

BQMS [ˌbiː kjuː em 'es] = BATTERY QUARTERMASTER SERGEANT

brace [breɪs] *verb* to prepare yourself for a crash or shock (usually by holding tightly onto something); **Brace ! Brace ! Brace !** = verbal warning given when an aircraft is about to crash *or* when a ship is about to be hit by a missile *or* torpedo

bracken ['brækən] *noun* plant with feather-like leaves, which grows extensively in woodland and heathland

bracket ['brækɪt] *verb* to correct artillery *or* mortar fire so that each adjusting round lands on the opposite side of the target to the last round, until the target is hit; ***he realised that his position was being bracketed***

> COMMENT: a competent FOO *or* MFC should be able to hit the target with his third correction

brackish ['brækɪʃ] *adjective; (of water)* unsuitable for drinking due to a high mineral content

Bradley ['brædli] *noun* American-designed 1980s-era infantry fighting vehicle (M2 or M3)

> COMMENT: the M2 is designed to carry a squad of infantry, while the M3 is an armored cavalry fighting vehicle carrying additional armament and equipment instead

bramble ['bræmbəl] *noun* common name for the plant of the wild blackberry, which grows as a thick thorny bush; ***it was impossible to get through the brambles***

brave [breɪv] *adjective* full of courage, able to control fear; ***it was brave of him to try to cross the street in front of the enemy positions***; *see also* COURAGEOUS

bravery ['breɪvri] *noun* ability to control fear; *see also* COURAGE

Bravo ['brɑːvəʊ] second letter of the phonetic alphabet (Bb)

BRDM [ˌbiː ɑː diː 'em] *noun* Soviet series of wheeled reconnaissance vehicles; **BRDM-2 =** late 1960s-era recce vehicle; **BRDM-3 =** late 1970s-era recce vehicle armed with ATGM

breach [briːtʃ] **1** *noun* point at which the enemy's line of defence is penetrated; ***the infantry poured through the breach in the enemy's defences*** **2** *verb* to break through an enemy's line of defence; ***after a heavy bombardment, they were still not able to breach the enemy's defences***

break [breɪk] **1** *noun* **(a)** place where something is broken; ***his leg had a clean break just above the ankle*** **(b)** period of rest taken during an activity; ***after two hours, we had a short break***; *GB* **NAAFI break** = break to have a cup of tea or coffee; **smoke break** = break to have a cigarette **2** *verb* **(a)** to cause damage to something; ***he broke the window***; ***he broke his leg*** **(b)** to stop being in a close group; ***the enemy has broken*** **(c)** to stop an activity (usually for a short period); ***they broke for lunch*** **(d) to break contact** = to stop fighting with the enemy and withdraw; *see also* DISENGAGE; **to break cover** = to come out into the open (NOTE: **breaking - broke - have broken**)

breakage ['breɪkɪdʒ] *noun* damage; ***soldiers have to pay for their breakages***

break down [ˌbreɪk 'daʊn] *verb* **(a)** *(people)* to suffer from a physical and mental collapse as a result of stress; ***he has broken down completely*** **(b)** *(machinery)* to stop working because of a malfunction; ***three of our tanks have broken down*** **(c)** to divide into separate components; ***a platoon can be broken down into sections***

breakdown ['breɪkdaʊn] *noun* **(a)** mechanical failure; ***a breakdown in communications with headquarters*** **(b)** physical and mental collapse; ***after three months on the front line he suffered a breakdown*** **(c)** analysis of an organization; ***I want a complete breakdown of the enemy force***

break in [ˌbreɪk 'ɪn] *verb* to use force to enter a building or vehicle

break out [ˌbreɪk 'aʊt] *verb* **(a)** to happen; ***fighting broke out along the front line*** **(b)** to fight your way out of an encirclement; ***they were encircled but managed to break out without much loss of life***

break through [ˌbreɪk 'θruː] *verb* to fight your way through a main line of defence; ***the enemy have broken through near Minden***

breakthrough ['breɪkθruː] *noun* act of fighting your way through a main line of defence; ***the enemy have made a breakthrough near Minden***

break up [ˌbreɪk 'ʌp] **(a)** to come apart, to fall to pieces; ***the aircraft broke up in mid-air*** **(b)** to disperse; ***the demonstration broke up when baton rounds were fired***; ***troops were sent in to break up the demonstration***

breastwork ['brestwɜːk] *noun* low field fortification constructed from earth, rocks, timber, etc.

breech [briːtʃ] *noun* rear part of a gun's barrel, into which a round is placed in order to be fired

Bren gun ['bren 'gʌn] *noun* type of light machine-gun

> COMMENT: the Bren was designed in Czechoslovakia and developed in Great Britain prior to World War II. It is still in use in many armies, including the British Army

brevet ['brevɪt] **1** *noun* commission which entitles an officer to take a higher rank without the appropriate pay; ***he was a brevet lieutenant-colonel*** **2** *verb* to confer a brevet rank on someone; ***he was breveted as a major*** (NOTE: **breveting - breveted**)

> COMMENT: brevet ranks are usually only conferred in wartime and are seen as temporary appointments

brew [bruː] **1** *noun* a cup of tea or coffee; ***they stopped for a brew*** **2** *verb* **(a)** to make beer **(b)** to make tea

brew up [ˌbruː 'ʌp] *verb* **(a)** to make a hot drink **(b)** *(vehicles)* to catch fire

brick [brɪk] *noun GB* team of four men, forming part of a multiple; ***Cpl Smith's brick captured the gunman***; *see also* MULTIPLE

bridge [brɪdʒ] **1** *noun* **(a)** structure built to carry a road or railway over a river, road or railway; **pontoon bridge** = temporary bridge supported by boats **(b)** control centre of a ship **2** *verb* to make a bridge over something; ***the enemy have bridged the river***

bridgehead ['brɪdʒhed] *noun* defensive position established on the enemy side of a river or other obstacle, which is used as a secure base for subsequent operations

bridgelayer ['brɪdʒleɪə] *noun* vehicle which carries and lays a portable bridge

brief [briːf] **1** *noun* **(a)** orders or instructions; ***that is not part of our brief***

(b) detailed summary or explanation; ***we received a brief on the enemy's organization*** **2** *verb* **(a)** to give orders or instructions; ***he briefed his platoon for the attack*** **(b)** to explain a situation in detail; ***he briefed the brigadier on the tactical situation***; *compare* DEBRIEF

briefing ['briːfɪŋ] *noun* **(a)** orders or instructions; ***we all assembled for the daily briefing*** **(b)** detailed explanation or summary; ***the press officer gave a briefing on the current situation to reporters*** **(c)** meeting where a briefing is given; **briefing room** = room where briefing and debriefing take place; *compare* DEBRIEFING

Brig = BRIGADIER

brig [brɪg] *noun US* military prison (especially on a warship)

brigade (Bde) [brɪ'geɪd] *noun* tactical army grouping of two or more battalions or regiments; **brigade administration area (BAA)** = operational location for the logistical elements of a brigade; **brigade major** = chief of staff of a brigade; **GB square brigade** = brigade, consisting of two armoured regiments and two battalions of armoured or mechanized infantry

> COMMENT: a British armoured brigade might consist of two armoured or mechanized infantry battalions and one armoured regiment or two armoured regiments and one infantry battalion, plus artillery and supporting arms. On operations, these units are broken down and combined into **battle groups.** As an example, an armoured infantry battle group might consist of two infantry companies and one squadron of tanks, which are organized into two **company and squadron groups** and a **squadron and company group** under the command of the infantry battalion HQ. The exact composition will vary according to the tactical requirement at the time. An air-portable infantry brigade might consist of three infantry battalions plus artillery and supporting arms. In the US Army, a battle group is known as a task force, while company and squadron groups and squadron and company groups are known as company teams.

brigadier (Brig) [brɪgə'dɪə] *noun GB* senior officer in the army or marines (usually in command of brigade) *US* **brigadier general** = senior officer in the army, marines or air force (junior to a major general and senior to a colonel, usually in command of a brigade)

British ['brɪtɪʃ] *adjective* relating to Great Britain (GB); **British Army of the Rhine (BAOR)** = obsolete title for British ground forces stationed in Germany (NOTE: Great Britain is formed of England, Scotland and Wales, and with Northern Ireland forms the **United Kingdom (UK)**)

brook [brʊk] *noun* small stream

BSM [ˌbiː es 'em] = *GB* BATTERY SERGEANT MAJOR

BTR [ˌbiː tiː 'ɑː] *noun* Soviet series of wheeled armoured personnel carriers (APCs); **BTR-60** = 1960s-era APC; **BTR-80** = 1980s-era APC; **BTR-90** = 1990s-era APC

Bty = BATTERY

BSM [ˌbiː es 'em] = BATTERY SERGEANT MAJOR

buckshee ['bʌkʃiː] *adjective GB (slang)* spare (and usually acquired unofficially or illegally); ***I've got a buckshee sleeping-bag***

buddy ['bʌdi] *noun US (informal)* comrade; **buddy-buddy system** = philosophy where comrades look after each other's welfare and protect each other in battle

buffer zone ['bʌfə ˌzəʊn] *noun* designated area between two groupings, which neither grouping can enter but in which enemy can be engaged by either grouping (designed to avoid fratricide between the two groupings)

bugle ['bjuːgl] *noun* musical instrument, similar to a trumpet, traditionally used to send signals or instructions in the form of music

bugler ['bjuːglə] *noun* person who plays the bugle

bug out [ˌbʌg 'aʊt] *verb (informal)* to abandon a position or location in a hurry

built-up ['bɪlt ˌʌp] *adjective* covered by buildings (ie cities, towns and other urban areas); ***we will have to move through a large built-up area***

bull [bʊl] *GB (slang)* **1** *noun* cleaning and polishing of kit; ***the RSM expects plenty of bull for this parade*** **2** *verb* to polish boots; ***he was bulling his boots***

bulldozer ['bʊldəʊzə] *noun* tracked vehicle designed to push obstructions out of the way

bullet ['bʊlɪt] *noun* projectile fired by a pistol, rifle or machine-gun; **armour-piercing bullet** = bullet designed to penetrate armour; **dum-dum bullet** = bullet modified to expand when it hits a person or animal, thereby causing a terrible wound; **high-velocity bullet** = bullet which travels faster than the speed of sound; **incendiary bullet** *or* **tracer bullet** = bullet which is designed to ignite after firing and burn in flight, so that the fall of shot can be observed; **plastic bullet** *or* **rubber bullet** = large projectile made of plastic or rubber which is fired from a special gun and is designed to knock a person over but not to cause a serious injury (NOTE: also called a **baton round**)

bulleted blank [ˌbʊlɪtɪd 'blæŋk] *noun* blank round designed for use with some automatic or semi-automatic weapons, containing a projectile which disintegrates upon leaving the muzzle of the weapon

bulletproof ['bʊlɪtpruːf] *adjective* designed to prevent penetration by bullets; **bulletproof vest** = vest fitted with panels of synthetic material (eg. Kevlar) designed to protect a soldier from shrapnel and low-velocity bullets; *see also* BODY ARMOUR, FLAK JACKET

bumf [bʌmf] *noun (slang)* written instructions, briefings, reports, etc.; ***I haven't had time to read all the bumf yet***

bunch [bʌntʃ] *verb (of a group of soldiers or vehicles)* to stand *or* move in close proximity to each other, thus presenting a good target for machine-guns and artillery; ***don't bunch ! Keep spread out !***

Bundeswehr ['bʊndəsveə(r)] *noun* German armed forces; ***the crossings are being held by units of the Bundeswehr***

bungee ['bʌndʒiː] *noun* elasticated cord used as a fastening

bunker ['bʌŋkə] *noun* **(a)** shelter with reinforced sides and roof, designed to withstand artillery and small-arms fire **(b)** reinforced underground shelter used for storage (especially of ammunition)

burial detail ['berɪəl ˌdiːteɪl] *noun* detachment of soldiers assigned to bury the dead

burlap ['bɜːlæp] *noun US* coarse fabric used as camouflage or to make sandbags (NOTE: British English is **hessian**)

burst [bɜːst] **1** *noun* **(a)** firing of a series of bullets rapidly; ***there was a burst of machine-gun fire from behind the wall*** **(b)** explosion; ***the burst of the shell deafened him*** **2** *verb* to explode; ***the shell burst next to the command post***

bury [berɪ] *verb* to place an object in a hole in the ground and then cover it with soil; ***there wasn't time to bury the dead***; ***the guerillas buried their weapons in the forest*** (NOTE: burial - buried - have buried)

busby ['bʌsbɪ] *noun* ceremonial fur headdress traditionally worn by cavalry soldiers

> COMMENT: the busby is sometimes confused with the *bearskin cap*, which is a similar but much taller headdress traditionally worn by guards infantry soldiers

bush [bʊʃ] *noun* **(a)** plant resembling a small tree **(b)** *(Australia and South Africa)* **the Bush** = wild uncultivated terrain (NOTE: no plural in this meaning)

bust [bʌst] *GB (slang)* **1** *adjective* damaged *or* broken; ***the radio is bust*** **2** *verb* **(a)** to damage *or* break; ***I've bust my binoculars*** **(b)** to demote; ***Cpl Hobbs has been busted***

butt [bʌt] *noun* **(a)** part of the rifle which a person places against his shoulder during firing; ***he killed the man with his rifle butt***; **butt salute** = salute made by slapping the butt or handguard of the rifle **(b) the butts** = the target end of a shooting range; ***he's in the butts***

butte [bjuːt] *noun US* small isolated hill, with a flat top and steep slopes

BVR [ˌbiː viː ˈɑːr] *adverb* = BEYOND VISUAL RANGE too far from an enemy aircraft to see it with the naked eye; ***we'll have to use our BVR missiles***; *compare* WVR

bypass [ˈbaɪpɑːs] **1** *noun* road which passes around the outside of a town (in order to avoid going through the centre); ***the bypass has been cratered*** **2** *verb* to move past an enemy position without engaging it; ***we've been ordered to bypass the village and continue our advance***

CHARLIE - Cc

C-130 [ˌsiː wʌn 'θɜːti] *noun* American-designed transport aircraft (NOTE: also known as the **Hercules**)

C-141 [ˌsiː wʌn fɔː 'wʌn] *noun* American-designed transport aircraft (NOTE: also known as the **Starlifter**)

C-17 [ˌsiː sevən 'tiːn] *noun* American-designed heavy-lift transport aircraft, which is capable of landing on short runways (NOTE: also known as **Globemaster**)

C3 [siː 'θriː] = COMMAND, CONTROL AND COMMUNICATIONS

C-601 [siː ˌsɪks əʊ 'wʌn] *noun* Chinese-designed anti-ship missile

C-801 [siː ˌeɪt əʊ 'wʌn] *noun* Chinese-designed anti-ship missile

CA = CRUISER (with guns)

CAB [ˌsiː eɪ 'biː *or* kɑb] = *US* COMBAT AVIATION BRIGADE

cab [kæb] *noun* driver's compartment of a lorry or truck

cabin ['kæbɪn] *noun* **(a)** room on an aircraft or ship (normally used as living quarters); ***the captain called a meeting in his cabin*** **(b)** *US* hut or simple shelter; ***they spent the night in a cabin in the mountains***

cable ['keɪbl] *noun* **(a)** thick metal wire which is used to convey electricity from one place to another **(b)** thick metal wire which is used to moor a ship, or to tow a ship or vehicle (NOTE: also called a **hawser**)

cache [kæʃ] **1** *noun* hidden store of ammunition *or* equipment *or* food; ***we found a cache of ammunition in a hollow tree*** **2** to put something in a cache; ***we cached our spare rations close to the track***

cadence ['keɪdəns] *noun* **(a)** standard time and pace for marching in step; ***they use a cadence of 95 paces to the minute*** **(b)** drum-beat *or* song designed to help maintain the cadence; ***I heard the squad chanting their cadence***

cadet [kə'det] *noun* schoolboy *or* girl who is a member of an official organization, which is designed to give young people a taste of life in the armed forces; ***a party of cadets will be visiting the barracks tomorrow***; **officer cadet** = rank held by a potential officer at an officer-training establishment

cadre ['kɑːdə(r)] *noun* small unit of trained *or* experienced personnel, which can be used to form the basis for a much larger unit consisting mainly of untrained *or* less experienced personnel (eg. recruits, reservists, territorials, etc)

cairn [keən] *noun* pile of stones *or* rocks, often built as a marker *or* monument; ***there's a cairn on the summit***

caisson ['keɪsən] *noun* trailer designed to carry ammunition

caliber *see* CALIBRE

calibre *US* **caliber** ['kælɪbə] *noun* **(a)** internal diameter of a gun barrel **(b)** external diameter of a projectile

call [kɔːl] **1** *noun* **(a)** shout or cry **(b)** radio message **2** *verb* **(a)** to speak loudly; ***we heard him calling*** **(b)** to summon; ***he called the man over*** **(c)** to request or order; ***he called for smoke***; **on call** = available on request; ***we have a section of mortars on call*** **(d)** to wake someone up; ***call me at 0600*** **(e)** to speak to someone on a radio or telephone; ***he called the duty officer to inform him of the incident***

call out [ˌkɔːl 'aʊt] *phrasal verb* to deploy a force in response to an incident *or* threat; ***the battalion has been called out***; ***call out the QRF !***

call-out ['kɔːl ˌaʊt] *noun* act of deploying a force in response to an incident *or* threat; ***the GOC was very unimpressed with our performance during the last call-out***

call-sign (C/S) ['kɔːl ˌsaɪn] *noun* name, letters or numbers used to identify a person or sub-unit on the radio

call up [ˌkɔːl 'ʌp] *verb* to summon for military service; ***all the young men have been called up***; ***he was called up in 1944 and immediately sent to the front***

call-up ['kɔːl ˌʌp] *noun* action of calling someone to join the armed forces; ***his call-up was deferred because he was still at university***

caltrops ['kæltrɒps] *plural noun* set of metal spikes designed to damage vehicle tyres

calvary ['kælvərɪ] *noun* small roadside religious monument, in the form of Jesus on the cross (common in Roman Catholic regions); ***there's a calvary 100 metres before the junction***

cam [kæm] *(informal)* **1** *noun* camouflage; ***put plenty of cam on this tank***; **cam-cream** = cosmetic face-paint for camouflage; **cam-net** = camouflage net **2** *verb* ***to cam up*** = to apply camouflage; ***they cammed up***

camouflage ['kæməflɑːʒ] **1** *noun* **(a)** use of natural and man-made materials to make something blend in with the surrounding area; ***camouflage is an essential military skill*** **(b)** materials used for camouflage (natural vegetation, camouflage-net, fabric, paint); ***put plenty of camouflage on this tank***; **camouflage cream** = cosmetic face-paint for camouflage; **camouflage net** = covering of knotted cord and pieces of fabric, used to conceal a vehicle, piece of equipment or structure **2** *verb* to conceal something by making it blend in with its surroundings; ***they were camouflaging their vehicles***

camp [kæmp] **1** *noun* **(a)** place where people are accommodated in temporary shelter (such as tents); ***the refugees are being housed in camps***; **POW camp** *or* **prison camp** = secure location where prisoners-of-war are accommodated; *see also* CONCENTRATION CAMP **(b)** place where troops are accommodated and trained; ***the recruits returned to camp*** *US*; *(informal)* **boot camp** = army training establishment for new recruits **2** *verb* to live outdoors in temporary shelter; ***we camped by the river***

campaign [kæm'peɪn] **1** *noun* prolonged period of military activity in a specific area or region; ***we are planning an autumn campaign at the end of the rainy season***; **the North Africa Campaign** = long period of warfare in North Africa; **campaign medal** = medal awarded for service during a certain campaign **2** *verb* to conduct or take part in a campaign; ***the army spent two years campaigning in North Africa***

camp-bed ['kæmp ˌbed] *noun* lightweight folding or collapsible bed used for camping (NOTE: American English for this is **cot**)

camp-follower ['kæmp ˌfɒləʊwə] *noun* civilian who attaches himself or herself to an army during a campaign

camp-site ['kæmp ˌsaɪt] *noun* place used for setting up a camp

camstick ['kæmstɪk] *noun* solid stick of face paint

can [kæn] *noun* **(a)** metal container for liquid (such as a petrol can) **(b)** *US* metal container in which food or drink is hermetically sealed for storage over long periods (NOTE: in British English, **tin** is more usual in this meaning when referring to containers for food)

canal [kə'næl] *noun* artificial waterway used for navigation or irrigation

Candid ['kændɪd] *noun* NATO name for Soviet-designed IL-76 transport aircraft

canister ['kænɪstə] *noun* **(a)** metal container for gas or aerosol; ***you should wear protective clothing when handling gas canisters***; **smoke canister** = metal container containing chemicals which produce smoke **(b)** direct-fire anti-personnel round for a tank gun or artillery piece, consisting of a fragile container filled with small projectiles; the container disintegrates as it leaves the muzzle of the gun and the projectiles spread out like shot from an enormous shotgun **(c)** *GB* disposable air filter for a respirator;

soldiers are trained how to change their canisters in NBC conditions

cannibalize ['kænəbəlaɪz] *verb* to use damaged or defective equipment as a source of spare parts

cannon ['kænən] *noun* **(a)** large calibre heavy machine-gun **(b)** large wheeled gun, often capable of firing explosive projectiles **(c)** *(historical)* outdated muzzle-loading artillery piece; **cannon-fodder** = soldiers who are seen merely as material to be used during a war

cannonball ['kænənbɔːl] *noun* large round stone or metal ball, fired from old cannons

canoe [kə'nuː] *noun* small hand-powered narrow boat with pointed ends

canopy ['kænəpi] *noun* **(a)** covering suspended over an object; ***we erected a canopy to protect the wounded from the sun*** **(b)** fabric part of a parachute; ***his canopy failed to open properly*** **(c)** cover provided by the leaves and branches of trees in a wood, forest, jungle, etc.; ***the jungle's canopy concealed the enemy's movements*** **(d)** transparent cover of an aircraft's cockpit; ***the aircraft needed a new canopy***

canteen [kæn'tiːn] *noun* **(a)** place where food and drink is sold **(b)** water-bottle **(c)** set of eating utensils

canvas ['kænvəs] *noun* very strong water-resistant fabric used to make tents, tarpaulins, etc; **under canvas** = accomodated in tents *or* in the field; ***we've spent the last six months under canvas***

CAOC [ˌsiː eɪ əʊ 'siː] = COMBINED AIR OPERATIONS CENTRE

CAP [ˌsiː eɪ 'piː] = COMBAT AIR PATROL

capability [keɪpə'bɪləti] *noun* being able to do something; **capability package (CP)** = forces, infrastructure and supplies which allow a commander to be able to do what is required; **required capability** = the resources that are necessary to do which is required

capable ['keɪpəb(ə)l] *noun* able to do something; ***I don't think he is capable of commanding a battalion***; ***he is a very capable NCO***; **night-capable** = able to be used in darkness; **special operations capable (SOC)** = having sufficient training and expertise to carry out specialized military tasks

cap-comforter ['kæp ˌkʌmfətə(r)] *noun* woollen hat, originally designed to be worn under a helmet in cold weather; ***the patrol will wear cap-comforters***

capital ['kæpɪtl] **1** *adjective* **(a)** *(legal)* punishable by death; ***treason is a capital offence***; **capital punishment** = execution of a convicted criminal **(b) capital letters** = letters of the alphabet written as A,B,C, instead of a,b,c **2** *noun* **(a)** most important city or town in a country or region; ***the army pressed on to the outskirts of the capital*** **(b) in block capitals** = written entirely in capital letters; ***this form should be completed in block capitals*** (NOTE: capital letters are also known as **upper case,** while small letters are known as **lower case**)

capitalism ['kæpɪtəlɪzm] *noun* economic system involving investment and profit-making by private individuals; *compare* COMMUNISM

capitalist ['kæpɪtəlɪst] **1** *adjective* favouring capitalism **2** *noun* someone who favours capitalism

capitulate [kə'pɪtjʊleɪt] *verb* to stop fighting and acknowledge the supremacy of an enemy; *see also* SURRENDER

capitulation [kəpɪtjʊ'leɪʃn] *noun* act of capitulating

capsize [kæp'saɪz] *verb* to overturn (a boat); ***the ship has capsized***

Capt = CAPTAIN

captain (Capt) ['kæptn] *noun* **(a)** *GB* senior officer in the navy (above a commander, and usually in command of a warship) **(b)** *GB* officer in the army or marines above the rank of lieutenant and below a major **(c)** *US* officer in the navy (usually in command of a warship) **(d)** *US* officer in the army, marines or air force (usually in command of company or equivalent-sized grouping); *see also* GROUP CAPTAIN

> COMMENT: British Army captains have enough experience to make them eligible for a variety of different roles. Within a unit, they might act as second-in-command of a company or

equivalent-sized grouping or have a specialist role in the unit's headquarters (as for example adjutant, intelligence officer, operations officer, etc.). Outside the unit, they might be employed as an aide-de-camp (ADC), or as a junior staff officer in a brigade or divisional headquarters. In the British armed forces, a captain in the marines is considered to be the equivalent of a major in the army. The rank of captain in the US Army requires more experience and higher qualifications than its equivalent in the British Army, and is therefore considered to have greater seniority. Captains in the US Army are eligible to command companies or equivalent-sized groupings

captive ['kæptɪv] *noun* someone who has been captured

captivity [kæp'tɪvəti] *noun* state of being a captive; **in captivity** = being held captive

captor ['kæptə(r)] *noun* someone who captures another person *or* holds them as a prisoner *or* hostage; ***none of our captors spoke English***

capture ['kæptʃə] **1** *noun* **(a)** act of taking someone prisoner; ***he evaded capture*** **(b)** act of taking possession of something by force; ***the capture of the town was a disaster*** **2** *verb* **(a)** to take someone prisoner; ***we captured two generals*** **(b)** to take possession of something by force; ***they captured an enemy supply dump***

car [kɑː] *noun* small motorized passenger vehicle; **car bomb** = terrorist bomb concealed in a vehicle; **car park** = area of ground or building used for parking cars; **car phone** = radio-telephone fitted to a motor vehicle; **staff car** = car used for official purposes by a senior officer

carabinier [kærəbɪ'nɪə] *noun (historical)* **(a)** elite light infantryman **(b)** elite heavy cavalryman

COMMENT: some modern armoured regiments retain their historical title as Carabiniers

carbine ['kɑːbaɪn] *noun* light short-barrelled rifle

card [kɑːd] *noun* small piece of stiff paper or plastic; **identity card (ID card)** = card issued by a government or organization as a means of identification; **range card** = card showing topographical features or targets and the distance to them from a specific location; **route card** = card showing the different stages of a journey, with locations, distances, bearings and other information

cargo ['kɑːgəʊ] *noun* goods or supplies carried by an aircraft or ship; *compare* FREIGHT

Carl Gustav [ˌkɑːl gʊ'stɑːv] *noun* Swedish-designed hand-held 84mm medium anti-tank weapon (MAW)

carpet bombing ['kɑːpɪt ˌbɒmɪŋ] *noun* dropping bombs *or* bomblets evenly over a wide area of ground

carrier ['kæriə] *noun* person or thing which carries something; **aircraft carrier** = large ship designed to carry aircraft and equipped with maintenance facilities and a runway for take-off and landing; **armoured personnel carrier (APC)** = armoured vehicle used to transport troops or police; **carrier air wing (CVW)** = US tactical grouping of naval aviation squadrons operating from a single aircraft carrier; **carrier battle group (CVBG)** = tactical grouping of warships which includes an aircraft carrier; **carrier pigeon** = bird used for carrying messages; **carrier wave** = electromagnetic wave used to carry a radio signal

COMMENT: a carrier air wing (CVW) usually consists of a fighter squadron, three ground-attack squadrons, an electronic-attack squadron, an airborne early warning squadron, a sea-strike squadron, a helicopter anti-submarine squadron, and a small detachment of logistics aircraft

carry ['kæri] *verb* **(a)** to hold; ***he was carrying a gun*** **(b)** to contain and transport goods; ***the trucks were carrying ammunition*** **(c)** to have in your possession; ***he was carrying drugs*** **(d)** to be infected with but not necessarily affected by a transmissible disease; ***he was***

carrying Hepatitis B **(e)** to bear the weight of; ***this bridge can carry tanks*** **(f)** *(weapons, radio signals, etc.)* to reach; ***this gun can carry up to five kilometers*** **(g)** to capture; ***they carried the enemy position***

carry on ['kæri 'ɒn] *verb* to continue doing something; ***he told the men to carry on with their tasks***; ***permission to carry on, Sir?***

carry out [ˌkæri 'aʊt] *verb* **(a)** to do something; ***he was carrying out a routine check*** **(b)** to complete an activity; ***we couldn't carry out our mission***

CAS [ˌsiː eɪ 'es] = CLOSE AIR SUPPORT

cart [kɑːt] *noun* wheeled vehicle pulled by a horse or other animal

carton ['kɑːtn] *noun* container made of cardboard or paper

cartridge ['kɑːtrɪdʒ] *noun* metal or plastic case containing the propellant for a projectile (and usually the projectile as well); **cartridge-belt** = belt fitted with loops or pouches to hold cartridges (NOTE: a cartridge together with its projectile are usually known as a **round**)

cas = CASUALTY

case [keɪs] *noun* **(a)** container; **empty case** = cartridge which has been fired; **shell case** = cartridge for an artillery or tank round **(b)** situation; ***in this case, we should not attack*** **(c)** matter under investigation or study; ***he is dealing with several cases*** **(d)** type of writing; **upper case** = capital letters written as A, B, C, etc.; **lower case** = small letters written as a, b, c, etc.

caseless ammunition *or* **rounds** ['keɪsləs] *noun* latest development in small arms ammunition, where the propellant is produced as a solid block which is formed around the projectile, thus removing the need for a metal cartridge case

> COMMENT: caseless rounds are lighter in weight and cheaper to produce than normal rounds and less likely to cause a stoppage, because there is no empty case to be ejected. They would normally be issued ready-packed in a disposable magazine

CASEVAC *or* **casevac** ['kæzɪvæk] **1** *noun* = CASUALTY EVACUATION movement of an injured person to a place where he/she can receive medical treatment; ***we must arrange a casevac*** **2** *verb (informal)* to move an injured person to a place where he/she can receive medical treatment; ***he has been casevacked*** (NOTE: **casevacking - casevacked**)

casket ['kɑːskɪt] *noun US* box in which a dead body is buried or cremated (NOTE: British English is **coffin**)

castle ['kɑːsl] *noun* large fortified building or complex

casualty ['kæʒjʊəlti] *noun* someone who is killed or injured; ***the enemy suffered heavy casualties***; ***newspapers carried reports of civilian casualties***; **casualty clearing-station** = place where casualties are assessed and given emergency medical treatment, before being evacuated to a place where they can receive proper medical treatment; **casualty evacuation (casevac)** = movement of an injured person to a place where he/she can receive medical treatment

catapult ['kætəpʌlt] *noun* **(a)** apparatus for helping planes take off from the deck of an aircraft carrier **(b)** weapon made of a Y-shaped piece of metal with a rubber attached, used to send stones and other small projectiles over long distances (NOTE: American English for this is **sling shot**)

Catholic ['kæθlɪk] *see* ROMAN CATHOLIC CHURCH

cattle ['kætl] *plural noun* collective word for bulls and cows; **cattle-grid** = obstacle in road, consisting of a shallow pit covered with a metal grid, which allows vehicles to pass freely but not cattle or other livestock

causeway ['kɔːzweɪ] *noun* raised road or path across water or wet ground

caution ['kɔːʃn] *noun* attention to safety; **caution signal** = warning signal that something is not safe (NOTE: the word **caution** is often used on signs warning of danger, for example: **Caution - Snipers!**)

cautious ['kɔːʃəs] *adjective* careful in regard to possible danger; ***he is a very cautious commander***

cavalry ['kævəlri] *noun* **(a)** *(traditional term)* troops mounted on horseback; **the Household Cavalry** = the Life Guards and the Blues and Royals, the elite troops who traditionally guard the British monarch **(b)**

tanks or armoured reconnaissance troops; *US* **air cavalry** = infantry equipped with their own integral transport helicopters and attack helicopters; *US* **armored cavalry** = highly mobile armoured troops specializing in the roles of reconnaissance and advance guard

> COMMENT: with the introduction of armoured fighting vehicles, the cavalry units of most armies were converted into armoured units. In general, these units have preserved their historical association with the horse, as well as retaining the traditional cavalry roles of reconnaissance and shock action on the battlefield; "The cavalry are there to add a touch of class to what would otherwise be a vulgar brawl." Anon

cave [keɪv] **1** *noun* natural chamber in the side of a hill; ***the deserters hid in a cave*** **2** *verb* ***to cave in*** = to collapse; ***the trench has caved in; as we advanced the opposition caved in***

CB [ˌsiː 'biː] = CITIZENS BAND, CONFINED TO BARRACKS

CBW [ˌsiː ˌbiː 'dʌb(ə)ljuː] *US* = CHEMICAL AND BIOLOGICAL WARFARE

cc [ˌsiː 'siː] *abbreviation* cubic centimetres; ***this vehicle has an 1800cc engine***

CCTV [ˌsiː siː tiː 'viː] *noun* = CLOSED-CIRCUIT TELEVISION surveillance system, consisting of cameras connected by cable to television receivers

Cdo = COMMANDO

cease [siːs] *verb* to stop; **to cease fire** = to stop shooting

ceasefire ['siːsfaɪə] *noun* agreement to stop fighting; ***the enemy have agreed to a ceasefire***; *see also* ARMISTICE, TRUCE

cell [sel] *noun* **(a)** small room used to hold a prisoner; ***he was found dead in his cell*** **(b)** small group which forms part of a larger organization; ***there are several terrorist cells operating in this area*** **(c)** department of a headquarters in the field; ***the G3 cell; the NBC cell***

cellar ['selə] *noun* part of a building below ground level (normally used for storage); *see also* BASEMENT

cemetery ['semətri] *noun* area of ground used for the burial of the dead; *see also* CHURCHYARD, GRAVEYARD

censor ['sensə] **1** *noun* person or organization authorized to examine letters, newspapers, books, radio or television broadcasts, etc., and to suppress any material which is judged to be subversive, obscene, a breach of security, or otherwise unsuitable for release to the general public; ***the censor had deleted most of the letter*** **2** *verb* to act as a censor; ***all reports from the battlefield have been heavily censored***

CENTCOM ['sentkɒm] *noun* *US* = CENTRAL COMMAND department of US forces responsible for defending American national interests in the Middle East (excluding Israel, Syria and Lebanon which are covered by EUCOM), parts of East Africa and south-west Asia

centre of gravity [ˌsentər əv 'grævɪtɪ] *noun* main source of an enemy's power and strength; ***the enemy's centre of gravity is formed by his elite armoured divisions***

Centurion [sen'tjʊəriən] *noun* British-designed late 1940s-era main battle tank (MBT)

CEP [ˌsiː iː 'piː] *noun* = CIRCULAR ERROR PROBABLE area surrounding an intended target within which a ballistic missile *or* stick of bombs might land; ***this missile has a CEP of around 3000 metres***

ceramic [sə'ræmɪk] *adjective* made of clay which has been hardened by heat; ***compound armour consists of steel and ceramic layers***

ceramic armour [sɪˌræmɪk 'ɑːmə(r)] *noun* combination armour which includes ceramic materials in its composition

ceremonial [serɪ'məʊniəl] **1** *adjective* relating to formal military occasions (such as a parade); **ceremonial uniform** = dress worn for special occasions, such as parades, which is more colourful than the normal khaki uniform; ***the guards were in their full ceremonial uniforms*** **2** *noun* **(a)** procedure carried out during formal military occasions; ***he is responsible for all the ceremonial*** **(b)** **ceremonials** = ceremonial uniform; ***the battalion was in full ceremonials***

cessation [sə'seɪʃn] *noun* the stopping of an activity or state of affairs; ***the UN has demanded a cessation of hostilities***

CET [ˌsiː iː 'tiː] = COMBAT ENGINEER TRACTOR

Cfn = CRAFTSMAN

CFV [ˌsiː ef 'viː] *noun* = CAVALRY FIGHTING VEHICLE M3 variant of the Bradley infantry fighting vehicle; *compare* BFV

CG [ˌsiː 'dʒiː] **(a)** = CRUISER (with guided missiles) **(b)** *US* = COMMANDING GENERAL **(c)** = CARBONYL CHLORIDE type of choking agent (NOTE: also known as **Phosgene**)

CGN = CRUISER (nuclear-powered, with guided missiles)

CH-47 [ˌsiː eɪtʃ ˌfɔːti 'sevən] *see* CHINOOK

CH-53 [ˌsiː eɪtʃ ˌfɪfti 'θriː] *noun* American-designed heavy transport helicopter (NOTE: its variants are known as **Sea Stallion, Super Stallion;** unofficially it is known as **Jolly Green Giant**)

chaff [tʃɑːf] *noun* strips of metal foil dropped by aircraft in order to confuse enemy radar or decoy radar-guided missiles

chain-gun ['tʃeɪn ˌgʌn] *noun* machine-gun, where the firing-mechanism is powered by a motor in order to produce a high rate of fire

chagul [tʃʌ'gʊl] *noun* *Arabic* water container made of course fabric, which is designed to keep water cool by the process of evaporation

chain of command [ˌtʃeɪn əv kə'mɑːnd] *noun* command structure within a grouping; ***demoralization is evident throughout the chain of command***

chalk [tʃɔːk] *noun* **(a)** soft white limestone rock, often found under a shallow covering of soil and grass; ***it's very hard work digging into chalk*** **(b)** writing instrument produced from chalk; ***the sign was written in chalk*** **(c)** group of passengers in an aircraft (especially helicopters); ***the first three chalks must be ready to move at 1400hrs***

challenge ['tʃæləndʒ] **1** *noun* **(a)** call to identify oneself; ***he didn't hear the sentry's challenge*** **(b)** invitation to take part in a contest or combat; ***we have received a challenge to a football match from B Company*** **(c)** difficult or demanding task; ***this mission will be a challenge for us all*** **(d)** opposition; ***your remarks were a challenge to my authority*** **2** *verb* **(a)** to call upon someone to identify himself; ***we crept up to the gate and were immediately challenged by a sentry*** **(b)** to invite someone to take part in a contest or combat; ***B Company have challenged us to a football match*** **(c)** to contradict or object to something; ***he challenged his platoon commander's report***

Challenger ['tʃæləndʒə] *noun* British-designed 1980s-era main battle tank (MBT)

challenging ['tʃæləndʒɪŋ] *adjective* difficult or demanding; ***the marines faced the challenging task of climbing up a 30m vertical cliff***

chamber ['tʃeɪmbə] *noun* part of a gun in which a round is placed for firing

channel ['tʃænəl] **1** *noun* **(a)** stretch of water between two seas; **English Channel** *or* **the Channel** = stretch of water between England and France **(b)** stretch of deep water through an area of shallow water; ***there are several navigable channels in the estuary*** **(c)** natural or man-made ditch or watercourse; **drainage channel** = ditch designed to remove surplus water; **storm channel** = ditch designed to receive water produced by seasonal rainstorms **(d)** band of radio frequencies; ***you are on the wrong channel*** **(e)** way in which information is passed from one place to another; **channels of communication** = ways of communicating; **official channels** = official ways of passing information; ***the complaint was sent to the ambassador by official channels*** **2** *verb* to make something move in a specific direction; ***the minefields will channel the enemy into our killing-zone*** (NOTE: channelling - channelled but American spelling is **channeling - channeled**)

chapel ['tʃæpl] *noun* small building (normally without a spire or tower) used for religious worship by Christians

chaplain ['tʃæplɪn] *noun* person authorized to lead religious worship; *see also* PADRE

character ['kærəktə] *noun* **(a)** personality; ***he has a strong character*** **(b)** letter, number or symbol; ***the message consists of one hundred and seventeen characters***

charge [tʃɑːdʒ] **1** *noun* **(a)** rapid and aggressive movement towards the enemy; ***our charge was stopped by a minefield***; **bayonet charge** = charge with the intention of using the bayonet; ***the colonel ordered a bayonet charge*** **(b)** official accusation of a crime or offence; ***you are on a charge of insubordination*** **(c)** measured quantity of propellant used to fire a projectile; ***the charges and shells are stored separately***; **bag charge** = fabric bag containing propellant for an artillery or tank round **(d)** explosive device; ***the engineers placed several charges on the bridge*** **(e)** electrical energy stored in a battery; ***none of these batteries have any charge left*** **2** *verb* **(a)** to move quickly and aggressively towards the enemy; ***the squad charged at the bunker***; ***we charged across the field towards the enemy positions*** **(b)** to make an official accusation against someone; ***you are charged with desertion*** **(c)** to put electrical energy into a battery or other device; ***he is charging radio batteries*** **(d) to charge bayonets** = to level the bayonet at an enemy prior to charging at him

Charlie ['tʃɑːli] third letter of the phonetic alphabet (Cc)

chart [tʃɑːt] *noun* map of an area of water (such as sea, river or lake)

chassis ['ʃæsɪ] *noun* base-frame of a vehicle; ***the Scorpion's chassis has been used for several other types of armoured vehicle***

check [tʃek] **1** *noun* examination to establish the accuracy, amount, condition or identity of something; ***he ordered a check of all the buildings***; **head check** = check to establish that everyone is present; ***he took a quick head check and found two men missing***; **radio check** = radio transmission to ensure that the radios are working and on the correct frequency **2** *verb* **(a)** to examine something in order to establish its condition; ***check your weapons***; ***he went round to check the sentries*** **(b)** to count; ***he checked his platoon*** **(c)** to look for; ***he checked the room for booby-traps*** **(d)** to stop doing something; **to check firing** = to stop firing **3** *adverb (informal)* **check!** = OK!, all right!

check-list ['tʃeklɪst] *noun* **(a)** list of things to be checked **(b)** list of tasks to be completed

check-point *or* **checkpoint** ['tʃekpɔɪnt] *noun* **(a)** place (usually on a road) where people or vehicles are stopped and inspected or searched **(b)** place or feature on the ground which is used as a navigational reference point; ***our next check-point is the track junction at grid 339648***

chemical ['kemɪkl] **1** *adjective* referring to chemistry; **chemical agent** = type of chemical weapon; **chemical attack** = attack using chemical weapons; **chemical shell** = artillery round used as a means of delivering a chemical agent; **chemical warfare** = warfare involving the use of chemical weapons; **chemical-warfare unit** = specialist unit trained to detect the presence of chemical weapons and to decontaminate persons, equipment and vehicles which have been affected *US*; **chemical and biological warfare (CBW)** = warfare using both chemical and biological weapons; **chemical warhead** = explosive part of a missile used as a means of delivering a chemical agent; **chemical weapon** = chemical substance used as a weapon (such as poisonous gas) **2** *noun* **(a)** substance formed by reactions between elements, obtained by or used in chemical processes; ***these are some of the most dangerous chemicals used in warfare*** **(b)** chemical weapon; ***the enemy are using chemicals***; *see also* AGENT, GAS, NBC

> COMMENT: during the Gulf War of 1990/91, chemical shells (believed to be of Soviet origin) found in Iraqi ammunition dumps were painted grey

chevron ['ʃevrən] *noun* V-shaped stripe worn on the sleeve and used to denote non-commissioned officer rank (NOTE: also simply called **stripes**)

COMMENT: in the British Army, as a general rule, a single chevron denotes a lance corporal, two chevrons a corporal, three chevrons a sergeant, and three chevrons surmounted by a crown a colour sergeant or staff sergeant

chief [tʃiːf] **1** *adjective* the most senior or important; ***he is the chief adviser*** GB & US; **chief petty officer (CPO)** = senior non-commissioned officer (SNCO) in the navy US; **chief master sergeant** = senior non-commissioned officer (SNCO) in the air force; **Commander in Chief (C-in-C)** = the most senior commander **2** *noun* **(a)** ruler of a tribal group or clan **(b)** head of a department; **chief of staff (COS)** = the most senior staff officer in a headquarters; **deputy chief of staff (DCOS)** = the second most senior staff officer in a headquarters (NOTE: in the British Army, the chief of staff of a brigade is referred to as the **Brigade Major**)

Chieftain ['tʃiːftən] *noun* British-designed 1960s-era main battle tank (MBT)

COMMENT: in a headquarters, the chief of staff (COS) coordinates operational matters, while the deputy chief of staff (DCOS) is responsible for logistics

chigger ['tʃɪgə(r)] *noun* tiny tropical insect, which burrows into an animal's skin (especially on the feet) in order to lay its eggs; *also written as* JIGGER

Chinagraph ['tʃaɪnəˌgrɑːf] *noun* wax pencil designed for writing on plastic (such as a map-case, overlay, talc, etc.)

Chinook [tʃɪ'nuːk] *noun* American twin-rotor CH-47 transport helicopter

chock [tʃɒk] *noun* metal or wooden block designed to stop a wheel moving

Choghi *or* **choggie** ['tʃɒgɪ] *noun* GB person of Asian origin, who runs a canteen *or* other shop on a British operational base *or* warship

COMMENT: the word Choghi is a legacy from the British Army in India, and is not supposed to be a derogatory term

choke [tʃəʊk] *verb* **(a)** to obstruct a person's airway so that he cannot breathe; ***I choked him with my belt*** **(b)** to be unable to breathe because of an obstruction to the airway; ***he is choking*** **(c)** to block a passage; ***the road was choked by refugees***

choke-point ['tʃəʊk ˌpɔɪnt] *noun* natural or man-made feature which restricts the movement of large numbers of people or vehicles; ***the valley is a potential choke-point for enemy armour***

choking agent ['tʃəʊkɪŋ ˌeɪdʒənt] *noun* chemical weapon designed to attack the lungs

cholera ['kɒlərə] *noun* a highly infectious disease of the intestine which causes vomiting and diarrhoea and is often fatal; it is caused by food and water infected by *Vibrio cholerae*

COMMENT: cholera frequently occurs during wartime due to a breakdown in sanitation

chopper ['tʃɒpə] *noun (informal)* helicopter

choppy ['tʃɒpi] *adjective (of water)* rough

chow [tʃaʊ] *noun* US *(slang)* food

chronometer [krə'nɒmɪtə] *noun* instrument used for measuring time

chuck [tʃʌk] *noun (informal)* to throw; ***he chucked a grenade into the dugout***

church [tʃɜːtʃ] *noun* **(a)** large building (usually with a tower or spire) used for religious worship by Christians **(b)** organized religious group of Christians (for example the Catholic Church, the Greek Orthodox Church)

churchyard ['tʃɜːtʃjɑːd] *noun* enclosed ground around a church used as a place to bury the dead; *see also* CEMETERY, GRAVEYARD

CIA [ˌsiː aɪ 'eɪ] *noun* = CENTRAL INTELLIGENCE AGENCY American secret service

CIC [ˌsiː aɪ 'siː] *noun* US = COMBAT INFORMATION CENTRE ops Room of an American warship; ***he's in the CIC***

CINC [ˌsiː ɪn 'siː *or* sɪŋk] = US COMMANDER IN CHIEF

C-in-C [ˌsiː ɪn 'siː] = COMMANDER IN CHIEF

cipher ['saɪfə] *noun* system of words, letters, numbers or other symbols, which is used to write secret messages; *see also* CODE

cirque [sɜːk] *noun* natural depression on the side of a mountain; *see also* CORRIE

city ['sɪti] *noun* large urban area

civil ['sɪvəl] *adjective* **(a)** relating to the ordinary citizens of a country; **the civil population** = the ordinary citizens of a country; **civil unrest** = breakdown of law and order, usually involving physical violence; **civil war** = war fought between groups of citizens of the same country **(b)** non-military; **civil defence** = organization and training of non-military personnel for the protection of life and property during wartime; **civil emergency planning** = defence planning by a government, not involving the Ministry of Defence (ie the fire service, police force, health services, civil defence, etc.)

civilian [sɪ'vɪljən] **1** *adjective* non-military; ***the enemy aimed at civilian targets***; ***newspapers reported many civilian casualties*** **2** *noun* someone who is not a member of the armed forces; ***hundreds of civilians were killed in the air raid***; *see also* CIVVY

civvy ['sɪvi] *GB (slang)* **1** *adjective* civilian; **Civvy Street** = civilian life; ***what did you do in Civvy Street?*** **2** *noun* **(a)** civilian; ***he shot a civvy*** **(b) civvies** = civilian clothing; ***he was in civvies***

CIWS [ˌsiː aɪ dʌb(ə)ljuː 'es] *noun* = CLOSE-IN WEAPONS SYSTEM radar-controlled naval anti-aircraft cannon, which automatically detects, tracks and engages targets (eg. Goalkeeper, Phalanx)

CJTF = COMBINED JOINT TASK FORCE

CK [ˌsiː 'keɪ] *noun* = CYANOGEN CHLORIDE type of blood agent

clan [klæn] *noun* group of families who share a common heritage and often, a strong sense of group identity

clap [klæp] *noun (slang)* gonorrhoea (a venereal disease)

clash [klæʃ] *noun* small engagement (usually indecisive); ***clashes reported between border security guards***

classification [klæsɪfɪ'keɪʃn] *noun* way of organizing things into categories; ***he is responsible for the classification of documents***

> COMMENT: the security classification of information varies according to its importance, eg: restricted, confidential, secret, top secret, etc.

classified ['klæsɪfaɪd] *adjective* secret; ***that information is classified***

classify ['klæsɪfaɪ] *verb* **(a)** to designate into classes or groups **(b)** to designate as secret; ***the report has been classified as top secret***

clay [kleɪ] *noun* wet, sticky type of soil, which can be used as a raw material for the manufacture of ceramics

Claymore ['kleɪmɔː] *noun* American anti-personnel device designed to fire a quantity of ball-bearings in a specific direction

> COMMENT: a Claymore can be initiated electronically or by means of a trip-wire

clear [klɪə] **1** *adjective* **(a)** free from obstructions; ***the road ahead is clear*** **(b)** free from hazards (such as chemical contamination, enemy troops, explosive devices, etc.); ***the area is clear of mines*** **(c)** unloaded; ***the weapon is clear*** **(d)** not close to; ***stand clear, please*** **(e)** *(of weather)* not cloudy or foggy; ***it's a clear day*** **(f)** easy to understand; ***is that clear?*** **2** *noun* uncoded radio transmission; ***he sent the message in clear*** **3** *verb* **(a)** to remove an obstruction; ***the road has been cleared*** **(b)** to remove a hazard; ***we cleared the enemy position*** **(c)** to unload a weapon; ***they cleared their rifles*** **(d)** to approve or authorize (where security is involved); ***he has not been cleared to read this document***

clearance ['klɪərəns] *noun* **(a)** act of clearing something; ***B Company is on route clearance*** **(b)** security approval; ***you do not have clearance to enter***

clearing ['klɪərɪŋ] *noun* **(a)** small area of ground in woodland where the trees have been removed; ***we came to a clearing*** **(b)** action of removing something; ***the clearing of the village took several hours***;

clearing patrol = patrol sent out from a patrol base or defensive position in order to check the surrounding area for enemy **(c)** authorization *or* permission; ***Range Control has given us clearance to start firing***

cleared hot [ˌklɪəd ˈhɒt] *adverb* permission for an aircraft to drop a bomb on a practice range; ***Foxhound 22, this is Range Control, you're cleared hot, over.***

cleared live [ˌklɪəd ˈlaɪv] *adverb* permission for an aircraft to drop a bomb on a real enemy target; ***Foxhound 22, this is Merlin, you're cleared live, over.***

clerk [klɑːk *US* klɜːrk] *noun* serviceman who carries out secretarial duties in a headquarters

click [klɪk] *noun* **(a)** short sharp noise made by a switch or lock; ***there was a loud click as he released his safety-catch*** **(b)** *US (informal)* kilometre; ***the bridge is three clicks down that road***

cliff [klɪf] *noun* steep wall of rock

climate [ˈklaɪmət] *noun* prevailing weather conditions of a region

climb [klaɪm] **1** *noun* upward movement **2** *verb* **(a)** *(person)* to move upwards (especially using hands and feet); ***the commandos had to climb a 50ft cliff*** **(b)** to move upwards; ***the planes climbed to 30,000ft***

clink [klɪŋk] *noun* noise made by something hitting metal; ***the sentry heard a clink***

clip [klɪp] **1** *noun* **(a)** spring-fitted device used to attach one object to another object, or to hold objects together **(b)** several rounds of ammunition held together by a clip for easy loading; ***he fired a whole clip at the man*** **2** *verb* to attach using a clip; ***he clipped on a new magazine*** (NOTE: **clipping - clipped**)

clock code [ˈklɒk ˌkəʊd] *noun* system used to indicate other aircraft in relation to your own aircraft; 12 o'clock is straight ahead; 6 o'clock is directly behind you; 3 o'clock is to your right; 9 o'clock is to your left.; ***enemy fighters at 3 o'clock !***

clog [klɒg] *verb* **(a)** to obstruct a mechanism or pipe with something; ***the fuel pipe was clogged with dirt*** **(b)** to obstruct a route; ***the road was clogged with refugees*** (NOTE: clogging - clogged)

close air support (CAS) [kləʊs ˈeə səˌpɔːt] *noun* attack by aircraft on a target which is close to friendly ground forces

> COMMENT: during the UN peacekeeping operation in Bosnia (1992-95), *close air support* meant limited attacks on individual positions *or* vehicles, which were actually firing at UN peacekeepers, while **air strikes** meant retaliatory attacks on multiple targets within a specified area. The difference between these two definitions was crucial during negotiations between the UNand the warring factions throughout this conflict.

cloud [klaʊd] *noun* visible mass of water, ice, gas, dust or other particles in the sky; **cloud cover** = area of sky covered by cloud; **mushroom cloud** = mushroom-shaped cloud of smoke and dust produced by the explosion of a nuclear weapon

clump [klʌmp] *noun* small group of trees *or* bushes; ***I saw something moving in that clump on the left***

cluster bomb [ˈklʌstə ˌbɒm] *noun* aircraft-dropped device containing a quantity of small bombs or bomblets which are released in mid-air over a target area

CMA [ˌsiː em ˈeɪ] = CONVOY MARSHALLING AREA

Cmd = COMMAND

Cmdr = COMMANDER (NAVY)

Cmdre = COMMODORE

CN [ˌsiː ˈen] *noun* = CHLOROACETOPHENONE type of tear agent

CO [ˌsiː ˈəʊ] *noun* = COMMANDING OFFICER officer commanding a battalion or equivalent-sized grouping; ***he was taken before the CO on a charge of being absent without leave***

coalition [kəʊəˈlɪʃn] *noun* temporary alliance formed as a result of an agreement rather than a formal treaty

coast [kəʊst] *noun* area of land where it meets the sea; ***the squadron sailed along the coast***; ***they planned a landing on the Normandy coast***

coastal ['kəʊstl] *adjective* relating to the coast; ***their coastal defences proved to be inadequate***

coastguard ['kəʊstgɑːd] *noun* **(a)** government organization responsible for the safety of shipping in coastal waters and the prevention of smuggling; ***the US Coastguard is on the lookout for drug smugglers*** **(b)** member of the coastguard; ***three coastguards boarded the ship***

coax ['kəʊæks] = COAXIAL MACHINE-GUN

coaxial [kəʊ'æksiəl] *adjective* having the same axis; **coaxial machine-gun** = machine-gun which is mounted alongside the main gun of an armoured fighting vehicle (AFV) and which shares its sighting systems

cobbled ['kɒbld] *adjective (of roads)* paved with an uneven surface of rounded stones

cobbler ['kɒblə(r)] *noun* person who repairs boots and shoes

cobbles *or* **cobblestones** ['kɒblz] *noun* rounded stones used to pave roads

Cobra ['kəʊbrə] *see* AH-1, HUEY COBRA

cock [kɒk] *verb* to pull back the firing mechanism of a firearm so that it is ready to fire; *see also* MAKE READY, LOCK AND LOAD

cocked [kɒkt] *adjective; (of firearms)* with the firing mechanism pulled back and ready to fire

cockpit ['kɒkpɪt] *noun* pilot's compartment in an aircraft; ***the aircraft crashed because the cockpit canopy had iced up***

cock-up ['kɒk ʌp] *noun GB (slang)* situation where everything is going wrong; ***this is turning into a right cock-up !***

code [kəʊd] *noun* **(a)** system of words, letters, numbers or other symbols, which is used to write secret messages; **codeword** = word or words which are used to convey a meaning, such as arrival at a destination, capture of an objective, order to withdraw, etc.; *see also* CIPHER **(b) code of conduct** *or* **code of honour** = the correct way to behave (honourably)

coded ['kəʊdɪd] *adjective* written in code; ***he received a coded message***

codename ['kəʊdneɪm] *noun* name which, for security purposes, is used instead of a real name; ***your codename will be Foxhound***

coerce [kəʊ'ɜːs] *verb* to persuade an unwilling person to do something by using force or threats; ***he was coerced into helping the soldiers***

coercion [kəʊ'ɜːʃn] *noun* use of force or threats to persuade an unwilling person to do something

coffin ['kɒfɪn] *noun GB* box in which a dead body is buried or cremated (NOTE: American English is **casket**)

CoH = CORPORAL OF HORSE

cohesion [kəʊ'hiːʒn] *noun* state of being organized and working together; ***the enemy's cohesion is starting to collapse***

COIN [kɔɪn] = COUNTER-INSURGENCY

Col = COLONEL

col [kɒl] *noun* high mountain pass

Cold War [ˌkəʊld 'wɔː] *noun* period between 1945 and 1989, when a state of near-hostility existed between the USA and its Western European allies (later NATO) on one side and the USSR and its Eastern European allies (later the Warsaw Pact) on the other

collaborate [kə'læbəreɪt] *verb* to assist the enemy

collaborator [kə'læbəreɪtə] *noun* someone who provides assistance to the enemy

collapse [kə'læps] **1** *noun* **(a)** falling down; ***the explosion caused the collapse of the building*** **(b)** loss of cohesion; ***the collapse of the enemy was due to a failure in logistics*** **(c)** failure; ***we were forced to watch the collapse of the whole plan*** **(d)** mental or physical breakdown; ***he suffered a collapse*** **2** *verb* **(a)** to fall down; ***the explosion caused the building to collapse*** **(b)** to lose cohesion; ***enemy resistance collapsed as soon as the allies entered the town*** **(c)** to fail; ***the plan collapsed as a result of poor planning*** **(d)** to suffer a mental or physical breakdown; ***he collapsed from exhaustion***

collapsible [kə'læpsəbl] *adjective* designed to fold up or be taken to pieces and then reassembled; ***we used a collapsible boat***

collate [kə'leɪt] *verb* to gather and analyze information

collateral damage [kəˌlætərl 'dæmɪdʒ] *noun* unintentional killing of civilians or destruction of civilian property as a result of military action; ***the Pentagon admitted that the bombing raids had caused some collateral damage***

collect [kə'lekt] *verb* to fetch or pick up

collection [kə'lekʃn] *noun* act of fetching or picking something up; **collection point** = place where personnel, equipment or supplies can be delivered and picked up

collide [kə'laɪd] *verb* to hit another object while moving; ***the helicopters collided in mid-air***

collision [kə'lɪʒn] *noun* act of colliding; ***a mid-air collision***

collocate ['kɒləkeɪt] *verb* to put in the same place; ***B Company was collocated with Battalion HQ***

colonel (Col) ['kɜːnl] *noun* **(a)** *GB* officer in the army or marines (ranking above a lieutenant-colonel and below a brigadier, usually employed as a senior staff officer); *see also* LIEUTENANT-COLONEL **(b)** *US* officer in the army, marines or air force (usually in command of a regiment or equivalent-sized grouping or employed as a senior staff officer) **(c)** *GB* **Colonel of the Regiment** = honorary position (usually held by a member of the Royal Family, a field marshal or general) **(d)** *GB* **half-colonel** = lieutenant-colonel

> COMMENT: in some regiments of the British Army, the lieutenant-colonel commanding a battalion or its equivalent is addressed as 'Colonel' and referred to as 'the colonel'

colonial [kə'ləʊniəl] *adjective* relating to colonies

colonist ['kɒlənɪst] *noun* settler in a colony

colony ['kɒləni] *noun* territory which is governed and exploited by a foreign power

colour *US* **color** ['kʌlə] *noun* ceremonial flag of a unit or sub-unit; **Trooping the Colour** = ceremonial parade where a unit's colours are displayed to the troops

> COMMENT: if you are watching a ceremonial parade in uniform, you should salute when the colours march past in front of you. If you are not in uniform, you should stand to attention, removing your hat if you are a man. If the parade includes guns from the artillery, you should salute the guns as they go past, since the guns are also the colours of the artillery.

colour sergeant (C/Sgt) ['kʌlə ˌsɑːdʒənt] *noun* *GB* senior non-commissioned officer (SNCO) in the infantry, usually employed as a company quartermaster sergeant (CQMS) (NOTE: the equivalent of colour sergeant in most other branches of the British Army is **staff sergeant (S/Sgt)**)

column ['kɒləm] *noun* **(a)** tactical formation consisting of several files moving forward together; ***two columns of infantry advanced across the desert*** **(b)** troops or vehicles moving in column formation; ***the column of tanks was strafed by enemy aircraft***

Comanche [kə'mæntʃi] *noun* American RH-66 light attack/reconnaissance helicopter

combat ['kɒmbæt] **1** *noun* fighting with the enemy; ***he has no experience of combat***; ***the unit was in combat three times***; **combat air patrol (CAP)** = patrol by fighter aircraft over a designated area; **combat-effective** = capable of fighting; ***only three of our tanks are still combat-effective***; **combat engineer tractor (CET)** = British designed armoured bulldozer; **combat fatigue** = mental and physical stress resulting from a long period in combat; **combat-loading** = loading ships with men and equipment in such a way that they are literally ready to fight the moment they disembark; **combat readiness** = degree to which a unit or sub-unit is considered capable of fighting effectively; ***the brigade is now at an advanced state of combat readiness***; **combat supplies** = ammunition, fuel and water; **combat vehicle reconnaissance (CVR)** light, fast-moving armoured vehicle designed for reconnaissance; **combat vehicle**

reconnaissance tracked (CVRT) = British series of light armoured reconnaissance vehicles (including the Scimitar and Scorpion light tanks); *see also* UNARMED **2** *verb* to take effective action against something; ***this oil will combat rust and corrosion***

combatant ['kɒmbətənt] *noun* person who is involved in fighting; ***combatants on both sides were affected by the chemical attack*** (NOTE: the opposite is **noncombatant**)

combination armour [kɒmbɪ'neɪʃn 'ɑːmə] *noun* armour composed of layers of steel and other substances (such as ceramics, plastics, other types of metal, etc.) (NOTE: also known as **composite armour** *or* **compound armour**)

combine [kəm'baɪn] *verb* **(a)** to bring or put together; ***the two battalions have been combined***; **combined joint task force (CJTF)** = multinational task force which can be used rapidly as a peacekeeping force; **combined logistic support** = support from various countries which is available for use by NATO; **combined operations (combined ops)** = **(i)** operations involving more than one arm (eg. aircraft, artillery, infantry, naval gunfire support, etc.); **(ii)** *US* operations carried out in conjunction with the armed forces of other states (NOTE: the Americans refer to combined arms operations as joint operations) **(b)** to come together; ***we must not allow the two enemy forces to combine***

Combo pen ['kɒmbəʊ ˌpen] *noun* automatic syrette of atropine

COMCEN ['kɒmsen] = COMMUNICATIONS CENTRE

Comd = COMMANDER

come-on ['kʌm ˌɒn] *noun* action designed to lure someone into an ambush or trap; ***the burning car was a come-on for a large bomb***

command [kə'mɑːnd] **1** *noun* **(a)** official instruction to do something; ***he gave the command to open fire*** **(b)** management and direction of troops, vehicles or equipment; ***he has taken command of B Company***; **command post (CP)** = place from which a unit or sub-unit is commanded; *US* **command sergeant major (CSM)** = most senior non-commissioned officer in an army unit; **command vehicle** = vehicle used as a command post; **chain of command** = command structure within a grouping; **in command** = holding a command; ***he is currently in command of D Troop***; **second in command (2IC)** = most senior person after the commander, nominated to take command in his absence **(c)** organization which manages and directs military forces at strategic level; **High Command** = senior command organization within a country's armed forces; **unified command** = placing all your military assets under one overall commander (commander in chief) and his headquarters **(d)** strategical grouping of armed forces (for example Bomber Command) **(e)** region or district under the command of a senior officer (for example Southern Command) **2** *verb* **(a)** to order someone to do something; ***I command you to arrest that man*** **(b)** to manage and direct troops, vehicles or equipment; ***he commands C Company*** **(c)** *(of ground)* to look down on; ***that hill commands the whole valley***

commandant [kɒmən'dænt] *noun* **(a)** officer commanding a military establishment (such as a prison camp, training depot, etc.) **(b) officer in the Irish army above the rank of captain and below a lieutenant-colonel (equivalent of a major in the British or US army)**

commandeer [kɒmən'dɪə] *verb* to take possession of something in order to use it for a military purpose; ***they commandeered our car***

commander [kə'mɑːndə] *noun* **(a)** someone who commands (NOTE: shortened to **Comd** in this meaning); **commander in chief (C-in-C)** = most senior commander; **platoon commander** = commander of a platoon; **component commander** = commander of one component of a combined arms force (eg. the air component commander coordinates the use of all aircraft in the force, regardless of whether they are provided by the air force *or* navy *or* army); **service commander** = most senior commander from one particular service (eg. army *or* navy *or* air force) in a combined arms force **(b)** *GB & US* rank of an officer in the navy (sometimes in command of a small warship) (NOTE: shortened to **Cmdr** in this meaning); *see also* LIEUTENANT-COMMANDER, WING COMMANDER

> COMMENT: under unified command, the service commander is responsible for the day-to-day management of the forces under his command, while the component commander decides how such forces will be used in a combined operation. Inevitably, differences of opinion will arise, in which case, it is the job of the commander in chief to mediate or make the final decision

commanding [kə'mɑːndɪŋ] *adjective* **(a)** holding a command; *US* **commanding general (CG)** = commander of a large tactical grouping (eg division, corps, army); **commanding officer (CO)** = officer commanding a battalion or equivalent-sized grouping; ***he was taken before the commanding officer on a charge of being absent without leave***; **officer commanding (OC)** = officer who commands a unit or sub-unit; *GB* **air officer commanding (AOC)** = commander of a large air-force grouping; *GB* **general officer commanding (GOC)** = commander of a large army grouping (usually a division) **(b)** *(of ground)* looking down on something; ***this position has a commanding view over the valley***

> COMMENT: in the British Army, the title **commanding officer (CO)** is only applied to an officer who commands a battalion or equivalent-sized grouping. Likewise, the title **officer commanding (OC)** is only applied to an officer who commands a company or equivalent-sized grouping

commando [kə'mɑːndəʊ] *noun* **(a)** *GB* battalion-sized grouping of the Royal Marines (such as 40 Commando, 45 Commando); ***40 Commando will lead the assault*** (NOTE: shortened to **Cdo** in this meaning) **(b)** *GB* member of the Royal Marines who has successfully completed basic training; ***he wants to be a commando*** **(c)** special forces unit or a unit of irregular troops; ***an enemy commando is operating in the area*** **(d)** a member of a special forces unit or a unit of irregular troops; ***the base was attacked by commandos***

commend [kə'mend] *verb* to praise (usually officially) an achievement; ***he was commended for his bravery***

commendation [kɒmen'deɪʃn] *noun* official recognition for an achievement; ***he received a commendation for leading the counter-attack***

Commie ['kɒmi] *noun (informal)* communist

commissar [kɒmɪ'sɑː] *noun* communist official responsible for political education and organization

commissariat [kɒmɪ'seəriət] *noun* official department responsible for the supply of food, clothing, etc.

commissary ['kɒmɪsəri] *noun* officer responsible for supply of food, clothing, etc.

commission [kə'mɪʃn] **1** *noun* authority by which an officer holds his rank in the armed forces; ***he resigned his commission;*** *GB* **quartermaster commission** = commission held by an officer who has been promoted from the ranks, instead of undergoing normal officer selection and training; *GB* **Queen's Commission** *or* **King's Commission** = commission held by an officer who has undergone normal officer selection and training **2** *verb* **(a)** to appoint someone as an officer; ***he was commissioned in 1980***; **non-commissioned officer (NCO)** = serviceman holding a supervisory rank which is not authorized by a commission (for example a lance corporal, corporal, sergeant, etc.) **(b)** to prepare a ship for operational duty; ***the ship was in action only two weeks after she was commissioned***; *compare* DECOMMISSION

commissioned officer [kəˌmɪʃnd 'ɒfɪsə] *noun* serviceman with a supervisory rank, who derives his authority from a commission (for example, a lieutenant, captain, major, etc.) (NOTE: a commissioned officer is normally referred to simply as an **officer**)

commissioning [kə'mɪʃənɪŋ] *noun* taking a ship into operational service in the navy; ***the ship was in action only two weeks after commissioning***; *compare* DECOMMISSIONING

commodore (Cmdre) ['kɒmədɔː] *noun GB & US* senior officer in the navy (usually in command of a naval squadron)

> COMMENT: the rank of commodore is temporary only, and is given to a captain when his job requires a greater degree of authority. When he finishes that job, he reverts to the rank of captain

commonality [kɒmə'næləti] *noun* state where various groups use common resources or have common aims

Commonwealth ['kɒmənwelθ] *noun* association consisting of the Great Britain and independent sovereign states which were once formerly ruled by Britain as colonies (such as Australia, Canada, New Zealand) (NOTE: the full title is: **the British Commonwealth of Nations**)

> COMMENT: the armed forces of many Commonwealth members are still modelled on those of Great Britain and have retained many of their traditions and customs)

comms [kɒmz] = COMMUNICATIONS

communal [kə'mjuːnəl] *adjective* for use by everyone; ***we have communal showers in this camp*** (NOTE: **communal** might refer to facilities that can be used by all ranks, or alternatively by both sexes)

communicate [kə'mjuːnɪkeɪt] *verb* to pass information to another person

communication [kəmjuːnɪ'keɪʃn] *noun* **(a)** the act of passing information to another person **(b)** ability to communicate; ***we have lost communication with B Company***; **lines of communication** = main roads used by an army to resupply its units, and along which its supply depots and reserve forces are located; ***the enemy's lines of communication have been cut*** **(c)** message; ***did you receive my communication?*** **(d) communications** = the means of passing information; ***our communications have broken down***; **to establish communications** = to carry out a radio check in order to ensure that all call-signs on the net are in radio contact

communism ['kɒmjunɪzm] *noun* **(a)** political theory developed by Karl Marx, which promotes the idea that all people should be considered equal and that all property should be owned by the state **(b)** political system based on the ideas of Karl Marx and others **(c)** any movement which favours communism; *compare* CAPITALISM

communist ['kɒmjʊnɪst] **1** *adjective* **(a)** relating to communism; **Communist Party** = international organization (with official status in some countries) which promotes communism **(b)** favouring communism **(c)** relating to a country with a communist government **2** *noun* **(a)** person who favours communism **(b) a Communist** = a member of the Communist Party

company (Coy) ['kʌmpni] *noun* tactical and administrative army grouping of three or more platoons; **company and squadron group** = combined arms grouping, based on an infantry company (equivalent of a company team in the US Army); **company quartermaster sergeant (CQMS)** = senior non-commissioned officer (SNCO), usually holding the rank of colour sergeant (C/Sgt) or staff sergeant (S/Sgt), responsible for the logistic support of a company; **company team** = US combined arms grouping based on a tank or mechanized infantry company (equivalent of a company and squadron group or a squadron and company group in the British Army); **squadron and company group** = combined-arms grouping based on a tank squadron (equivalent of a company team in the US Army)

> COMMENT: in the British army, company-sized groupings of tanks and certain supporting arms (for example engineers) are known as **squadrons,** while artillery companies in many armies, including Great Britain and the USA, are known as **batteries.** American armored cavalry companies are known as **troops,** although normal armored units use the term **company.** Companies or equivalent-sized groupings are usually commanded by majors (although companies in the Royal

Marines are commanded by captains). In the US Army, companies or equivalent-sized groupings are usually commanded by captains. A British armoured brigade might consist of two armoured or mechanized infantry battalions and one armoured regiment or alternatively, two armoured regiments and one infantry battalion, plus artillery and supporting arms. On operations, these units are broken down and combined into **battle groups.** As an example, an armoured infantry battle group might consist of two infantry companies and one squadron of tanks, which are organized into two **company and squadron groups** and a **squadron and company group** under the command of the infantry battalion HQ. The exact composition will vary according to the tactical requirement at the time In the US Army, a battle group is known as a task force, while company and squadron groups and squadron and company groups are known as company teams

compass ['kʌmpəs] *noun* instrument designed to calculate direction by indicating magnetic north; **compass bearing** = magnetic bearing obtained by using a compass; **Silva™ compass** = compass which is designed to be placed onto a map in order to calculate bearings (without the need for a protractor)

Compass Call ['kʌmpəs ˌkɔːl] *noun* *US* air-force role, involving the use of EW aircraft to jam enemy communications

compassionate leave [kəm'pæʃənət ˌliːv] *noun* leave granted when a serviceman has problems at home (such as the death of a relative)

compatibility [kəmpætə'bɪləti] *noun* being able to fit in or work with other types of equipment

compatible [kəm'pætəbl] *adjective* able to fit in or work with other types of equipment

compatriot [kɒm'pætriət] *noun* someone of the same nationality

comply (with) [kəm'plaɪ ˌwɪð] *verb* to carry out an instruction, order or request; ***the troops refused to comply with the order to withdraw***

compo ['kɒmpəʊ] *noun* *GB* *(informal)* *(short for 'composite rations')* tinned or dehydrated food supplied to the British Army

component [kəm'pəʊnənt] *noun* **(a)** part of something (especially machinery *or* instruments); ***we'll need to replace many of the components of this radio*** **(b)** part of a combined arms force; **air component** = all aircraft, regardless of whether they are from the air force *or* navy *or* army; **ground** *or* **land component** = all ground forces, including marines (which are actually part of the navy)

composite ['kɒmpəzɪt] *adjective* made up of several parts; **composite armour;** *see* COMBINATION ARMOUR; **composite rations;** *see* COMPO

compound ['kɒmpaʊnd] *noun* **(a)** secure area enclosed by a fence; **ammunition compound** = place where ammunition is stored **(b)** mixture of two or more substances; **compound armour;** *see* COMBINATION ARMOUR

compromise ['kɒmprəmaɪz] *verb* **(a)** to settle a dispute by agreeing to accept some of the other party's demands; ***we will have to compromise on this issue*** **(b)** to reveal or to allow your intentions, location or secrets to become known (usually unintentionally); ***the ambush has been compromised***; ***our security codes have been compromised***

computer [kəm'pjuːtə] *noun* electronic device used for storing and processing data

comrade ['kɒmreɪd] *noun* fellow soldier, worker, etc.

COMMENT: **Comrade** was a common form of address among communists, and was often used to express the idea of social equality among people of different rank or status (for example addressing someone as 'Comrade General')

conceal [kən'siːl] *verb* **(a)** to hide something; ***the bomb was concealed in a suitcase*** **(b)** to keep something secret; ***we need to conceal our intentions from the enemy***

concealment [kən'siːlmənt] *noun* act of concealing something

concentrate ['kɒnsəntreɪt] *verb* **(a)** to bring together; ***he concentrated his forces for the attack*** **(b)** to come together; ***the division will concentrate around Bocksheim*** **(c)** to focus your attention, energy or resources on something; ***they concentrated on destroying the enemy rail network***

concentrated ['kɒnsəntreɪtɪd] *adjective* intense or strong; ***a concentrated barrage***

concentration [ˌkɒnsən'treɪʃn] *noun* **(a)** act of bringing something together; **concentration camp** = camp where people are interned in harsh conditions for political reasons or because they belong to a certain ethnic or religious group **(b)** act of coming together; **concentration area** = area where the units of a large tactical grouping (such as a brigade or division) come together to reorganize, before starting the next phase of an operation **(c)** act of focusing your attention, energy or resources on something; **concentration of firepower** = utilization of all weapons available aimed at the same target

concept of operations [ˌkɒnsept əv ɒpə'reɪʃnz] *noun* general outline of how an operation is intended to proceed

concertina wire [kɒnsə'tiːnə 'waɪə] *noun* barbed wire rolled into a series of loops, which can be compressed for storage and transportation, but easily extended for use as an obstacle

conchie ['kɒnʃi] *noun (slang)* conscientious objector

concrete ['kɒŋkriːt] *noun* building material composed of cement, gravel, sand and water; ***concrete blockhouses were built along the frontier***

concurrent [kən'kʌrənt] *noun* happening at the same time as something else

concussed [kən'kʌst] *adjective* suffering from concussion

concussion [kən'kʌʃn] *noun* **(a)** temporary incapacity caused by a blow to the head; ***he is suffering from concussion*** **(b)** shock; ***he was killed by the concussion from an exploding shell***

cone [kəʊn] *noun* shape which is round at the base, tapering to a point; **cone of fire** = shape like a triangle made when guns in two positions fire at the same target

confidential [kɒnfɪ'denʃl] *adjective* secret

> COMMENT: the security classification of information varies according to its importance eg: restricted, confidential, secret, top secret, etc.

confine [kən'faɪn] *verb* **(a)** to keep within specific limits; **confined to barracks (CB)** = punishment by which a soldier is not allowed to leave the barracks; ***he was awarded 10 days CB***; ***he was confined to barracks for 10 days*** **(b)** to imprison; ***they were confined in a barn***

confirm [kən'fɜːm] *verb* to say that something is true or correct; ***aerial photographs confirmed the enemy's movements***

confirmation [kɒnfə'meɪʃn] *noun* statement that something is true or correct; ***we need confirmation of the chemical attack***

confiscate ['kɒnfɪskeɪt] *verb* to take something away from someone, with authority; ***all privately-owned radios were confiscated***

conflict ['kɒnflɪkt] *noun* state of hostility or war; ***the whole region is in a state of conflict***; ***we are trying to settle the conflict by diplomatic means***

confluence ['kɒnfluəns] *noun* place where two rivers join; ***Koblenz lies at the confluence of the Rhine and the Mosel***

confront [kən'frʌnt] *verb* **(a)** to take a hostile attitude towards someone or something; ***we were confronted by the 7th Infantry Regiment*** **(b)** to deal with a problem or difficulty; ***we need to confront the lack of discipline in this battalion***

confrontation [kɒnfrən'teɪʃn] *noun* aggressive or hostile behaviour; ***this confrontation could lead to war***

confusion [kən'fjuːʒn] *noun* **(a)** situation where no one knows what is happening; ***the whole headquarters is in a state of confusion*** **(b)** loss of order and cohesion; ***the enemy retreated in confusion***

conifer ['kɒnɪfə] *noun* tree which does not lose its leaves in winter (such as fir, pine, spruce, etc.); *see also* EVERGREEN, FIR

coniferous [kə'nɪfərəs] *adjective* relating to trees which do not lose their leaves in winter (such as fir, pine, spruce, etc.); *compare* DECIDUOUS

conning tower ['kɒnɪŋ ˌtaʊwə] *noun* vertical construction on a submarine, which houses the periscope and is used as an observation platform

CONPLAN ['kɒnplæn] = US CONTINGENCY PLAN

conquer ['kɒŋkə] *verb* to gain control over an enemy's territory by defeating his armed forces; ***the victorious army conquered one state after another***; ***I came; I saw; I conquered - Caesar***

conqueror ['kɒŋkrə] *noun* someone who conquers; ***Julius Caesar was the conqueror of Gaul***

conscientious objector [kɒnʃɪˌenʃəs əb'dʒektə] *noun* someone who, for moral reasons, refuses to serve in the armed forces when required to do so (NOTE: also called, rudely, a **conchie**)

conscript 1 ['kɒnskrɪpt] *noun* person who joins the armed forces because he is forced to do so by law, rather than because he wants to; *compare* VOLUNTEER, PROFESSIONAL SOLDIER **2** [kən'skrɪpt] to select someone for compulsory military service; ***all men over the age of 18 were conscripted***; *see also* DRAFT

conscription [kən'skrɪpʃn] *noun* compulsory enlistment for military service; *see also* DRAFT

conserve [kən'sɜːv] *verb* to avoid unnecessary waste; ***conserve your ammunition!***

consolidate [kən'sɒlɪdeɪt] *verb* to strengthen or make more secure; ***the enemy is consolidating his bridgehead***

constable ['kʌnstəbl] *noun* policeman or policewoman

constrain [kən'streɪn] *verb* to restrict someone's actions

constraint [kən'streɪnt] *noun* something which restricts someone's actions; ***NATO forces in the area are operating under a number of constraints***

consul ['kɒnsəl] *noun* official appointed to live in a foreign city in order to protect the interests of his own countrymen

consulate ['kɒnsjʊleɪt] *noun* building used by a consul and his staff

contact ['kɒntækt] *noun* **(a)** ability to communicate with another person or grouping; **radio contact** = ability to communicate with another person or grouping over the radio; ***we were in radio contact with the base this morning***; ***the HQ has lost radio contact with the platoon***; **visual contact** = situation where two or more people or groupings can see each other **(b)** first sighting of the enemy (usually resulting in an exchange of fire); ***'hullo 2, this is 22, contact, wait out!'***; ***D Company have just had a contact***; **in contact** = state in which you and the enemy are within effective range of each others' weapons (and usually, shooting at each other); ***B Company are in contact with the enemy***; **to break contact** = to stop fighting with the enemy and withdraw; *see also* DISENGAGE; **to make contact** = to see the enemy

> COMMENT: To avoid causing unnecessary excitement at headquarters, you should only use the word "contact" on the radio when talking about contact with the enemy. When you are talking about your ability to communicate with another person, use an alternative term **I can't raise 22, get in touch with 33B**

contain [kən'teɪn] *verb* **(a)** to hold; ***this box contains live ammunition*** **(b)** to prevent or restrict the movement of a group of people or vehicles; ***we have managed to contain the enemy in the western end of the town***

container [kən'teɪnə] *noun* **(a)** anything which is used to contain something; ***the room was full of ammunition containers*** **(b)** very large metal case of a standard size for loading and transporting goods on trucks, trains and ships

contaminate [kən'tæmɪneɪt] *verb* to infect or pollute (for example through the use of biological or chemical weapons);

the whole area has been contaminated with anthrax; *compare* DECONTAMINATE

contaminated [kən'tæmɪneɪtɪd] *adjective* infected or polluted; ***the contaminated clothing was burnt***

contamination [kəntæmɪ'neɪʃn] *noun* **(a)** act of contaminating something; ***the enemy was not responsible for the contamination of the water supply*** **(b)** something which contaminates (such as a biological weapon, chemical agent, radioactive fallout, etc.); ***there are still traces of contamination***

continent ['kɒntɪnənt] *noun* **(a)** one of the major land areas in the world (Africa, North America, South America, Asia, Australia, Antarctica, Europe) **(b)** *(in Britain)* **the Continent** = the rest of Europe, as opposed to Britain itself which is an island; **on the Continent** = in Europe; **to the Continent** = to Europe; ***when you drive on the Continent remember to drive on the right***

contingency [kən'tɪndʒənsi] *noun* action or situation which is considered possible or likely and which could affect another action or situation; **contingency planning** = planning for a possible future operation

contingent [kən'tɪndʒənt] *noun* small military force which forms part of a larger grouping; ***the British contingent is made up of marines and reconnaissance units***

> COMMENT: **contingent** is normally used to describe the different members of a multinational force, eg: the British contingent, the German contingent, etc.

contour ['kɒntʊə] *noun* line on a map connecting points of equal altitude

contract soldier [ˌkɒntrækt'səʊldʒə(r)] *noun* ex-serviceman serving in the armed forces of a foreign state

contrail ['kɒntreɪl] *noun* = CONDENSATION TRAIL white trail of vapour given off by an aircraft in flight (normally at high altitudes); ***hello 2, this is 22, contrails heading south-west, over.***

contravene [kɒntrə'viːn] *verb* to act contrary to a law or code of conduct; ***your actions contravene the Geneva Convention***

contravention [kɒntrə'venʃn] *noun* act of contravening a law or code of conduct

control [kən'trəʊl] **1** *noun* **(a)** power to direct the actions of people or things; ***you must keep your men under strict control***; ***the men are out of control***; ***he lost control of his vehicle***; **in control of** = having power over something; ***the rebels are in control of the southern part of the country***; **control tower** = observation tower on an airfield or airport used to direct the landing and take-off of aircraft; **fire control** = direction of a unit or sub-unit's weapons in battle; **radio control** = means of operating a device or machine by radio signals (for example radio-controlled aircraft, radio-controlled bomb); **remote control** = means by which a device or machine can be operated from a distance (eg. command wire, radio signals); ***the bomb was detonated by remote control***; *see also* SELF-CONTROL **(b) the controls** = the instruments by which a machine or device is operated; ***I wasn't familiar with the controls of the aircraft*** **2** *verb* to direct the actions of people or things; ***he was unable to control his platoon***

controller [kən'trəʊlə] *noun* someone who controls; **mortar-fire controller (MFC)** = soldier (usually an NCO) who directs mortar fire

convalesce [kɒnvə'les] *verb* to recover your health after illness or injury

convention [kən'venʃn] *noun* formal agreement; **Geneva Convention** = international agreement concerning the conduct of military personnel in war, and dealing with subjects such as treatment of prisoners, care of the wounded, protection of civilian lives and property, etc.; ***your actions contravene the Geneva Convention***

conventional [kən'venʃən(ə)l] *adjective* non-nuclear; ***we will be defeated if we have to rely on purely conventional weapons***

convoy ['kɒnvɔɪ] *noun* group of ships or vehicles travelling together

cookhouse ['kʊkhaʊs] *noun* place where food is cooked

cook off [ˌkʊk 'ɒf] *verb (of ammunition)* to explode prematurely in the breech of the weapon because it is too hot

coolant ['kuːlənt] *noun* fluid designed to stop an engine from overheating

cooperative logistics [kəʊˌɒpərətɪv lə'dʒɪstɪks] *noun* logistics involved in the manufacture, procurement, and storage of supplies

coordinate 1 *noun* [kəʊ'ɔːdɪnət] **(a)** one of a series of two-digit numbers shown on a map grid in order to produce grid references **(b)** grid reference; ***what are your coordinates?*** (NOTE: the coordinates running from left to right are known as **eastings,** while the coordinates running from the bottom to the top are known as **northings)** **2** *verb* [kəʊ'ɔːdɪneɪt] to manage the actions of two or more people or groups so that they work towards a common goal; ***the attack was not properly coordinated***; **coordinating authority** = authority given to a NATO commander to coordinate the work of various agencies and forces from different countries

cop [kɒp] *noun (informal)* policeman or policewoman

copilot ['kəʊpaɪlət] *noun* second pilot of an aircraft crew

Copperhead ['kɒpəhed] *noun* American-designed laser-guided anti-tank artillery round

copse [kɒps] *noun* small wood

copy [kɒpi] **1** *noun* **(a)** thing produced to be the same as something else; ***local gunsmiths are producing good copies of the AK-47 assault weapon*** **(b)** one specimen of a document or publication, where several specimens have been produced; ***I need two copies of the report*** **2** *verb* **(a)** to produce a copy; ***the Chinese are trying to copy the latest Russian tank*** **(b)** US *(radio terminology)* to receive a radio transmission; **copy that** = I have received (and understood) your message; **do you copy?** = are you receiving me?; *see also* AFFIRMATIVE, ROGER

cord [kɔːd] *noun* thick rope of twisted fibres, normally used for tying things together; **det-cord** = fast-burning explosive fuse; **rip-cord** = device which is pulled by hand in order to open a parachute

> COMMENT: cord is thicker than string, and thinner than rope

cordite ['kɔːdaɪt] *noun* smokeless explosive used as propellant for bullets and other projectiles

cordon ['kɔːdən] **1** *noun* line of men or series of outposts designed to control, monitor or prevent movement into or out of an area; ***the cordon went in at first light***; ***he broke through the cordon*** **2** *verb* to place a cordon around an area; ***the village was cordoned off***

corridor ['kɒrɪdɔːr] *noun* **(a)** passage between rooms; ***he is waiting in the corridor*** **(b)** strip of territory *or* airspace along which one can move; ***our mission is to clear a corridor through the enemy's forward defences***

cornet ['kɔːnət] *noun* second lieutenant in certain cavalry regiments

corporal (Cpl) ['kɔːprəl] *noun* **(a)** GB junior non-commissioned officer (NCO) in the army, marines or air force; *see also* LANCE-CORPORAL; GB **corporal major** = rank used by the Household Cavalry as an equivalent to staff sergeant and sergeant major; GB **corporal of horse (CoH)** = rank used by the Household Cavalry as an equivalent to sergeant **(b)** US junior non-commissioned officer (NCO) in the army or marines

> COMMENT: British infantry sections are usually commanded by corporals. In the British army, a corporal in the Brigade of Guards is known as a **lance-sergeant,** while a corporal in the artillery is known as a **bombardier**

corps [kɔː] *noun* **(a)** tactical army grouping of two or more divisions **(b)** administrative grouping used by certain specialist troops (for example the Army Air Corps, the Royal Corps of Signals, the US Marine Corps, etc.); **Corps of Drums** = band of drummers and fife-players belonging to a battalion or regiment; *(also* *known as the Drums or the Fifes and Drums)*

corpse [kɔːps] *noun* dead body

corpsman ['kɔːmən] *noun US* specialist soldier trained to give first aid on the battlefield

correct [kə'rekt] **1** *adjective* **(a)** true or accurate; ***that is correct*** **(b)** *(of artillery or mortar fire)* on target **2** *verb* **(a)** to amend or put right; ***he corrected the timetable*** **(b)** to make calculations and issue instructions in order to bring artillery or mortar fire onto a target; ***he corrected onto the second enemy position***

correction [kə'rekʃn] *noun* **(a)** act of correcting something **(b)** calculation made by a forward observer and sent to an artillery or mortar unit in order to bring fire onto a target **(c)** process of directing artillery or mortar fire onto a target

correspondent *see* WAR CORRESPONDENT

corrie ['kɒri] *noun (Scotland)* natural depression on the side of a mountain; *see also* CIRQUE

corrugated iron [ˌkɒrəgeɪtɪd 'aɪən] *noun* wrinkled metal sheets used in the construction of field fortifications and shelters

Corsair ['kɔːseə] *see* A-7

corvette [kɔː'vet] *noun* small ocean-going warship

COS [ˌsi ː əʊ 'es] = CHIEF OF STAFF

COSCOM ['kɒskɒm] *noun US* = CORPS SUPPORT COMMAND organization responsible for the resupply of a corps

cot [kɒt] *noun US* camp-bed

cottage ['kɒtɪdʒ] *noun* small rural house

counter ['kaʊntə] **1** *adjective* contrary to; ***your attack was counter to my orders***; **counter-battery fire** = artillery attack on an enemy artillery fire-position; ***we lost three guns through counter-battery fire***; **counter-espionage** = action taken to impede the activities of enemy spies; **counter-insurgency (COIN) operation** = military operation mounted to destroy armed resistance to the established government or foreign domination **2** *verb* to take action in order to prevent or impede another action taking place; ***we countered the tank threat by mining all the likely approaches***

counter-air [ˌkaʊntə 'eə] *adjective* relating to operations directed against the enemy air force; **offensive counter-air operations** = attacks on enemy airfields, surface-to-air missile sites, radar sites and other facilities associated with the enemy air force; **defensive counter-air operations** = use of fighter aircraft and air defence weapons to protect one's own territory and forces

counter-attack ['kaʊntə əˌtæk] **1** *noun* retaliatory attack on an enemy force which is in the process of attacking or has just completed an attack; ***the enemy counter-attack was successfully beaten off*** **2** *verb* to mount a counter-attack; ***B Company counter-attacked while the enemy were reorganizing***

counter-concentrate [ˌkaʊntə 'kɒnsəntreɪt] *verb* to bring forces together to repel an enemy attack

counterfire ['kaʊntəˌfaɪə(r)] *noun* attack on the enemy's artillery assets

countermand [kaʊntə'mɑːnd] *verb* to cancel an order or instruction (usually made by someone else); ***the general countermanded the brigade commander's order to withdraw***

countermeasure ['kaʊntəˌmeʒə] *noun* action or procedure designed to neutralize a danger or threat; **electronic countermeasures (ECM)** = standard procedures designed to minimize a unit's chances of being located by the enemy through emissions given off by its electrical equipment; **electronic counter-countermeasures (ECCM)** = procedures used to defeat the enemy's electronic countermeasures

countersign ['kaʊntəˌsaɪn] *noun* words, letters or numbers used as a verbal recognition signal, usually in the form of a challenge and a reply; ***a sentry should use the countersign when he does not recognize a person***; *see also* PASSWORD

country ['kʌntri] *noun* **(a)** land forming the territory of a nation or state **(b) the country** = rural district or region (consisting of agricultural land, villages and small towns as opposed to large towns and cities); ***guerillas are still operating in***

the country; **country house** = large ornate dwelling, usually built by an aristocrat or rich person **(c)** terrain; ***we will have to cross some difficult country***

countryside ['kʌntrisaɪd] *noun* **(a)** rural district or region **(b)** terrain consisting of agricultural land, woodland and villages

county ['kaʊnti] *noun* rural region or district with its own local administration

coup [kuː] *noun France* **(a)** significant (and usually successful) action; **coup de grâce** = act of killing a wounded person or animal in order to prevent further suffering; **coup de main** = surprise attack; **coup d'oeil** = assessment of terrain or a situation simply through observation **(b)** coup d'état, a sudden seizure of power by use of force; ***the army has staged a coup***

courage ['kʌrɪdʒ] *noun* ability to control fear (also known as physical courage); **moral courage** = ability to disagree with or reprimand other people; ***although he is very brave under fire, he lacks the moral courage to control his NCOs***; *see also* BRAVERY

courageous [kə'reɪdʒəs] *adjective* able to control your fear; *see also* BRAVE

courier ['kʊrɪə(r)] *noun* person who takes something (eg. message, weapon, etc) from one place to another; ***the gunman escaped, but we captured the courier and the weapon***

course [kɔːs] *noun* **(a)** series of lessons, lectures and practical exercises in a specific subject; ***I am going on a survival course*** **(b)** series of obstacles or practical tasks forming part of a test or competition; **assault course** = series of obstacles used by infantry training establishments to practise obstacle-crossing **(c)** direction taken by a ship or aircraft; ***the ship was heading on a course of 220°***; **off course** = going in the wrong direction

court-martial [ˌkɔːt 'mɑːʃl] **1** *noun* trial of someone serving in the armed forces by the armed forces authorities under military law; ***the court-martial was held in the army headquarters***; ***he was found guilty by the court-martial and sentenced to imprisonment***; **drumhead court-martial** = court martial held in the field (NOTE: plural is **courts-martial**) **2** *verb* to try someone who is serving in the armed forces; ***he was court-martialled for cowardice*** (NOTE: **court-martialling - court-martialled**; US **court-martialing - court-martialed**)

cove [kəʊv] *noun* small sheltered bay or inlet

cover ['kʌvə] **1** *noun* **(a)** something that conceals or protects another object; **cover from fire** = anything which provides protection from bullets, shrapnel or other projectiles; **to break cover** = to move out of a place of concealment; **cloud cover** = area of sky covered by cloud; **overhead cover** = roof of a trench or other field fortification, which is designed to withstand bullets and shrapnel; **to take cover** = to hide or to seek protection from enemy fire **(b)** support for another person or unit; **air cover** = aircraft which are in the air or on call in order to provide air support if required; **medical cover** = medical personnel, ambulances, etc., which are available in the event of casualties **(c)** false identity or explanation; ***his business trip was just a cover for meeting the rebel leaders*** **2** *verb* **(a)** to put something over another object in order to conceal *or* protect it; ***he covered the body with a blanket*** **(b)** to provide fire support for another person or unit; ***cover me while I move forward!*** **(c)** to point a weapon at a person or group; ***he covered the prisoners while they were being searched*** **(d)** to be able to observe or shoot into a specific area; ***6 Platoon is covering the main road***; **covered by fire** = having a clear field of fire over an area of ground or, if this is not possible, having the area registered as an artillery or mortar target; ***all obstacles must be covered by fire*** **(e) to cover for someone** = to carry out the duties of another person; ***I am covering for Sgt Hobbs tonight***

coveralls ['kʌvərɔːlz] *plural noun* garment combining jacket and trousers, which is worn over other clothes in order to protect them from dirt, mud, oil, etc

covering fire [ˌkʌvrɪŋ 'faɪə] *noun* shooting at an enemy, to make him keep his head down, so that another person or unit can move; ***my platoon provided covering fire while the rest of the company withdrew***

cover-up ['kʌvə ˌʌp] *noun* attempt to conceal true facts by lying or concealing information; ***the media are saying that there has been a cover-up***

covert ['kəʊvɜːt] *adjective* concealed or secret; ***this is a covert operation***

coward ['kaʊəd] *noun* someone who cannot control his fear

cowardice ['kaʊədɪs] *noun* **(a)** inability to control your fear; **moral cowardice** = reluctance to disagree with or reprimand other people, even when you think that you are right **(b)** military offence of running away from the enemy; ***he was shot for cowardice*** (NOTE: the opposite is **bravery**)

cowardly ['kaʊədli] *adjective* **(a)** *(of a person)* unable to control fear (NOTE: the opposite is **brave**) **(b)** *(of an action)* done against someone who cannot retaliate

coxswain ['kɒksweɪn *or* 'kɒks(ə)n] *noun* **(a)** person who steers a boat *or* ship **(b)** senior petty officer on a small ship

Coy = COMPANY

CP [ˌsiː 'piː] = COMMAND POST

Cpl = CORPORAL

CPO = CHIEF PETTY OFFICER

CPX [ˌsiː piː 'eks] *noun* = COMMAND POST EXERCISE radio exercise involving only the command elements of a grouping

CQB [ˌsiː kjuː 'biː] *noun* = CLOSE - QUARTER BATTLE art of fighting at very close range (eg. FIBUA, jungle fighting, etc); ***we've set up a CQB range in the wood***

CQMS [ˌsiː kjuː em 'es] = COMPANY QUARTERMASTER SERGEANT

CRA [ˌsiː ɑː 'reɪ] *noun GB* = COMMANDER ROYAL ARTILLERY brigadier in command of a division's artillery assets

> COMMENT: although regiments and batteries are nominally allocated in support of specific brigades and battle groups respectively, artillery is considered to be a divisional asset and batteries may be tasked to support other groupings as the tactical situation dictates

crab [kræb] *noun GB (army slang)* member of the Royal Air Force

crack [kræk] *adjective* elite, of very high quality; ***he served in a crack cavalry regiment***

craft [krɑːft] *noun* boat or ship; **assault craft** = small boat designed for amphibious operations; **landing craft** = small flat-bottomed boat designed to move troops and vehicles from a transport ship to a beach (NOTE: the word **craft** is used for both singular and plural)

craftsman (Cfn) ['krɑːftsmən] *noun GB* private in the Royal Electrical and Mechanical Engineers (REME)

crag [krɑg] *noun* rocky summit of a hill; ***we lost a lot of men taking that crag***

crampon ['kræmpɒn] *plural noun* set of spikes which are fitted to the sole of a boot for climbing on snow *or* ice; ***we'll need crampons for this operation***

crane [kreɪn] *noun* tall machine with a long arm, designed for lifting heavy objects

crap-hat ['kræp ˌhæt] *noun GB* derogatory term used by members of the Parachute Regiment to describe any soldier who is not a trained paratrooper

crash [kræʃ] **1** *noun* violent collision; ***he was killed in a car crash*** **2** *verb* to collide violently with something else; ***the truck crashed into a tank***

crash out [ˌkræʃ 'aʊt] *phrasal verb informal* **(a)** to deploy at short notice from a base *or* position in response to an incident *or* threat; ***the QRF crashed out as soon as they heard the explosion*** **(b)** to go to sleep (but not in a bed); ***we'll just crash out here***

crate [kreɪt] *noun* large container made of wood or metal

crater ['kreɪtə] **1** *noun* hole in the ground made by an explosion **2** *verb* to make craters (as an obstacle); ***the engineers cratered the road***

crawl [krɔːl] *verb* to move on your hands and knees; ***he crawled up to the bunker***

creek [kriːk] *noun* **(a)** small stream **(b)** narrow inlet

creep [kriːp] *verb* to move slowly and cautiously; ***he crept up to the bunker*** (NOTE: **creeping - crept**)

creeping barrage *or* **bombardment** ['kriːpɪŋ] *noun* artillery bombardment which is constantly adjusted, so that the shells continue to land in front of friendly troops as they advance; ***we will advance behind a creeping barrage***

crest [krest] *noun* top of a hill or ridge; **military or tactical crest** = highest point of the slope from which there is an unrestricted view down to the bottom.; **topographical crest** = actual crest (from which the bottom of the slope may not be visible)

crevasse [krɪ'væs] *noun* large crack in the surface of a glacier *or* mass of snow; ***he fell down a crevasse***

crew [kruː] *noun* **(a)** team of people who man a ship, aircraft or vehicle; ***the crew of the helicopter which was brought down have all been rescued*** **(b)** team of people who operate a weapon or equipment

crewman *or* **crewmember** ['kruːmən *or* 'kruːmembə] *noun* member of a crew

crime [kraɪm] *noun* **(a)** illegal act which is punishable by law; **war crime** = act which violates international rules of war **(b)** illegal activity in general

criminal ['krɪmɪnəl] **1** *adjective* relating to crime **2** *noun* someone who commits a crime

crippled ['krɪpld] *adjective* **(a)** *(of people)* physically disabled **(b)** *(of vehicles)* badly damaged and unable to function properly

crisis ['kraɪsɪs] *noun* very difficult or dangerous situation; **crisis management** = taking rapid decisions to deal with a crisis (NOTE: plural is **crises** 'kraɪsiːz])

critical point ['krɪtɪkl 'pɔɪnt] *noun* **(a)** location or position which could influence the outcome of an operation **(b)** *(for aircraft)* the point midway between two airbases from which a plane will take the same time to reach either base another name for WAYPOINT

cross [krɒs] *verb* to move from one side of a feature to the other; ***the enemy are crossing the river***

cross-attachment [ˌkrɒs ə'tætʃmənt] *noun* attachment to a different arm *or* service (eg. a platoon of armoured infantry attached to a tank squadron)

crossfire ['krɒsfaɪə] *noun* fire directed at a target from two or more different locations; ***we were caught in a crossfire***

cross-grain [krɒs'greɪn] *adverb* across the grain of the country; ***we'll have to move cross-grain***

crossing ['krɒsɪŋ] *noun* **(a)** place where a railway line, river, road or other feature can be crossed **(b)** bridge or ford; ***the enemy have captured several crossings*** **(c)** act of crossing a feature or obstacle; ***the enemy is preparing a river crossing***; **assault river crossing** = act of crossing a river while in contact with the enemy

crossroads ['krɒsrəʊdz] *noun* place where two or more roads cross over each other

cross-servicing [ˌkrɒs 'sɜːvɪsɪŋ] *noun* work done by one service for another

crosswind ['krɒswɪnd] *noun* wind which blows across your direction of travel

Crotale [krəʊ'tɑːl] *noun* French-designed short-range surface-to-air missile (SAM)

crowd [kraʊd] *noun* large group of people

crown [kraʊn] *noun* **(a)** ceremonial head-dress worn by a king or queen **(b)** insignia used in some badges of rank of the British Army (for example a single crown denotes major, while a crown and a star denotes lieutenant-colonel)

cruise [kruːz] *verb* to make a long journey by sea; **cruise missile** = American-designed low-flying missile which is capable of navigating itself to a target; *see also* TOMAHAWK

cruiser ['kruːzə] *noun* large ocean-going warship, armed with missiles or guns, which is capable of spending long periods at sea without support

CS [ˌsiː 'es] *noun* = ORTHOCHLOROBENZYLIDENE MALONONITRILE type of tear agent (NOTE: commonly known as **CS gas**)

> COMMENT: CS gas is normally used by riot police for crowd control

C/S = CALL-SIGN

C/Sgt = COLOUR SERGEANT

CSAR [ˌsiː es eɪ 'ɑː(r)] *noun* = COMBAT SEARCH AND RESCUE operation mounted to locate and rescue aircrew who have been shot down over enemy territory

CSM [ˌsiː es 'em] **1** = COMPANY SERGEANT MAJOR **2** *US* = COMMAND SERGEANT MAJOR

CSS [ˌsiː es 'es] *noun* = COMBAT SERVICE SUPPORT resupply of ammunition, food, fuel and other necessities on the battlefield; ***we need to practise CSS on the move***

cul-de-sac ['kʌldɪˌsæk] *noun* street *or* road which suddenly ends, so that the only way out is to go back the way you came; *also known as a* ***DEAD END***

culminating point ['kʌlmɪneɪtɪŋ ˌpɔɪnt] *noun* point at which an attacking force is unable to continue its attack *or* even defend itself (because of casualties, shortages of fuel, ammunition, and rations, and sheer physical exhaustion); ***the enemy attack had reached its culminating point***

culvert ['kʌlvɜːt] *noun* small tunnel used to carry drainage water under a road; ***the bomb was placed in a culvert***

cupola ['kjuːpələ] *noun* revolving turret housing a gun or machine-guns, which is fitted to a warship, aircraft or fighting vehicle

curfew ['kɜːfjuː] *noun* regulation requiring people to be off the streets during a specified period, usually at night; ***the military commander ordered a dusk-to-dawn curfew***; ***soldiers patrolled the streets during the curfew***

custody ['kʌstədi] *noun* imprisonment or being held under close arrest; ***he was taken into custody***

cutlass ['kʌtləs] *noun* short sword with a curved blade, formerly used in the navy and now only used on ceremonial occasions

cut off [ˌkʌt 'ɒf] *verb* **(a)** to prevent someone from retreating or from rejoining his comrades; **cut-off group =** small group of soldiers positioned on the likely approaches to or exits from the killing area of an ambush, in order to prevent any of the survivors from escaping **(b)** to surround a unit so that it can neither retreat, nor be reinforced or supported; ***the platoon was cut off when the rest of the company withdrew*** **(c)** to stop the supply of food, power, water, etc.; ***the electricity has been cut off*** **(d)** to prevent movement to or from a location; ***the village has been cut off by snow***; *see also* ISOLATE

cutter ['kʌtə] *noun* **(a)** tool used for cutting; **wire cutters =** special type of scissors used for cutting through barbed wire **(b)** small armed naval boat

cutting ['kʌtɪŋ] *noun* man-made channel allowing a road or railway to pass through an area of high ground; *compare* EMBANKMENT

CV = AIRCRAFT CARRIER

CVBG [ˌsiː viː biː 'dʒiː] = CARRIER BATTLE GROUP

CVN = AIRCRAFT CARRIER (nuclear powered)

CVR [ˌsiː viː 'ɑː] *noun* = COMBAT VEHICLE RECONNAISSANCE light, fast-moving armoured vehicle designed for reconnaissance

CVR (T) [ˌsiː viː ɑː 'tiː] *noun* = COMBAT VEHICLE RECONNAISSANCE TRACKED armoured reconnaissance vehicle fitted with tracks (especially the British-designed Scimitar and Scorpion light tanks)

CVR (W) [ˌsiː viː ɑː 'dʌb(ə)ljuː] *noun* = COMBAT VEHICLE RECONNAISSANCE WHEELED armoured reconnaissance vehicle fitted with wheels (as opposed to tracks)

CVW [ˌsiː viː 'dʌb(ə)ljuː] = CARRIER AIR WING

CX [ˌsiː 'eks] *noun* = DICHLOROFORMIXIME type of blister agent (NOTE: also known as **Phosgene oxime**)

DELTA - Dd

D-30 [ˌdiː 'θɜːti] *noun* Soviet-designed 121.92mm light artillery piece

DA [ˌdiː 'eɪ] *noun* = DIPHENYLCHLOROARSINE type of vomiting agent

dagger ['dægə] *noun* long knife with a thin blade, designed for stabbing

dam [dæm] **1** *noun* barrier designed to restrict the flow of water, in order to make a reservoir or to prevent flooding; ***the valley was flooded when the dam was destroyed*** **2** *verb* to construct a dam; ***the river has been dammed*** (NOTE: damming - has dammed)

damage ['dæmɪdʒ] **1** *noun* harm done to something; ***the bomb caused extensive damage to civilian property***; **collateral damage** = unintentional killing of civilians or destruction of civilian property as a result of military action **2** *verb* to cause harm to something; ***the rebel radio station has been damaged***

danger ['deɪnʒə] *noun* **(a)** situation where people may be killed or injured; **danger area** = area within which casualties may occur; ***the danger area of this grenade is 90 metres***; ***there is a danger area behind the shooting range***; **danger close** = *US* deliberately calling friendly artillery fire down on top of your own positions; ***the platoon commander brought the rounds in danger close*** **(b)** something which may cause harm or injury; ***landmines are a constant danger in this area*** **(c)** possibility of failure or unfortunate consequences; ***there is a danger that the enemy will outflank us***

dangerous ['deɪnʒrəs] *adjective* **(a)** likely to cause harm or injury; ***this vehicle is in a dangerous condition*** **(b)** likely to have unfortunate consequences; ***the international situation is now extremely dangerous***

dannart wire ['dænət ˌwaɪə(r)] *noun* *GB* barbed wire

dare [deə] *verb* to do something in the knowledge that it is dangerous or risky; ***the sergeant dared his men to follow him up the hill*** (NOTE: **dare** is followed by the infinitive, and can be used with or without **to**)

`Who dares wins`

motto of the Special Air Service

---

daring ['deərɪŋ] *adjective* involving risks; ***it was a daring plan***; ***they carried out a daring escape from the POW camp***

dash [dæʃ] **1** *noun* **(a)** movement at high speed; ***the enemy made a dash for the coast*** **(b)** longer signal in Morse code (the shorter signal is the dot) **2** *verb* to move at high speed; ***he dashed to the latrine***

dashboard ['dæʃbɔːd] *noun* instrument panel of a vehicle or aircraft

data ['deɪtə] *noun* information

database ['deɪtəbeɪs] *noun* information stored on a computer; ***hackers tried to get into the Pentagon database***

date [deɪt] *noun* number of a particular day, month and year; **date of birth (DOB)** = number of the day, month and year when a person was born

date-time group (DTG) [deɪt 'taɪm ˌgruːp] *noun* timing, consisting of day of the month, time, time zone and, if necessary, month and even year; ***231645ZSeptember 2001***= 1645hrs (GMT) on 23 September 2001

COMMENT: if the month and year are are not included, then you can assume that the date-time group applies to the current month and year

davit ['dævɪt] *noun* small crane on a ship for suspending and lowering a lifeboat

dawn [dɔːn] *noun* time of day when it is starting to get light; *see also* DAYBREAK, FIRST LIGHT; *compare* DUSK

daybreak ['deɪbreɪk] *noun* time of day when it is starting to get light; *see also* DAWN, FIRST LIGHT

daysack ['deɪsæk] *noun* small rucksack for carrying supplies or clothing during an operation

DC [ˌdiː 'siː] *noun* = DIPHENYLCYANOARSINE type of vomiting agent

DCOS ['diːkɒs] = DEPUTY CHIEF OF STAFF

DD = DESTROYER (with guns)

D-Day ['diː ˌdeɪ] *noun* day on which an operation starts; **D-minus-two** = two days before D-Day; **D-plus-three** = three days after D-Day

COMMENT: other letters are also used to mark the start of an operation. In the Gulf War of 1991, the start of the ground operation was designated as G-day

DDG = DESTROYER (with guided missiles)

dead [ded] **1** *adjective* **(a)** no longer alive; ***after the attack, 50% of the battalion were left dead or wounded***; **dead on arrival (DOA)** = found to be already dead on reaching a hospital or casualty clearing station **(b)** *(of radios, etc.)* not working **(c)** **dead end** = street *or* road which suddenly ends, so that the only way out is to go back the way you came **(d)** **dead ground** = area of ground which provides cover from view (eg. the reverse slope of a hill); ***the company formed up in dead ground to the enemy position*** **(e)** **dead letter-box** = secret location used to deposit and collect messages, equipment, etc. **(f)** **dead reckoning** = calculating your position using a compass, speed of travel, etc., but not using the sun, moon or stars **2**; **the dead** = people who have died or have been killed; ***after the battle troops were sent to collect and bury the dead***

deadfall ['dedfɔːl] *noun* dead branch which falls away from a tree as a result of wind *or* simply because of its own weight

COMMENT: deadfall is a serious and constant hazard in jungle regions

dead reckoning [ded 'rekənɪŋ] *noun* method of calculating your position entirely from the compass bearing on which you have been travelling and the distance which you have covered

COMMENT: dead reckoning is used at sea *or* in terrain where there are no obvious features (eg. desert *or* jungle). However, as a result of satellite navigation and other modern technology, it is rapidly becoming a forgotten skill.

death [deθ] *noun* act of dying or being killed; **death's head** = traditional insignia consisting of a human skull over two crossed bones; **death squad** = group who assassinate or execute people

débâcle *or* **debacle** [deɪ'bɑːkl] *noun* complete failure (for example a decisive defeat)

debark [diː'bɑːk] *verb* to land from a ship; *compare* EMBARK (NOTE: also **disembark**)

debarkation [diːbɑː'keɪʃn] *noun* act of landing from a ship; *compare* EMBARKATION (NOTE: also **disembarkation**)

debouch [dɪ'baʊtʃ] *verb* to come out of a re-entrant, valley or wood; ***we engaged the enemy as they were debouching from the valley***

debrief [diː'briːf] *verb* to question people who have taken part in a mission or operation; *compare* BRIEF

debriefing [diː'briːfɪŋ] *noun* **(a)** act of debriefing someone **(b)** meeting where debriefing is carried out; *compare* BRIEFING

debris ['debri *US* də'briː] *noun* pieces of something which has been destroyed or badly damaged; ***we found debris from the crashed plane***

debus [diː'bʌs] *verb* to get out of a bus or other vehicle; ***the platoon debussed on the objective***

decamp [diː'kæmp] *verb* to leave suddenly; ***the enemy had already decamped***

deception [dɪ'sepʃn] *noun* activity intended to give the enemy a false idea of your location or intentions

deciduous [dɪ'sɪdjʊəs] *adjective* relating to trees which lose their leaves in winter (such as ash, beech, oak, etc.); *compare* CONIFEROUS

decimate ['desɪmeɪt] *verb* to kill a large number of people; ***the division has been decimated***

> COMMENT: this word is derived from the ancient Roman military punishment of killing every tenth man in a legion

decipher [dɪ'saɪfə] *verb* to convert from code into normal language; *compare* ENCIPHER; *see also* DECODE, DECRYPT

decisive [dɪ'saɪsɪv] *adjective* **(a)** *(of an event)* settling an issue (such as a campaign or war); ***it was a decisive victory*** **(b)** *(of a person)* capable of making a decision quickly; ***he's not very decisive***

decisively [dɪ'saɪsɪvli] *adverb* in a decisive way; ***we need to act decisively before the enemy recovers***

deck [dek] *noun* floor or level in a ship; **to hit the deck** = to take cover (literally, to lie down); *see also* FLIGHT DECK

declaration [deklə'reɪʃn] *noun* formal announcement; **declaration of war** = formal announcement by one country to another, that they are now at war with each other

declare [dɪ'kleə] *verb* to make a formal announcement; **to declare war on someone** = to inform the government of another country that a state of war exists; ***Britain had declared war on Germany***

decode [diː'kəʊd] *verb* to convert from code into normal language; *compare* ENCODE; *see also* DECIPHER, DECRYPT

decommission [diːkə'mɪʃn] *verb* to take a ship out of operational service with the navy; *compare* COMMISSION

decommissioning [diːkə'mɪʃnɪŋ] *noun* taking a ship out of operational service with the navy; *compare* COMMISSIONING

decontaminate [dikən'tæmɪneɪt] *verb* to remove or neutralize contamination (such as chemical agent, radioactive fallout, etc.); ***we had to decontaminate our vehicles***; *compare* CONTAMINATE

decorate ['dekəreɪt] *verb* to award a medal to someone (usually for bravery or outstanding achievement); ***a much-decorated pilot***

decoration [dekə'reɪʃn] *noun* medal (usually for bravery or outstanding achievement); ***he wore his decorations to the ceremony***

decoy 1 ['diːkɔɪ *US* dɪ'kɔɪ] *noun* **(a)** something which is used to draw the enemy into an ambush or to make him reveal his location **(b)** something which draws a guided missile away from its intended target (for example chaff or flares) **2** [dɪ'kɔɪ] *verb* **(a)** to present the enemy with a target in order to draw him into an ambush or to make him reveal his location; ***a small patrol was used to decoy the enemy into the ambush*** **(b)** to draw a guided missile away from its intended target; ***the aircraft dropped chaff to decoy surface-to-air missiles***

decrypt [diː'krɪpt] *verb* to convert from code into normal language; *compare* ENCRYPT; *see also* DECIPHER, DECODE

deer [dɪə(r)] *noun* large brown four-legged wild animal (often with horns on its head); ***several deer came running out of the wood*** (NOTE: plural form is also **deer**)

defaulter [dɪ'fɒltə] *noun* someone who has been accused or found guilty of a military offence; ***defaulters will parade at 0745hrs***

defeat [dɪ'fiːt] **1** *noun* the act of losing a battle, campaign or war; ***the enemy will not recover from this defeat***; ***messengers brought back news of the naval defeat*** **2** *verb* to destroy or drive off an enemy force; ***we won a great victory and defeated the***

enemy; Napoleon's army was defeated by the Russian winter

defeatist [dɪ'fi:tɪst] **1** *adjective* believing that defeat or failure is inevitable; ***that is a very defeatist attitude*** **2** *noun* someone who believes that defeat or failure is inevitable; ***the last president was a defeatist***

defect 1 [dɪ'fekt] *verb* to abandon your country in favour of an enemy power; ***he has defected to the Russians*** **2** ['di:fekt] *noun* fault or imperfection; ***this equipment has several defects***

defective [dɪ'fektɪv] *adjective* not working properly; ***this equipment is defective***

defector [dɪ'fektə] *noun* someone who abandons his own country in favour of an enemy power

defence *US* **defense** [dɪ'fens] *noun* **(a)** act of resisting an attack; ***the enemy put up a stubborn defence***; **all-round defence** = having all the approaches to your position covered by fire, including those from the flanks and rear; **area defence** = naval anti-air warfare (AAW) term for warships' use of their long-range surface-to-air missiles (SAM) for the mutual defence; **civil defence** = organization and training of non-military personnel for the protection of life and property during wartime; **home defence** = defence of a state's own territory in the event of war (as opposed to territory belonging to another state); ***the division will be used for home defence***; **defence stores** = material used in the construction of field fortifications (for example barbed wire, corrugated iron, sandbags); **point defence** = naval anti-air warfare (AAW) term for a warship's use of its short-range surface-to-air missiles (SAM) and other weapons (eg. CIWS) for self-defence **(b)** military power; ***the Government has cut its spending on defence***; *GB* **Ministry of Defence (MOD)** = British government department in charge of the armed forces; **Secretary of State for Defence** *or* **Defence Secretary** = government minister in charge of the armed forces **(c) defences** = fortifications built to protect a place; ***the enemy easily breached our defences***

defend [dɪ'fend] *verb* **(a)** to resist an attack; ***the town was defended by a small group of soldiers*** **(b)** *(legal)* to represent an accused person in a court of law or court-martial; ***he was defended by a major*** **(c)** to justify an action or opinion; ***he found it difficult to defend the CO's decision***

defended locality [dɪ'fendɪd ləʊ'kæləti] *noun* area containing several defensive positions, which are able to provide each other with mutual support

defense [dɪ'fens] *see* DEFENCE *US* **Department of Defense (DOD)** *or* **Defense Department** = US government department in charge of the armed forces (NOTE: also called the **Pentagon**); **Secretary for Defense** *or* **Defense Secretary** = US government minister in charge of the armed forces

defensive [dɪ'fensɪv] *adjective* relating to defence; **defensive fire task (DF)** = pre-determined artillery target, which has been registered and given a target number; **defensive position** = area prepared for defence; *compare* OFFENSIVE

deficiency [dɪ'fɪʃənsɪ] *noun* item of equipment which is reported lost *or* mislaid; ***platoon commanders are to report all deficiencies ASAP***

deficient [dɪ'fɪʃənt] *adjective; (relating to equipment issued to a serviceman)* lost *or* mislaid; ***he is deficient his helmet***; **to go deficient** = to report the loss of an item of equipment; ***you'll have to go deficient on your helmet***

defilade [defɪ'leɪd] **1** *adjective* shielded from observation and direct fire by a natural or man-made obstacle; ***the anti-tank missiles were sited in defilade positions*** **2** *noun* defilade position; ***find yourself a defilade on the right and prepare to give us covering fire***

defile [dɪ'faɪl] **1** *verb* to move in file formation **2** *noun* pass or valley which forces an advancing body of soldiers or other tactical grouping to move in a narrow column

defoliate [di:'fəʊlieɪt] *verb* to remove the leaves from, and usually kill, vegetation; **defoliating agent** = chemical designed to kill vegetation (such as Agent Orange)

defoliant [diː'fəʊliənt] *noun* something which kills vegetation by removing leaves

defuse [diː'fjuːz] *verb* **(a)** to remove the fuse from an explosive device **(b)** to reduce tension between people; ***the troop withdrawal has defused the situation***

degrade [dɪ'greɪd] *verb* **(a)** to make something smaller or weaker **(b)** *(current term)* to destroy military assets belonging to another state, in order to reduce that state's ability to mount offensive operations; ***the object of the airstrikes was to degrade the country's offensive capability***

degree [dɪ'griː] *noun* **(a)** unit of measurement for angles or bearings **(b)** unit of measurement for temperature (NOTE: there are 360 degrees in a circle. 360 degrees are the equivalent of 6,400 mils. The symbol for degrees is ° (214 °))

> COMMENT: many armies use **mils** instead of degrees in order to measure bearings, because they offer greater precision

delay [dɪ'leɪ] **1** *noun* **(a)** act of arranging or causing an action to take place later than originally planned; ***he was not responsible for the delay in implementing the airstrikes*** **(b)** act of making someone late or slowing someone down; **aggressive delay =** tactic involving the aggressive use of small units to slow down an advancing enemy force so that a main line of defence can be prepared or strengthened **(c)** period of time lost as a result of a delay; ***there will be a delay of forty minutes***; **without delay** = immediately **2** *verb* **(a)** to arrange or cause an action to take place later than originally planned; ***the attack has been delayed*** **(b)** to make someone late or to slow someone down; ***we were delayed by the weather***; **delaying force =** small force used to slow down an enemy advance while the main force prepares or strengthens a line of defence

Delta ['deltə] fourth letter of the phonetic alphabet (Dd)

delta ['deltə] *noun* triangular area of land or marsh at the mouth of a river; ***the Danube Delta***; **delta wing =** triangular aircraft wing

demilitarized zone (DMZ) [diː'mɪlɪtəraɪzd ˌzəʊn] *noun* area or region in which the presence of military forces is forbidden under the terms of a treaty or other international agreement

demo ['deməʊ] *noun (informal)* demonstration

demob [diː'mɒb] *verb (informal)* to demobilize

demobilize [dɪ'məʊbɪlaɪz] *verb* to return conscripted servicemen to civilian life

demolish [dɪ'mɒlɪʃ] *verb* to destroy a structure (such as a bridge or building)

demolition [demə'lɪʃn] *noun* act of demolishing something; **demolition gun =** large calibre gun, which is fitted to an armoured engineer vehicle for the purpose of demolishing buildings or destroying obstructions; **prepared for demolition =** fitted with explosive charges

demonstrate ['demənstreɪt] *verb* **(a)** to show someone how something is done; ***he demonstrated the use of the respirator*** **(b)** to take part in a public assembly or procession in order to express an opinion or grievance; ***they were demonstrating against the invasion of their country***

demonstration [demən'streɪʃn] *noun* **(a)** act of showing someone how something is done; ***we were given a demonstration of how to load the mortar*** **(b)** public assembly or procession in order to express an opinion or grievance; ***the demonstration was organized to protest against the invasion of their country*** **(c)** show of military force intended to intimidate the enemy or to divert the enemy's attention; ***our battalion made a demonstration to the enemy's front while the rest of the brigade moved round to attack the flank***

demonstrator ['demənstreɪtə] *noun* **(a)** someone who demonstrates something **(b)** someone who takes part in a demonstration to protest against something

demoralization [ˌdiːmɒrəlaɪ'zeɪʃən] *noun* loss of morale (usually as a result of defeat *or* high casualties); ***demoralization is evident throughout the entire chain of command***

demoralize [dɪ'mɒrəlaɪz] *verb* to destroy someone's morale

demote [diː'məʊt] *verb* to reduce to a lower rank (usually as a punishment); ***he was demoted for being drunk on duty***; *compare* PROMOTE; *see also* BUST

denial [dɪ'naɪəl] *noun* act of denying something

dense [dens] *adjective* thick or crowded; ***dense undergrowth***; ***a dense crowd***

deny [dɪ'naɪ] *verb* **(a)** to say that something is untrue; ***he denied the accusation*** **(b)** to prevent someone from using or having access to; ***we must deny the river crossings to the enemy***

depart [dɪ'pɑːt] *verb* to leave a location

department [dɪ'pɑːtmənt] *noun* **(a)** part of an organization; ***which department do you work in?*** **(b) (i)** major section of the British government headed by a Secretary of State; **(ii)** major section of the US government headed by a Secretary; ***a spokesman for the US Department of Defense***

departure [dɪ'pɑːtʃə] *noun* act of leaving a location; **line of departure** = real or imaginary line which marks the start of an advance or attack; *see also* START LINE

deplane [diː'pleɪn] *verb* to get off an airplane; ***the force will begin to deplane at 0600hrs***

depleted uranium (DU) [dɪˌpliːtɪd juː'reɪniəm] *noun* uranium with its harmful radioactive properties reduced; used in the manufacture of some long-rod penetrators (NOTE: in the US Army, depleted uranium is known as **Staballoy**)

deploy [dɪ'plɔɪ] *verb* **(a)** *(strategical)* move to a war zone or area of operations; ***7 Armoured Brigade deployed to the Gulf in October*** **(b)** *(tactical)* to adopt a battle formation; ***the platoon deployed into extended line***

deployment [dɪ'plɔɪmənt] *noun* movement of troops to a war zone or area of operations; ***the deployment to Germany was completed in 72 hours***

depot ['depəʊ *US also* 'diːpəʊ] *noun* **(a)** location where equipment and supplies are stored; ***the bomb hit an oil storage depot*** **(b)** military training establishment; ***recruits were ordered to report to the Guards' Depot***

depth [depθ] *noun* **(a)** vertical distance in water; ***the depth is ten metres***; **depth charge** = anti-submarine bomb which can be set to explode at a specified depth **(b)** extent of a force's position from front to rear; **deployed in depth** = deployed with units or sub-units behind the forward units or sub-units, in order to provide support and to deal with any enemy breakthroughs

deputize ['depjʊtaɪz] *verb* to do someone else's job (on a temporary basis)

deputy ['depjʊti] *noun* person authorised to act in support of or instead of another official; **deputy chief-of-staff (DCOS)** = second-most senior staff officer in a headquarters

derail [di'reɪl] *verb* to make a train come off the rails

derelict ['derəlɪkt] **1** *adjective* **(a)** *(of buildings and ships)* abandoned and no longer maintained **(b)** *US* negligent; ***he was derelict in his duty*** **2** *noun* abandoned building or ship which is no longer maintained

dereliction of duty [derəˌlɪkʃn əv 'djuːti] *noun* failure to carry out your duty

descend [dɪ'send] *verb* to go down

descent [dɪ'sent] *noun* act of going down

desert 1 ['dezət] *noun* region where there is very little water and therefore hardly any life or vegetation; **Desert Storm** = operation mounted by an international coalition to recapture Kuwait in 1991, following its invasion by Iraq; *see also* THE GULF **2** [dɪ'zɜːt] *verb* to leave a military unit without permission; ***he was accused of deserting his post***

deserted [dɪ'zɜːtɪd] *adjective* with no people present; ***the village was deserted***

deserter [dɪ'zɜːtə] *noun* serviceman who leaves his unit without permission

desertion [dɪ'zɜːʃn] *noun* military offence of leaving a unit without permission; ***the punishment for desertion was execution by firing squad***

> COMMENT: desertion implies an intention to absent oneself

> permanently, while temporary absence is usually classified as **absent without leave (AWOL)**

designator ['dezɪgneɪtə] *see* LASER TARGET DESIGNATOR

destination [ˌdestɪ'neɪʃn] *noun* location to which a person or thing is going; ***our destination is Hamburg***

destroy [dɪ'strɔɪ] *verb* to damage something completely; ***the factory has been destroyed***

destroyer [dɪ'strɔɪjə] *noun* medium-sized high-speed warship used to support amphibious or strike forces

> COMMENT: in the British Navy, the destroyer's primary role is air defence (AD)

destruction [dɪ'strʌkʃn] *noun* act of destroying something; ***he was responsible for the destruction of the village***

detach [dɪ'tætʃ] *verb* to remove a soldier or sub-unit from their parent unit, in order to assign them to a separate mission or task; ***6 Platoon has been detached to guard the hospital***; *compare* ATTACH

detachment [dɪ'tætʃmənt] *noun* **(a)** act of detaching a soldier or sub-unit; ***he is on detachment to the air force*** **(b)** small administrative or tactical grouping (normally attached to or supporting another unit); ***two detachments of sappers arrived on the scene***

detail ['diːteɪl] **1** *noun* **(a)** one of several items of information which relate to the same subject; ***he gave me some details on the tactical situation***; **to go into detail** = to give all the information available; **in detail** = item by item; **personal details** = name, date of birth, occupation, address, etc. **(b)** part of a diagram, photograph or picture which is magnified for closer examination; ***we studied a detail showing the bridge*** **(c)** small detachment of soldiers assigned to carry out a specific task; ***a detail of Marines was sent to clear the mines*** **(d)** written order or instruction; ***have you read the company detail today?*** **2** *verb* **(a)** to give a piece of information item by item; ***he detailed the duties for the day***; **detailed support arrangements** = *see* TECHNICAL ARRANGEMENTS **(b)** to assign a soldier or unit to a specific task; ***he was detailed to guard the prisoners***

detain [dɪ'teɪn] *verb* to confine someone or restrict his movements

det-cord ['det kɔːd] *noun* = DETONATING CORD explosive substance contained in a thin length of plastic tube, which is used as an explosive or to detonate a larger explosive charge

detect [dɪ'tekt] *verb* to indicate the presence of an object or substance

detector [dɪ'tektə] *noun* device designed to indicate the presence of something; **detector paper** = specially treated paper, which is designed to detect the presence of chemical agents; **mine-detector** = device designed to locate mines

detention [dɪ'tenʃn] *noun* confinement of a serviceman who has been found guilty of a military offence

> COMMENT: **detention** normally refers to a period of confinement at a person's unit location, whereas **imprisonment** usually refers to confinement in a military prison

deter [dɪ'tɜː] *verb* to discourage someone from doing something through fear of unpleasant consequences

deterrent [dɪ'terənt] *noun* something which deters; **nuclear deterrent** = possession of nuclear weapons in order to deter an attack by a foreign power

detonate ['detəneɪt] *verb* to make an explosive charge explode; **detonating cord** = *see* DET-CORD

detonation [detə'neɪʃn] *noun* **(a)** act of detonating an explosive charge **(b)** explosion

detonator ['detəneɪtə] *noun* small explosive device used to detonate an explosive charge

detour ['diːtʊə] *noun* alteration to a planned route; ***we had to make a detour to avoid the minefield***

detrain [diː'treɪn] *verb* to get out of a train

devastate ['devəsteɪt] *verb* to cause great destruction

de-turf [diː'tɜːf] *verb* to carefully remove the turf from the ground, so that it can be replaced and will continue growing; ***we***

came under fire before we had even finished de-turfing the trenches

devastation [devəs'teɪʃn] *noun* **(a)** act of devastating an area **(b)** widespread destruction

device [dɪ'vaɪs] *noun* instrument or machine which performs a function; **improvised explosive device (IED)** = home-made bomb, booby-trap or mine

DF [ˌdiː 'ef] *noun* = DEFENSIVE FIRE TASK pre-determined artillery target, which has been registered and given a target number

dhobi ['dəʊbɪ] *noun GB* laundry; ***my kit is still at the dhobi, he's doing his dhobi***

dhow [daʊ] *noun* traditional Arab sailing ship

diamond ['daɪəmənd] *noun* **(a)**; *(of vehicles or dismounted infantry)* tactical formation in the form of a square, with one corner pointing in the direction of advance **(b)**; *(of groupings)* tactical formation, with one sub-unit leading as point, followed by two sub-units abreast of each other, followed by one sub-unit centre rear

DIBUA ['dɪbʊə(r)] = DEFENCE IN BUILT-UP AREAS

die [daɪ] *verb* to stop living; ***thousands of soldiers died in the trenches during the First World War***

diesel ['diːzl] *noun* liquid fuel, made from petroleum, used in certain motor vehicles, especially buses, vans, trucks, etc.

diffy [dɪfɪ] *noun GB (slang)* deficient *or* deficiency; ***he's diffy his helmet***

dig [dɪg] *verb* to make a hole in the ground; **to dig in** = to dig trenches or prepare other field fortifications; ***the enemy is digging in*** (NOTE: **digging - dug**)

digging tool ['dɪgɪŋ ˌtuːl] *noun* lightweight tool (such as a pickaxe or shovel) carried by infantrymen in order to dig trenches

dike *see* DYKE

Diphosgene ['daɪfɒzdʒiːn] *see* DP

direct [daɪ'rekt] **1** *verb* **(a)** to control or guide the actions of subordinates or supporting arms; ***a troop commander directs the fire of all the tanks in his troop***; **direct support** = assistance from another unit or arm in which the unit being assisted has control over how the assistance is used; ***the battalion had a battery of guns in direct support for the entire attack*** **(b)** to tell someone the way to a destination; ***a military policeman directed us to the Brigade RV*** **2** *adjective* without deviation or by the shortest way; ***he was ordered to find the most direct route to the bridge***; **direct fire** = fire from weapons which are pointed directly at their targets (eg. rifle, anti-tank gun, guided missile); **direct weapon** = weapon which is pointed directly at its target (for example a rifle, anti-tank gun, guided missile) **3** *adverb* in a straight line or by the shortest route; ***the squadron moved direct to the bridge***

directing staff (DS) [daɪ'rektɪŋ ˌstɑːf] *noun* officers and non-commissioned officers (NCOs) who act as instructors on a course

direction [daɪ'rekʃn] *noun* **(a)** line or course along which anything moves or looks, or along which anything lies; ***the enemy tanks were moving in a south-easterly direction***; ***I looked in the direction of the church*** **(b)** bearing (usually a grid reference) to an artillery or mortar target **(c)** control or guidance; ***he was expected to work without direction from his superiors*** **(d)** instructions on how to go to a destination; ***the sergeant gave us directions to the fuel dump***

direction-finding [daɪ'rekʃn ˌfaɪndɪŋ] *adjective* relating to equipment which is designed to locate radio sets *or* radar by intercepting their emissions; ***the enemy has good direction-finding equipment***

directive [daɪ'rektɪv] *noun* order or instruction which indicates an intended result but does not specify how that result should be achieved; **directive command** = doctrine of command and control where commanders at all levels are informed of the intended result of an operation, but are then free to exercise their own initiative in order to achieve that result, with minimum interference from higher command; *compare* RESTRICTIVE CONTROL

directly [daɪ'rektli] *adverb* **(a)** immediately, without any delay; ***move to the bridge directly*** **(b)** in a straight line, without deviation; ***move directly to the bridge***

dirt road *or* **track** [dɜːt] *noun* US unmetalled road *or* track

disable [dɪs'eɪbl] *verb* **(a)** *(of people)* to injure a person so that he is deprived of the use of one or more of his faculties (such as movement, sight, etc.); ***he was disabled by a sniper's bullet*** **(b)** *(of machines)* to do something to a machine so that it does not work properly; ***the mine disabled the tank's steering system***

disabled [dɪs'eɪbld] *adjective* **(a)** *(of people)* deprived of the use of one or more of your faculties (such as movement, sight, etc.) **(b)** *(of machines)* unable to work properly; ***they towed the disabled tanker into the harbour***

disarm [dɪs'ɑːm] *verb* **(a)** to take a person's weapon away from him; ***we disarmed the enemy forces as they surrendered*** **(b)** to do something to a weapon so that it cannot be fired; ***the gun was disarmed by removing the firing mechanism*** **(c)** to do something to a bomb or other explosive device so that it cannot explode; ***they managed to disarm the bomb before it exploded***

disarmament [dɪs'ɑːməmənt] *noun* reduction of a state's military resources; **nuclear disarmament** = removal or destruction of a country's nuclear weapons

disaster [dɪz'ɑːstə] *noun* situation where a lot of people are killed or injured , or where a lot of damage is caused

disband [dɪs'bænd] *verb* to break up a group or organization; ***the regiment has been disbanded***

disc [dɪsk] *see* IDENTITY DISC

discharge 1 ['dɪstʃɑːdʒ] *noun* **(a)** release of a person from duty; **discharge papers** = document proving that a person has been discharged from the armed forces; *see also* DISHONOURABLE DISCHARGE **(b)** act of carrying out a duty; ***he was accused of obstructing the sergeant in the discharge of his duty*** **(c)** act of firing a weapon; **accidental discharge** *or* **negligent discharge (ND)** = unintentional firing of a weapon **2** [dɪs'tʃɑːdʒ] *verb* **(a)** to release a person from duty; ***he was discharged from the army*** **(b)** to carry out a duty; ***he has discharged his duties satisfactorily*** **(c)** to fire a weapon; ***he discharged his weapon into the crowd***

discharger [dɪs'tʃɑːdʒə] *noun* **(a)** device which fires or releases a projectile or other object; **smoke discharger** = device which releases smoke or smoke canisters **(b)** device which releases the electrical charge from a battery

disciplinary [dɪsɪ'plɪnəri] *adjective* designed to enforce discipline; **disciplinary offence** = offence which is punishable under military law

discipline ['dɪsɪplɪn] **1** *noun* **(a)** control which an army has over its soldiers' actions and behaviour; ***the British Army is famous for its discipline*** **(b)** rules and regulations which maintain control; ***your actions were contrary to good order and military discipline*** **(c)** self-control; **fire discipline** = personal judgement preventing unnecessary wastage of ammunition **2** *verb* to punish; ***he was disciplined under Section 69 of the Army Act 1955***

DISCOM ['dɪskkɒm] *noun* US = DIVISIONAL SUPPORT COMMAND organization responsible for the resupply of a division

disembark [ˌdɪsɪm'bɑːk] *verb* to land from a ship (NOTE: also **debark**)

disembarkation [ˌdɪsembɑː'keɪʃn] *noun* act of landing from a ship (NOTE: also **debarkation**)

disengage [ˌdɪsɪn'geɪdʒ] *verb* to stop fighting with the enemy and withdraw; *compare* ENGAGE; *see also* BREAK CONTACT

disengagement [ˌdɪsɪn'geɪdʒmənt] *noun* action of stopping fighting with the enemy; *compare* ENGAGEMENT

disguise [dɪs'gaɪz] **1** *noun* anything which alters the appearance of something in order to conceal its true identity; ***he was wearing a disguise*** **2** *verb* to alter the appearance of something in order to conceal its true identity; ***the chemical weapons factory was disguised as a hospital***

dishonor *see* DISHONOUR

dishonorable *see* DISHONOURABLE

dishonour *US* **dishonor** [dɪs'ɒnə] **1** *noun* state of disgrace resulting from an action or failure; ***your actions have***

brought dishonour to the regiment **2** *verb* to do something which causes dishonour

dishonourable *US* **dishonorable** [dɪs'ɒnrəbl] *adjective* causing dishonour; **dishonourable discharge** = dismissal of a person from the armed forces after being found guilty of a civil or military offence

dislodge [dɪs'lɒdʒ] *verb* to remove from a firm or secure position; ***the gun became dislodged from its mounting***; ***we were unable to dislodge the enemy from the village***

disintegrate [dɪs'ɪntəˌgreɪt] *noun* to come apart; ***the plane started to disintegrate in mid-air***

disk *see* IDENTITY DISC

dismiss [dɪs'mɪs] *verb* **(a)** to send someone away; ***he dismissed the clerk*** **(b)** to remove someone from their job; ***the brigade commander has been dismissed*** **(c)** to release servicemen at the end of a parade; ***Company, dismiss!***

dismissal [dɪs'mɪsl] *noun* act of dismissing someone

dismount [dɪs'maʊnt] *verb* to get out of a vehicle; ***the infantry dismounted 100 metres from the objective***; *see also* DEBUS

dismounted [ˌdɪs'maʊntɪd] *adjective; (of armoured or mechanized infantry)* on foot; ***this will be a dismounted attack***

disobedience [dɪsə'biːdiəns] *noun* failure or refusal to carry out an order or command

disobey [dɪsə'beɪ] *verb* to fail or refuse to carry out an order or command

disorder [dɪs'ɔːdə] *noun* **(a)** lack of order or cohesion; ***the enemy retreated in disorder*** **(b)** breakdown of law and order; ***the police are unable to deal with the disorder in the capital***

disorient [dɪs'ɔːriənt] *verb* *US* to destroy a person's awareness of his exact location (NOTE: **disorient - disoriented - disorientation;** British English is **disorientate - disorientated - disorientation**)

disorientate [dɪs'ɔːriənteɪt] *verb* *GB* to destroy a person's awareness of his exact location (NOTE: **disorientate - disorientated - disorientation;** American English is **disorient - disoriented - disorientation**)

disorientated [dɪs'ɔːriənteɪtɪd] *adjective* *GB* unsure of your exact location; ***the squad became disorientated and headed towards the minefield***

disoriented [ˌdɪs'ɔːriəntɪd] *adjective* *US* unsure of your exact location; ***when we debussed, we were completely disoriented*** (NOTE: British English is **disorientated**)

dispatch [dɪs'pætʃ] **1** *noun* **(a)** written message; **dispatch-rider** = army motorcyclist used for delivering messages **(b)** official military report; **Mentioned in Dispatches (MID)** = British award in recognition of achievement on operational service **2** *verb* to send someone or something; ***messengers were dispatched to HQ***

dispersal [dɪs'pɜːsl] *noun* act of dispersing; **dispersal point** = location where the sub-units of a grouping divide and go off in different directions

disperse [dɪs'pɜːs] *verb* **(a)** *(of a crowd or group)* to split up and go off in different directions; ***the crowd dispersed when baton rounds were fired*** **(b)** to make something split up and go in different directions; ***the soldiers fired into the air to disperse the crowd*** **(c)** to send information or instructions to several different locations; ***orders were dispersed to the units*** **(d)** *(of a chemical agent)* to become weaker and eventually disappear; ***this gas takes about five minutes to disperse***

displaced person [ˌdɪspleɪst 'pɜːsən] *noun* person who is forced to leave his or her home as a result of war or some other disaster; *see also* REFUGEE

displacement [dɪs'pleɪsmənt] *noun* amount of water moved when a solid object is placed in it; ***this ship has a displacement of 17,000 tons***

disposal [dɪs'pəʊzl] *noun* act of getting rid of something; **bomb disposal** = disarming and safe destruction of unexploded bombs; **bomb-disposal unit** = small group of soldiers trained to make unexploded bombs safe; **explosive ordnance disposal (EOD)** = disarming and safe destruction of explosive ordnance (such as booby-traps, misfires, captured ammunition)

dispose of [dɪs'pəʊz ɒv] *verb* **(a)** to get rid of; ***I disposed of the contaminated***

clothing **(b)** to kill; ***he used a knife to dispose of the sentry*** **(c)** to destroy; ***we used a missile to dispose of the tank***

disposition [dɪspə'zɪʃn] *noun* **(a)** positioning of troops on the ground **(b)** **dispositions** = orders for the positioning of troops

disregard [dɪsrɪ'gɑːd] *verb* to ignore; ***disregard my last order***

disrupt [dɪs'rʌpt] *verb* to cause disorder , to interrupt an activity in progress; ***our mission is to disrupt the enemy's lines of communication***

disruption [dɪs'rʌpʃn] *noun* act of disrupting something

dissemination [dɪˌsemɪ'neɪʃn] *noun* act of sending information or instructions throughout a grouping or other organization; ***the dissemination of the orders took longer than expected***

dissident ['dɪsɪdənt] **1** *noun* **(a)** person who opposes the established government of his own country (where such opposition is illegal) *or* who opposes the system of government itself (especially totalitarian forms of government such as communism *or* fascism); ***the police have been arresting known dissidents*** **(b)** person who actively opposes the leadership of his own political party *or* group; ***dissidents are trying to sabotage the peace talks*** **2** *adjective* being a dissident; ***the bombing was the work of dissident nationalists***

distance ['dɪstəns] *noun* **(a)** space between two locations; ***a distance of five kilometres*** **(b)** **the distance** = area at the limit of a person's vision; ***we saw them in the distance***; **the middle distance** = area half way between an observer's location and the horizon

distant ['dɪstənt] *adjective* far away; ***we are aiming at a distant target***

Distilled Mustard [dɪˌstɪld 'mʌstəd] *see* HD

distress [dɪs'tres] *noun* **(a)** great unhappiness or fear; ***the regulations caused great distress to the civilian population*** **(b)** danger; **distress signal** = signal signifying that a person, ship or aircraft is in danger; **in distress** = in danger; (ship) likely to sink

district ['dɪstrɪkt] *noun* area (normally defined for administrative purposes)

ditch [dɪtʃ] **1** *noun* man-made channel used for drainage; **anti-tank ditch** = ditch dug as an obstacle to tanks and other armoured vehicles **2** *verb (of aircraft)* to make an emergency landing in the sea; ***we were forced to ditch in the sea***

Div = DIVISION

dive [daɪv] **1** *noun* the act of diving **2** *verb* **(a)** to throw oneself head first into water **(b)** to operate underwater (usually with breathing apparatus) **(c)** *(of submarines)* to submerge **(d)** *(of aircraft)* to make a steep descent

dive-bomb ['daɪv ˌbɒm] *verb* to make a steep descent in order to drop a bomb; ***they tried to dive-bomb the cruiser***

dive-bomber ['daɪv ˌbɒmə] *noun* aircraft which makes a steep descent in order to drop a bomb

diver ['daɪvə] *noun* person who operates underwater (usually with breathing apparatus); *see also* FROGMAN

diversion [daɪ'vɜːʃn] *noun* **(a)** attack or raid intended to distract the enemy while another operation is carried out elsewhere; ***the attack was just a diversion***; *see also* FEINT **(b)** alternative route when the road ahead is closed; ***the convoy was late because of a diversion***

diversionary [daɪ'vɜːʃnəri] *adjective* relating to a diversion; ***this is a diversionary attack***

divert [daɪ'vɜːt] *verb* **(a)** to change the direction in which something is heading **(b)** to distract someone

division (Div) [dɪ'vɪʒn] *noun* tactical army grouping of two or more brigades

divisional [dɪ'vɪʒnl] *adjective* relating to a division; ***he reported to divisional headquarters***

dixie ['dɪksɪ] *noun* large rectangular metal cooking-pot, used for cooking in the field; ***as a punishment, you can clean all the dixies***

DM [ˌdiː 'em] *noun* = DIPHENYLAMINOCHLOROARSINE type of vomiting agent (NOTE: also known as **Adamsite**)

DMPI ['dɪmpi] *noun* = DIRECT MEAN POINT OF IMPACT) exact grid reference of a target for an air attack

Dmr = DRUMMER

DMZ [ˌdiː em 'zed *US* ˌdiː em 'ziː] = DEMILITARIZED ZONE

DNBI [ˌdiː en biː 'aɪ] = *US* DISEASE NON-BATTLE INJURY

DOA [ˌdiː əʊ 'eɪ] = DEAD ON ARRIVAL

DOB [ˌdiː əʊ 'biː] = DATE OF BIRTH

doc [dɒk] *noun (informal)* doctor

dock [dɒk] **1** *noun* small area of water enclosed by wharves, where a ship can be loaded and unloaded; **dry dock** = dock from which the water can be removed in order to allow repairs to the hull of a ship **2** *verb (of ships)* to go into a dock

dockyard ['dɒkjɑːd] *noun* place where ships are built and repaired

doctrine ['dɒktrɪn] *noun* standard teaching on a subject; standard principles which guide an action

document ['dɒkjuːmənt] *noun* **(a)** any piece of written material (but not a book *or* pamphlet *or* newspaper, etc), which provides information *or* identification *or* evidence *or* instructions; ***we found a lot of documents in the enemy command post*** **(b)** serviceman's personal records; ***his documents haven't arrived from the depot yet***

DOD [ˌdiː əʊ 'diː] *US* = DEPARTMENT OF DEFENSE

dog [dɒg] *noun* intelligent meat-eating animal with four legs which can be trained to work with man; **dog-handler** = person trained to work with dogs; **guard dog** = dog trained to attack intruders; **sniffer dog** = dog trained to detect explosives or weapons by their smell; **tracker dog** = dog trained to follow the smell of a person

dogfight ['dɒgfaɪt] *noun* battle between aircraft

dogleg ['dɒgleg] *noun* movement to the side at an angle to the normal direction of advance; ***we made a dogleg to avoid the village***

dog-tag ['dɒg ˌtæg] *noun (informal)* metal or plastic disc or lozenge, bearing a soldier's personal details, which is worn round the neck; *see also* IDENTITY DISC

dogwatch ['dɒgwɒtʃ] *noun (naval terminology)* one of two short periods of duty which alternate each day, in order to change a person's daily routine; **first dogwatch** = period of duty from 1600-1800hrs; **second dogwatch** = period of duty from 1800-2000hrs; *see also* WATCH

dominate ['dɒmɪneɪt] *verb* **(a)** to have control over someone or something **(b)** *(of ground)* to look down on; ***this hill dominates the entire valley***

donga ['dɒŋgə] *noun (South Africa)* dry riverbed

doolally [ˌduː'lælɪ] *adjective GB slang* insane; ***he's gone completely doolally***

DOP [ˌdiː əʊ 'piː] = DROP-OFF POINT

DOR [ˌdiː əʊ 'ɑː(r)] *verb US* = DISCHARGE ON REQUEST to ask to be removed from a training course; ***he was DOR'd from the SEAL programme***

dose [dəʊs] *noun* **(a)** amount of medicine given to a person **(b)** amount of radiation received by a person **(c)** *(informal)* infection with a venereal disease

dosimeter [dəʊ'sɪmɪtə] *noun* instrument which measures radiation

dossier ['dɒsiə] *noun* set of documents containing information about someone or something

dot [dɒt] *noun* shorter signal in Morse code (the longer signal is the dash)

down draught ['daʊn ˌdrɑːft] *noun* strong downward current of air given off by a helicopter's rotors

downstream [ˌdaʊn'striːm] *adverb* in the direction in which a river or stream is flowing; ***we moved downstream***; ***the enemy are crossing downstream of the town***; *compare* UPSTREAM

downwind [ˌdaʊn'wɪnd] *adverb* in a position where the wind is blowing from another location towards your own location; ***B Company was downwind of the nuclear explosion***; *compare* UPWIND

DP [ˌdiː 'piː] *noun* = TRICHLOROMETHYL CHLOROFORMATE type of choking agent (NOTE: also known as **Diphosgene**)

DPICM [ˌdiː piː ˌaɪ siː 'em] *noun US* = DUAL-PURPOSE IMPROVED CONVENTIONAL MUNITION one of several small bomblets, which are realeased by airburst from an artillery shell

DPM [ˌdiː piː 'em] *noun GB* = DISRUPTIVE PATTERN MATERIAL camouflage combat uniform; ***he was wearing DPMs*** (NOTE: American English is **BDU**)

draft [drɑːft] **1** *noun* **(a)** *especially US* method of selecting men for compulsory military service; ***he went to Canada to avoid the draft***; *see also* CONSCRIPTION; **draft-dodger** = someone who tries to avoid doing compulsory military service **(b)** group of newly conscripted recruits **(c)** group of reinforcements **2** *verb* to select men for compulsory military service; ***all men over 18 were drafted into the armed forces***; *see also* CONSCRIPT

drag [dræg] **1** *noun* natural force which slows down a flying object ***2*** *verb* to pull along the ground; ***the guns were dragged into position***

Dragon ['drægən] *noun* American hand-held anti-tank guided missile (ATGM)

dragon's teeth ['drægənz ˌtiːθ] *plural noun* concrete pillars used as an obstacle for tanks

dragoon [drə'guːn] *noun (historical)* heavy cavalryman who could also fight as an infantryman

> COMMENT: some modern armoured regiments retain their historical title as Dragoons

Dragunov ['drægu:ˌnɒv] *noun* Soviet-designed 7.62mm sniper rifle

Draken ['drɑːkən] *see* Saab-35

draw [drɔː] *verb* to collect or be issued with; ***you will draw rations at 1500 hours*** (NOTE: **drawing - drew - have drawn**)

dress [dres] **1** *noun* clothing; **barrack dress** = everyday uniform consisting of a sweater and service-dress trousers; **service dress** = smart khaki uniform worn on formal duties and parades **2** *verb* **(a)** to put on clothing; ***for operations in winter, the troops are dressed in white uniforms*** **(b)** to apply a dressing to a wound; ***he went to the RAP to have his wound dressed*** **(c)** to correct the alignment of soldiers on parade; *(command)* **right dress!** = form a straight line, aligned on the soldier at the right end of the line

dressing ['dresɪŋ] *noun* **(a)** absorbent pad and bandage used to cover a wound; **dressing station** = place where battle casualties receive emergency medical treatment before being moved back to a field hospital; **field dressing** = camouflaged dressing designed to treat serious wounds **(b)** alignment of soldiers on parade; ***the dressing is terrible***

drift [drɪft] **1** *noun* **(a)** effects of a current or wind on the course of a ship or aircraft; ***the convoy was slowed down by the strong drift*** **(b)** bank of snow formed by the wind; ***the mountain road was blocked by snow drifts*** **(c)** *(South Africa)* ford; ***we can cross the river at Rorke's Drift*** **2** *verb* to be moved by a current or wind; ***the ship's steering broke and she drifted into a minefield***

drill [drɪl] **1** *noun* **(a)** routine procedure; ***the unloading drill is designed to prevent accidents***; **drill round** = inert round used to practise weapon-handing drills **(b)** the practising of ceremonial movements; ***there was one hour of drill every morning*** **2** *verb* **(a)** to teach a routine procedure through repeated practice; ***the sergeant drilled his recruits in the use of the mortar*** **(b)** to teach ceremonial movements through repeated practice; ***we spent two hours drilling on the parade ground***

drink [drɪŋk] *verb* **(a)** to swallow liquid; ***the men are drinking up their water too quickly*** **(b)** to drink alcohol; ***he was charged with drinking on duty***

drunk [drʌŋk] *adjective* under the influence of alcohol; ***he was drunk on parade***

drip [drɪp] *noun* apparatus, consisting of a bottle *or* other container and a tube, which is designed to introduce liquid gradually into a person's body, either through a needle inserted into a vein *or* through an orifice (eg. mouth, nose, rectum); ***every man was taught how to insert a drip***

drive [draɪv] **1** *noun* **(a)** energy and motivation; ***he has plenty of drive*** **(b)** move forwards; ***the enemy's drive towards the coast was halted*** **2** *verb* **(a)** to operate and steer a vehicle; ***he drives a tank*** **(b)** to travel by vehicle; ***she drove to the hospital***

(c) to make someone do something; ***he drove his men on to take the position*** **(d)** to push in a certain direction; ***they drove on to take the rebel stronghold***; **to drive back** *or* **off** = to force an enemy to retreat; ***we drove back repeated enemy attacks*** (NOTE: **driving - drove - have driven**)

drive-by ['draɪv baɪ] *noun* terrorist assassination method, where the gunman drives up to the victim in a vehicle, shoots him and then drives off.; ***there's been a drive-by in the town square***

driver ['draɪvə] *noun* person who operates and steers a vehicle

drone [drəʊn] *noun* small unmanned radio-controlled aircraft designed to carry surveillance equipment; *see also* REMOTE PILOTED VEHICLE (RPV) (NOTE: also known as an **unmanned aerial vehicle (UAV)**)

drop [drɒp] **1** *noun* **(a)** act of going down; ***a drop in temperature*** **(b)** vertical distance downwards; ***there is a sheer drop of 90m into the sea*** **(c)** deployment by parachute; ***he broke his leg in the last drop*** **(d)** act of leaving something where it can be collected by someone else; ***we are making a drop of ammunition tonight*** **(e)** small portion of liquid (such as blood, rain, etc.); ***there were drops of blood on the floor*** **2** *verb* **(a)** to let something fall to the ground; ***he dropped his rifle***; **drop tank** = additional fuel tank for an aircraft, which can be jettisoned when empty **(b)** to fall or throw oneself onto the ground; ***the platoon dropped when the enemy fired a flare*** **(c)** to make a vertical descent (usually under control); ***he dropped from the window*** **(d)** to deploy troops by parachute; ***the enemy have dropped two airborne divisions in the area of Mensdorf***; **to drop in** = to deploy by parachute **(e)** to deliver supplies by helicopter or parachute; ***the enemy is dropping supplies at night*** **(f)** to offload men or supplies from a vehicle; ***we dropped the patrol at the RV***; ***they dropped the ammunition at the collection point*** **(g)** to correct artillery or mortar fire so that the rounds land closer to the observer; ***Drop 20! (ie. 20 metres)*** *compare* ADD **(h)** *(informal)* to shoot someone; ***drop him!***

droplet ['drɒplət] *noun* tiny particle of liquid (such as chemical agent)

drop-off point (DOP) ['drɒp ɒf ˌpɔɪnt] *noun* **(a)** place where soldiers leave their vehicles to continue an operation on foot **(b)** pre-selected location where men or supplies can be offloaded from vehicles, in order to be collected by another unit

drop zone (DZ) ['drɒp ˌzəʊn] *noun GB* area of ground selected for the landing of troops by parachute (NOTE: the American army uses the phrase **landing zone (LZ)** in this meaning)

drum [drʌm] **1** *noun* **(a)** musical instrument consisting of a cylinder, closed at each end with skin or plastic, which the player beats with two sticks **(b) the Drums** = band of drummers and fife-players belonging to a battalion or regiment (NOTE: also known as **the Fifes and Drums**); **drum major** = senior non-commissioned officer (NCO) in charge of the Drums **(c)** cylindrical container designed to contain liquid (such as oil, petrol); ***the weapons were hidden in an oil drum*** **(d)** cylindrical magazine for certain types of machine-gun; ***this weapon can use belts or drums*** **2** *verb* to beat a drum; *(informal)* **to drum someone out of the forces** = to dismiss a person from the armed forces

drummer (Dmr) ['drʌmə] *noun* **(a)** musician who plays the drums **(b)** rank held by a private soldier in the Drums (NOTE: shortened to **Dmr** in this meaning: **Dmr Jones**)

drumhead court martial [ˌdrʌmhed kɔːt 'mɑːʃl] *noun* court martial held in the field

drumhead service ['drʌmhed ˌsɜːvɪs] *noun* religious service held in the field or on the parade ground

dry dock ['draɪ ˌdɒk] *noun* dock from which the water can be removed in order to allow repairs to the hull of a ship

dry run [ˌdraɪ 'rʌn] *noun* rehearsal for an operation; ***there will be a dry run at 1400hrs***

dry season ['draɪ ˌsiːzn] *noun* time of the year when there is no rain; *compare* MONSOON, RAINY SEASON

DS [ˌdiː 'es] = DIRECTING STAFF

DSO ['diːsəʊ] *noun US* = DEFENSIVE SYSTEMS OFFICER aircrew member on a bomber, who detects and locates threats to the aircraft (eg. enemy aircraft, radar, missiles, etc); *compare* OSO

DTG [ˌdiː tiː ˈdʒiː] = DATE-TIME GROUP

DU [ˌdiː ˈjuː] = DEPLETED URANIUM

duck [dʌk] *verb* to lower one's head and upper body instinctively to avoid a projectile; ***he ducked at the sound of the explosion***

duckboard [ˈdʌkbɔːd] *noun* strips of wood nailed together, in order to provide a dry path across muddy ground

dud [dʌd] *noun* **(a)** shell or other projectile which fails to fire or explode **(b)** battery without any electric charge

dug in [ˌdʌg ˈɪn] *adverb* protected by field fortifications; ***the enemy are well dug in***

dugout [ˈdʌgaʊt] *noun* shelter dug into the side of a trench

dumb bomb [ˈdʌm ˌbɒm] *noun (informal)* bomb without a guidance system which is simply dropped by an aircraft onto its target; *see also* GP BOMB, IRON BOMB; *compare* SMART BOMB

dum-dum bullet [ˈdʌmdʌm ˌbʊlɪt] *noun* bullet modified to expand when it hits a person or animal, thereby causing a terrible wound

dummy [ˈdʌmɪ] *adjective* imitation (for the purposes of deception); ***the engineers have been constructing dummy positions on the ridge***

dummy run [ˌdʌmi ˈrʌn] *noun* act of practising an operation before doing it for real

dump [dʌmp] **1** *noun* temporary store in the field; ***the bomb scored a direct hit on an ammunition dump***; ***we must try to locate the enemy's supply dump*** **2** *verb* **(a)** to leave ammunition, fuel, etc., in a temporary store; ***the ammunition has been dumped at grid 341632*** **(b)** to abandon a vehicle; ***the truck was dumped at the side of the road***

dune [ˈdjuːn] *noun* bank or small hill of loose sand formed by the wind

duplicate 1 [ˈdjuːplɪkət] *noun* a second copy of a document; **in duplicate** = in two copies **2** [ˈdjuːplɪkeɪt] *verb* **(a)** to do something twice **(b)** to produce two copies of a document

duration [djuˈreɪʃn] *noun* the length of time that an activity continues; **for the duration** = until an activity finishes; ***for the duration of the war***

dusk [dʌsk] *noun* period between sunset and when it is fully dark; *compare* DAWN

dust [dʌst] *noun* fine particles of sand, soil or any other material; **dust storm** = strong wind producing dense clouds of dust

dust-off [dʌst ˈɒf] *noun US (radio terminology)* evacuation of casualties by helicopter; ***we require dust-off at grid 342659***

duty [ˈdjuːti] *noun* **(a)** moral or legal obligation; ***it is your duty to obey orders*** **(b)** specified tasks which a person is required to do as part of his job; ***that is not one of my duties***; **duty officer** = officer assigned by his unit to deal with incidents and carry out various routine tasks during a specified period; **off duty** = not at work; **on duty** = at work; ***England expects every man to do his duty - Nelson***

dwelling [ˈdwelɪŋ] *noun* building used as a home

dyke [ˈdaɪk] *noun* **(a)** embankment built to prevent flooding **(b)** drainage ditch (NOTE: can also be spelled **dike**)

dynamite [ˈdaɪnəmaɪt] *noun* explosive material made from nitroglycerine

dysentery [ˈdɪsəntri] *noun* disease which inflames the intestines, causing severe diarrhoea

DZ [ˌdiː ˈzed *US* ˌdiː ˈziː] = DROP ZONE

ECHO - Ee

E-2 [ˌiː ˈtuː] *noun* American-designed airborne early warning aircraft, with a large disc-like antenna (radome) mounted on the fuselage, which is capable of being used from an aircraft carrier (NOTE: also known as **Hawkeye**)

E-3 [ˌiː ˈθriː] *noun* American-designed airborne warning and control system (AWACS) aircraft, which is based on a passenger airliner, and has a large disk-like antenna (radome) mounted on the fuselage (NOTE: also known as **Sentry**)

EA [ˌiː ˈeɪ] = ENGAGEMENT AREA

EA-6 [ˌiː eɪ ˈsɪks] *noun* American-designed electronic warfare aircraft, based on the A-6 Intruder (NOTE: also known as **PROWLER**)

Eagle [ˈiːgəl] *see* F-15

ear-defenders [ˈɪə dɪˌfendəz] *plural noun* device which is placed in *or* over the ears, in order to protect them from noise; ***ear-defenders must be worn on the range***

earphone [ˈɪəfəʊn] *noun* part of the headset for a radio or other audio equipment, which is put on the ear to listen to a transmission or signal

earpiece [ˈɪəpiːs] *noun* part of a radio or telephone handset, which is put in the ear to listen to a transmission

earthmover [ˈɜːθmuːvə] *see* ARMORED COMBAT EARTHMOVER

earthwork [ˈɜːθwɜːk] *noun* man-made

ease springs [iːz ˈsprɪŋz] *verb GB* final action of the unloading drill for an automatic *or* semi-automatic weapon; after checking that the breech is clear of ammunition, let the working-parts go forward and pull the trigger

east [iːst] **1** *noun* **(a)** one of the four main points of the compass, corresponding to a bearing of 90 degrees or 1600 mils **(b)** area to the east of your location; ***the enemy are approaching from the east*** **(c) the East** = part of the world to the east of Europe; **the Far East** = region consisting of China, Japan and neighbouring countries; **the Middle East** = region consisting of Arab countries (such as Egypt, Iran, Iraq, Saudi Arabia, the United Arab Emirates) and Israel; **the Near East** = region consisting of countries of the eastern Mediterranean (such as Cyprus, Lebanon, Turkey) **(d)** the eastern part of a country **2** *adjective* relating to east; ***the East Gate***; **east wind** = wind blowing from the east **3** *adverb* towards the east; ***the enemy is moving east***

eastbound [ˈiːstbaʊnd] *adjective* moving or leading towards the east; ***an eastbound convoy***

easterly [ˈiːstəli] *adjective* **(a)** towards the east; **to move in an easterly direction** = to move towards the east **(b)** *(of wind)* from the east

eastern [ˈiːstən] *adjective* relating to the east; ***the eastern part of the country***; **Eastern Bloc** = name sometimes given to the Warsaw Pact; **Eastern Europe** = region consisting of countries which were allied to the USSR during the Cold War (such as Bulgaria, the Czech Republic, Hungary, Poland, Romania, Slovakia, the former Yugoslavia)

easting [ˈiːstɪŋ] *noun* **(a)** vertical line of a map grid **(b)** one of the coordinates running from left to right across a map; *compare* NORTHING

eastward [ˈiːstwəd] **1** *adjective* towards the east; ***a eastward direction*** **2** *adverb US* towards the east; ***they are moving eastward***

eastwards [ˈiːstwədz] *adverb* towards the east; ***they are moving eastwards***

eavesdrop ['iːvzdrɒp] *verb* to listen secretly to a conversation between other people

ebb tide ['eb ˌtaɪd] *noun* tide which is moving out to sea

Ebola [iː'bəʊlə] *noun* virus which causes Ebola haemorrhagic fever (Ebola HF), a severe and often fatal disease affecting man and monkeys

> COMMENT: the Ebola virus is believed to have originated in Africa. It is extremely rare and as yet, noone knows how it is initially transmitted to humans. However, once people are affected, the disease can then be transmitted through contact with blood *or* body fluids from an infected person. Its symptoms are fever, headaches, vomiting, diarrhoea, massive internal bleeding and in most cases death. There is no known cure at present, although some people seem to have a natural immunity to the disease and do recover. Research has shown that the virus could also be spread through airborne particles (aerosols) and there is a strong possibility that it might be developed for use as a biological weapon

ECCM [ˌiː siː siː 'em] *noun* = ELECTRONIC COUNTER-COUNTER MEASURES procedures used to defeat an enemy's electronic counter measures (ECM)

echelon ['eʃlɒn] *noun* **(a)** tactical formation in which troops, vehicles or aircraft are deployed in a series of parallel lines, each of which is longer than the one in front; **A1 Echelon** = combat supplies; **A2 Echelon** = rations, spares, clothing, etc; **echelon attack** *or* **attack in echelon** = attack made by several units deployed side by side, where one unit sets off first, followed after an interval by the second, followed after another interval by the third, and so on **(b)** part of a tactical grouping; **A Echelon** = logistical elements of a tactical grouping; **B Echelon** = administrative elements of a tactical grouping; **F Echelon** = fighting elements of a tactical grouping

Echo ['ekəʊ] fifth letter of the phonetic alphabet (Ee)

echo ['ekəʊ] *noun* sound which is reflected by an object back towards the listener; **echo location** = method of finding objects under water by sending sound signals down and listening for the echo

ECM = ELECTRONIC COUNTERMEASURES

ECMM [ˌiː siː em 'em] = EUROPEAN COMMUNITY MONITORING MISSION

ECOMOG ['ekəʊˌmɒg] = ECONOMIC COMMUNITY OF WEST AFRICAN STATES CEASE-FIRE MONITORING GROUP

economy of force [ɪˌkɒnɪmɪ əv 'fɔːs] *noun* not wasting your military assets and, if practicable, using fewer forces than normal on your initial tasks, so that you will have the maximum forces avaliable for your main effort

ECP [ˌiː siː 'piː] = EQUIPMENT COLLECTION POINT

ED [ˌiː 'diː] *noun* = ETHYLDICHLOROARSINE type of blister and vomiting agent

EF-111A [ˌiː ef wʌn ɪˌlevən 'eɪ] *noun* American-designed electronic warfare (EW) aircraft (NOTE: also known as **Raven**)

effective enemy fire [ɪˌfektɪv ˌenəmi 'faɪə] *noun* situation where the enemy's fire starts to cause casualties amongst your own troops

EFP [ˌiː ef 'piː] = EXPLOSIVELY FORMED PROJECTILE

eject [ɪ'dʒekt] *verb* **(a)** to throw or drive out; ***the enemy have been ejected from the village***; ***my rifle is not ejecting the empty cases properly*** **(b)** to use an ejector seat; ***the pilot ejected over the sea***

ejection [ɪ'dʒekʃn] *noun* escape from an aircraft using an ejector seat

ejector [ɪ'dʒektə] *noun* something which ejects (such as the part of a firearm which ejects the empty cases); **ejector seat** = seat designed to eject a pilot or crew member from a damaged aircraft

élan [eɪ'læn] *noun* great enthusiasm and drive

electromagnetic pulse (EMP) [ɪˌləktrəʊmægˌnetɪk 'pʌls] *noun* surge of electromagnetic radiation given off by a nuclear explosion, which causes electrical equipment (such as radios, vehicle batteries, etc.) to stop working

electronic [ɪlek'trɒnɪk] *adjective* relating to the use of electricity; **electronic counter-countermeasures (ECCM)** = procedures used to defeat the enemy's electronic countermeasures; **electronic countermeasures (ECM)** = standard procedures designed to minimize a unit's chances of being located by the enemy through emissions given off by its electrical equipment; **electronic silence** = state when all radios and other transmitting equipment (such as radar) must be switched off; **to impose electronic silence** = to start electronic silence; **to lift electronic silence** = to end electronic silence; **electronic warfare (ELW** *or* **EW)** = location and suppression of an enemy's electronic equipment; **electronic warfare officer (EWO)** = crewman of an EW aircraft who navigates the aircraft and operates its electronic warfare equipment

elephant ['elɪfənt] US *to see the elephant*; = to experience combat for the first time; ***you don't know how you'll react until you've seen the elephant***

element ['elɪmənt] *noun* part of a grouping; ***elements of the enemy advance guard have been sighted***

elevate ['elɪveɪt] *verb* **(a)** to put something in a higher position **(b)** to raise the barrel of an artillery piece or mortar

elevation [elɪ'veɪʃn] *noun* **(a)** act of elevating something; ***the elevation is too high*** **(b)** angle at which the barrel of an artillery piece or mortar is raised in order to engage a target; ***elevation: 51 degrees!*** **(c)** area of high ground; ***an enemy battalion is dug in on that elevation***

elevator ['elɪveɪtə(r)] *noun* **(a)** machine which takes people up *or* down from one floor to another in a building **(b)** moving part of the tailplane of an aircraft, which is used to control pitch

eliminate [ɪ'lɪmɪneɪt] *verb* **(a)** to get rid of; ***that option has been eliminated*** **(b)** to kill; ***we must eliminate the sentries before the main assault goes in***

ELINT ['i:lɪnt] *noun* = ELECTRONIC INTELLIGENCE information on the enemy, which is obtained through the monitoring of his electronic transmissions by specially equipped aircraft

elite [eɪ'li:t] *adjective* of very high quality; ***he commands an elite regiment of Presidential Guards***

ELW [ˌi: el 'dʌb(ə)lju:] = ELECTRONIC WARFARE

embankment [ɪm'bæŋkmənt] *noun* man-made bank of soil or stone used as a barrier, or to carry a railway or road; *compare* CUTTING

embark [ɪm'bɑ:k] *verb* to go on board an aircraft or ship (in order to travel somewhere); *compare* DEBARK

embarkation [ˌembɑ:'keɪʃn] *noun* act of going on board an aircraft or ship; *compare* DEBARKATION

embassy ['embəsi] *noun* building used by an ambassador and his or her staff

embrasure [ɪm'breɪʒə] *noun* opening in a wall or parapet, through which a weapon can be fired; *compare* FIRING PORT

embus [ɪm'bʌs] *verb* to get into a bus, in order to travel somewhere; *compare* DEBUS

emergency [ɪ'mɜ:dʒənsi] *noun* situation where immediate action is required in order to prevent injury or damage or some other serious misfortune; **emergency rendezvous (ERV)** = location where people assemble in the event of an emergency

EMCON ['emkɒn] *noun* = EMISSION CONTROL measures to reduce emissions which can be detected by the enemy; ***we have a strict EMCON policy***

emission [ɪ'mɪʃən] *verb* **(a)** act of emitting something; ***this engine has been designed to produce a lower emission of heat*** **(b)** something emitted; ***we must reduce our emissions***

emit [ɪ'mɪt] *verb* to give off something (eg. heat, radiation, radar, noise, etc)

EMP [ˌi: em 'pi:] = ELECTROMAGNETIC PULSE

emplacement [em'pleɪsmənt] *noun* prepared firing position for an artillery piece or other large weapons system; ***the camp is surrounded with gun emplacements***

emplane [ɪm'pleɪn] *verb* to go on board an aircraft (in order to travel somewhere); ***we emplane at 0600hrs***

encamp [en'kæmp] *verb* to set up a camp; ***they were encamped by the river***

encampment [en'kæmpmənt] *noun* place where troops are camped

encipher [en'saɪfə] *verb* to convert from normal language into code; *compare* DECIPHER; *see also* ENCODE, ENCRYPT

enclave ['enkleɪv] *noun* a piece of territory, belonging to one state or occupied by one ethnic group, which is surrounded by territory belonging to another state or occupied by a different ethnic group; ***a Muslim enclave, surrounded by Orthodox territory***

encode [ɪn'kəʊd] *verb* to convert from normal language into code; *compare* DECODE; *see also* ENCIPHER, ENCRYPT

encounter [ɪn'kaʊntə] **1** *noun* **(a)** meeting which happens by chance; ***he did not report the encounter*** **(b)** military engagement which happens by chance; ***three of our men were killed in the encounter*** **2** *verb* to meet or make contact by chance; ***we didn't encounter any enemy***; ***the invading force encountered only light resistance***

encrypt [ɪn'krɪpt] *verb* to convert from normal language into code; *compare* DECRYPT; *see also* ENCIPHER, ENCODE

endemic [en'demɪk] *adjective; (of disease)* regularly affecting a large proportion of the population of an area *or* region; ***venereal disease is virtually endemic amongst the local population***

ENDEX ['endeks] = END OF EXERCISE

enemy ['enəmi] **1** *adjective* referring to a state which is at war with your own country; ***enemy snipers attacked the convoy***; ***she listened to enemy propaganda on the radio***; *compare* FRIENDLY; *see also* HOSTILE **2** *noun* **(a)** national of a state which is at war with your own country; ***we consider the French as allies, not enemies*** **(b)** state which is at war with your own country **(c) the enemy** = enemy forces; ***the enemy is withdrawing*** (NOTE: use the third person singular, when referring to the enemy in general)

enfilade [ˌenfɪ'leɪd] **1** *adjective* capable of engaging the entire frontage or length of a formation or position; ***our advance was halted by heavy enfilade fire*** **2** *verb* to engage the entire frontage or length of a formation or position; ***the enemy managed to enfilade our forward trenches***

enforce [ɪn'fɔːs] *verb* to use force or the law in order to make someone do something

enforcement [ɪn'fɔːsmənt] *noun* act of enforcing something

engage [ɪn'geɪdʒ] *verb* to start to fight or shoot at someone; ***we started to engage the enemy infantry at around 600 metres***; *compare* DISENGAGE; ***You engage, and then you see what happens - Napoleon***

engagement [ɪn'geɪdʒmənt] *noun* exchange of fire between opposing forces; ***the engagement lasted just over an hour***; *compare* DISENGAGEMENT; *see also* BATTLE, FIREFIGHT, SKIRMISH; **rules of engagement** = set of rules governing the firing of weapons and use of force by soldiers (usually in a peacekeeping *or* counter- insurgency role)

engineer [endʒə'nɪə] *noun* **(a)** specialist soldier trained in the construction and demolition of bridges, field fortifications, obstacles, roads, etc.; **Royal Engineers (RE)** = engineers of the British Army (NOTE: traditionally called **Sappers**) **(b)** mechanic on a ship; **engineer officer** = officer in the navy who specializes in ship's engines

engineering [endʒə'nɪərɪŋ] *noun* **(a)** construction and use of engines and other mechanical devices **(b)** construction or demolition of buildings, installations, roads, etc.; **field engineering** = tasks carried out by engineers in support of ground forces (such as the construction, repair and demolition of bridges, construction of field fortifications, construction and removal of obstacles, etc.)

enlist [ɪn'lɪst] *verb* to join the armed forces as a career; ***he enlisted at the age of 18***

enlisted man [ɪn'lɪstɪd mæn] *noun US* any serviceman who is not an officer (NOTE: British English is **other rank (OR)**)

enquiry [ɪŋ'kwaɪəri] *noun* official investigation into the cause of an incident

ensign ['ensaɪn] *noun* **(a)** flag; **White Ensign** = flag displayed by ships of the British Royal Navy (it is white, with a red cross and the Union Jack in one corner); **Red Ensign** = flag displayed by ships of the British Merchant Navy (it is red, with the Union Jack in one corner) **(b)** officer who carries a flag on parade **(c)** *GB* second-lieutenant in the Brigade of Guards **(d)** *US* lowest officer rank in the navy

entanglement [ɪn'tæŋglmənt] *noun* obstacle to infantry constructed from barbed wire; **low-wire entanglement** = obstacle, consisting of a lattice of barbed wire, which is set at ankle height, in order to trip up infantry as they assault a position

entrain [ɪn'treɪn] *verb* to get on a train (in order to travel somewhere); ***we entrained at Dover***

entrench [ɪn'trentʃ] *verb* to dig a trench (as a field fortification); *see also* DIG IN

entrenched [ɪn'trentʃt] *adjective* protected by trenches and other field fortifications; ***the enemy was well entrenched to the north of the hill***; *see also* DUG IN

entrenching tool [ɪn'trentʃɪŋ ˌtu:l] *noun* light-weight tool carried by infantrymen in order to dig trenches; *see also* DIGGING TOOL

envelop [ɪn'veləp] *verb* to manoeuvre against the flanks or rear of an enemy force in order to surround them

envelopment [ɪn'veləpmənt] *noun* attack made on one or both of the enemy's flanks or rear, and usually accompanied by a frontal attack

environs [ɪn'vaɪərənz] *noun* surrounding area; ***they concentrated the bombing on the environs of the city***

EOD [ˌi: əʊ 'di:] *noun* = EXPLOSIVE ORDNANCE DISPOSAL disarming and safe destruction of explosive ordnance (such as booby-traps, misfires, captured ammunition); *see also* BOMB DISPOSAL

epaulette *US* **epaulet** [epə'let] *noun* **(a)** shoulder decoration on a uniform jacket; ***aides-de-camp wear gold epaulettes*** **(b)** semi-detachable flap on the shoulders of a uniform jacket, designed to carry badges of rank or unit insignia

epidemic [ˌepɪ'demɪk] *noun* rapid spreading of an infectious disease through a community; ***we've got an epidemic of typhus in the town***

EPW [ˌi: pi: 'dʌb(ə)lju:] = ENEMY PRISONER OF WAR

Equator [ɪ'kweɪtə] *noun* imaginary line running around the earth, exactly halfway between the North and South Poles; *see also* THE LINE

equerry ['ekwəri] *noun* officer who acts as personal assistant to a member of the British Royal Family; ***he has been appointed an equerry to the Prince of Wales***

equip [ɪ'kwɪp] *verb* to provide someone with equipment; ***we equipped the platoon with shovels***

equipment [ɪ'kwɪpmənt] *noun* any article which a person needs in order to carry out a task (for example clothing, radios, tools, weapons, vehicles, etc.)

ERA [ˌi: ɑ: 'eɪ] = EXPLOSIVE REACTIVE ARMOUR

era ['ɪərə] *noun* a distinct period of time; **from the 1960s-era** = introduced during the years 1960-1969; ***the M-60 is an American 1960s-era main battle tank***

ERFB [ˌi: ɑ: ef 'bi:] = EXTENDED RANGE FULL BORE *noun* advanced aerodynamic design for artillery shells, which increases their range

ERFBB = EXTENDED RANGE FULL BORE BASE BLEED

ERV [ˌi: ɑ: 'vi:] = EMERGENCY RENDEZVOUS

escape [ɪs'keɪp] **1** *noun* act of escaping; ***his daring escape from the POW camp*** **2** *verb* **(a)** to get away from captivity; ***three prisoners escaped during the night*** **(b) to escape capture** = to avoid being captured **(c)** to survive a potentially lethal situation; ***the pilot of the crashed helicopter escaped with minor injuries***; **escape capsule** = enclosed box with seats and survival equipment, in which crewmembers can escape from an aircraft **(d)** *(of a chemical agent, gas, liquid, etc.)* to leak; ***gas escaped into the cabin***

escarpment [ɪs'kɑ:pmənt] *noun* steep slope along the edge of a plateau; ***we saw a***

column of vehicles moving along the base of the escarpment

escort 1 ['eskɔːt] *noun* person, vehicle or aircraft or ship which accompanies an individual or group in order to protect them **2** [ɪs'kɔːt] *verb* to act as an escort; ***the convoy was escorted by two destroyers***

ESDI = EUROPEAN SECURITY AND DEFENCE IDENTITY

espionage ['espiənɑːʒ] *noun* use of spies, surveillance equipment, etc., in order to collect information about the enemy; **counter-espionage** = action taken to impede the activities of enemy spies

esprit de corps [eˌspriː də 'kɔː] *French phrase* feeling of devotion to and pride in your unit or grouping

establish [ɪs'tæblɪʃ] *verb* **(a)** to set something up; ***we have established an OP on the ridge*** **(b)** to consolidate your position; ***the enemy is now established on the western bank of the river*** **(c)** to find out or verify a piece of information; ***we haven't managed to establish the full strength of the enemy***; **to establish communications** = to carry out a radio check in order to ensure that all call-signs on the net are in radio contact

establishment [ɪs'tæblɪʃmənt] *noun* **(a)** act of establishing something **(b)** number of men, vehicles and equipment which should to be held by a grouping at full strength; ***this platoon has an establishment of 28 men***

estimate 1 ['estɪmət] *noun* approximate idea of distance, size, time, etc., obtained by judgement rather than by accurate measurement; ***we think there are 3,000 enemy troops in the camp, but that is only an estimate*** **2** ['estɪmeɪt] *verb* to judge distance, size, time, etc., instead of counting or making accurate measurements; ***he estimated the distance at 1,500m***; **estimated time of arrival (ETA)** = time when a vehicle, group of soldiers, etc., is expected to arrive

ETA [ˌiː tiː 'eɪ] = ESTIMATED TIME OF ARRIVAL

ethnic ['eθnɪk] *adjective* relating to race or cultural background; **ethnic cleansing** = systematic attempt by the main population to drive members of an ethnic minority away from their homes by the use of force (including intimidation, destruction of property, physical violence and even murder); **ethnic minority** = smaller group of people who have a different racial or cultural background to the main population

ETR [ˌiː tiː 'ɑː(r)] *noun* = ELECTRONIC TARGET RANGE shooting range where the targets are raised and lowered by electricity

EUCOM ['juːkɒm] *noun* *US* = EUROPEAN COMMAND department of US forces responsible for defending American national interests in Europe and also Israel, Syria and Lebanon

Eurofighter ['jʊərəʊˌfaɪtə] *noun* European multirole fighter aircraft, produced by a consortium of companies from the UK, Germany, Italy and Spain (NOTE: now called the **Typhoon**)

European Security and Defence Identity (ESDI) *noun* feeling of common aims and responsibilities with is developed among members of NATO

evacuate [ɪ'vækjʊeɪt] *verb* **(a)** to remove people from their homes because of danger and make them stay elsewhere until that danger is over; ***the civilian population was evacuated across the river*** **(b)** to leave a place of danger; ***the platoon evacuated the position when it became too dangerous***

evacuation [ɪˌvækjʊ'eɪʃn] *noun* act of evacuating

evacuee [ɪˌvækjʊ'iː] *noun* person who has been evacuated

evade ['ɪveɪd] *verb* to take avoiding action; ***we managed to evade the enemy patrols***

evasion [ɪ'veɪʒn] *noun* the art of avoiding enemy forces (usually after escape from capture or encirclement)

evasive [ɪ'veɪzɪv] *adjective* intended to evade a danger or threat; ***we were forced to take evasive action***

evergreen ['evəgriːn] *noun* tree which does not lose its leaves in winter (such as fir, pine, spruce, etc.); *compare* DECIDUOUS; *see also* CONIFER, FIR

EW [ˌiː 'dʌb(ə)ljuː] = ELECTRONIC WARFARE

EWO ['i:wəʊ] = ELECTRONIC WARFARE OFFICER

exclude [ɪk'sklu:d] *verb* **(a)** to keep someone or something out; ***the aim is to exclude enemy ships from our territorial waters*** **(b)** to not include; ***the plane was carrying 215 men, excluding the crew***

exclusion zone [ɪk'sklu:ʒn ˌzəʊn] *noun* area or region, defined by a state or by international agreement, which the armed forces or shipping of another state are not allowed to enter; ***in 1982, the British Government declared a 400 mile exclusion zone around the Falkland Islands***

exclusive [ɪk'sklu:sɪv] *adjective* not including; ***our sector is exclusive of the main road***; *compare* INCLUSIVE

execute ['eksɪkju:t] *verb* **(a)** to kill a person who has been found guilty of an offence for which the punishment is death; ***he was executed for cowardice*** **(b)** to carry out a planned task; ***we were unable to execute our mission***

execution [ˌeksɪ'kju:ʃn] *noun* **(a)** act of killing a person who has been found guilty of an offence for which the punishment is death **(b)** method by which a planned task is carried out; **to put a plan into execution** = to carry out a plan; ***the art of war is a simple art and all in execution; there is nothing vague about it; it is all common sense - Napoleon***

executive officer (XO) [ɪgˌzekjʊtɪv 'ɒfɪsə] *noun* *US* officer responsible for coordinating staff functions within a headquarters

> COMMENT: in many groupings, the XO is also the second in command (2IC)

exercise ['eksəsaɪz] *noun* **(a)** act of practising the skills that a unit or sub-unit will be required to carry out on operational service **(b)** **exercises** = military training plan; ***the fleet is taking part in NATO exercises in the Mediterranean***; ***joint Anglo-Spanish exercises will be held next week***; *see also* MANOEUVRES, WAR GAMES **(c)** physical activity designed to improve or maintain fitness; ***you should take more exercise***; ***these exercises are designed to improve your arm muscles***

exfiltrate ['eksfɪltreɪt] *verb* to break down a grouping into smaller groups which can withdraw by different routes through territory controlled by the enemy; *compare* INFILTRATE

exfiltration [eksfɪl'treɪʃn] *noun* act of exfiltrating; *compare* INFILTRATION

exhaust [ɪg'zɔ:st] **1** *noun* smoke and waste gases expelled from a running engine or motor; **exhaust pipe** = pipe through which the exhaust is expelled from an engine or motor **2** *verb* to use up a resource completely; ***we have exhausted our fuel supply***

exhausted [ɪg'zɔ:stɪd] *adjective* **(a)** *(of resources)* completely used up; ***our ammunition is exhausted*** **(b)** *(of people)* very tired and weak, as a result of strenuous physical activity; ***after two weeks in the jungle, the commandos returned to base completely exhausted***

exhaustion [ɪg'zɔ:stʃn] *noun* total loss of strength (as a result of strenuous physical activity); ***the three escaped prisoners were picked up by one of our patrols in a state of complete exhaustion***

Exocet ['eksəset] *noun* French-designed short-range radar-guided anti-ship missile, usually launched from a ship or aircraft

expedite ['ekspədaɪt] *verb* to carry out an action or task

expeditionary force [ekspəˌdɪʃənri 'fɔ:s] *noun* military grouping sent on a special mission overseas; ***the expeditionary force landed under cover of darkness***

explode [ɪk'spləʊd] *verb* to burst outwards due to a release of internal energy; ***the bomb exploded at five o'clock***

exploit [ɪk'splɔɪt] *verb* to take advantage of something; ***the general failed to exploit the breakthrough***

exploitation [ˌeksplɔɪ'teɪʃn] *noun* continuation of a successful attack after the objective has been taken, in order to destroy the enemy's ability to conduct an orderly withdrawal or organize a defence or counter attack; **limit of exploitation** = point on the ground beyond which the exploitation of a successful attack should not continue; ***our limit of exploitation is the rear edge of the wood***

explosion [ɪk'spləʊʒn] *noun* act of exploding; ***the bombing raid set off a series of explosions at the munitions factory***

explosive [ɪk'spləʊsɪv] **1** *adjective* **(a)** designed or liable to explode; ***nitrogen-based fertilizers can be used to make an explosive substance***; **explosive ordnance** = general term for any projectile or device which contains an explosive substance or which uses an explosive substance as its propellant; **explosive ordnance disposal (EOD)** = disarming and safe destruction of explosive ordnance (such as booby-traps, misfires, captured ammunition); *see also* BOMB DISPOSAL; **high explosive (HE)** = powerful explosive substance used in bombs, grenades, shells, etc. **(b)** liable to cause an outburst of violent behaviour; ***the Chief of Police described the situation as 'explosive'*** **2** *noun* explosive substance; ***a large quantity of explosive was found in the house***; **explosive reactive armour (ERA)** = secondary armour, containing a thin layer of explosive, fitted to the outside of an armoured vehicle; designed to counter the effect of an anti-tank projectile by exploding outwards when hit; **home-made explosive** = explosive substance manufactured by terrorists or criminals from readily available ingredients (such as fertilizer); **improvised explosive device (IED)** = home-made bomb or mine; **plastic explosive** = soft explosive substance which can be moulded into a required shape by hand

explosively formed projectile (EFP) [ɪkˌspləʊzɪvli ˌfɔːmd prə'dʒektaɪl] *noun* anti-tank warhead where high explosive is packed around a shallow hemispherical metal plate (on impact, the plate forms itself into a solid metal projectile, which is capable of penetrating armour)

ex-serviceman [ˌeks 'sɜːvɪsmən] *noun* man who formerly served in the armed forces; ***the parade was attended by ex-servicemen*** (NOTE: American English is **veteran**)

extract [ɪk'strækt] *verb* **(a)** to remove one object from another; ***he extracted the empty case from the gun*** **(b)** to move out of an area of operations; ***we will extract by helicopter*** **(c)** to move someone out of an area of operations; ***the patrol was extracted by helicopter***; *compare* INSERT

extraction [ɪk'strækʃn] *noun* act of extracting; *compare* INSERTION

FOXTROT - Ff

F-111 [ˌef ˌwʌn ɪˈlevn] *noun* American-designed attack aircraft (NOTE: also called the **Aardvark;** plural is **F-111s** [ˌef ˌwʌnɪ ˈlevnz])

F-117A [ˌef wʌn sevəntiːn ˈeɪ] *noun* American-designed stealth attack aircraft (NOTE: also called the **Nighthawk;** plural is **F-117As** [ˌef wʌn sevəntiːn ˈeɪz])

F-14 [ˌef fɔːˈtiːn] *noun* American-designed multirole fighter, designed to operate from an aircraft carrier (NOTE: also known as the **Tomcat;** plural is **F-14s** [ˌef fɔːˈtiːnz])

F-15 [ˌef fɪfˈtiːn] *noun* American-designed fighter aircraft with a secondary attack role (NOTE: also known as the **Eagle;** plural is **F-15s** [ˌef fɪfˈtiːnz])

F-16 [ˌef sɪksˈtiːn] *noun* American-designed multirole fighter aircraft, with advanced fighter-ground-attack (FGA) capability (NOTE: also known as the **Fighting Falcon;** plural is **F-16s** [ˌef sɪksˈtiːnz])

FA-18 [ˌef eɪ eɪˈtiːn] *noun* American-designed lightweight multirole fighter aircraft (NOTE: also known as the **Hornet;** plural is **FA-18s** [ˌef eɪ eɪˈtiːnz])

F-22 [ˌef twentiˈtuː] *noun* American-designed stealth fighter aircraft (NOTE: also known as **Raptor**)

F-4 [ˌef ˈfɔː] *noun* American-designed multirole fighter aircraft, suitable for use from aircraft carriers (NOTE: also known as the **Phantom;** plural is **F-4s** [ˌef ˈfɔːz])

FAA [ˌef eɪ ˈeɪ] = FLEET AIR ARM

FAARP [fɑːp] *noun US* = FORWARD ARMING AND REFUELING POINT place where forward units can replenish ammunition and fuel during an advance

Fablon™ [ˈfæblɒn] *noun* clear adhesive plastic sheet, which is used to make maps waterproof and allow them to be marked with Chinagraph *or* Lumocolor

FAC [ˌef eɪ ˈsiː] = FORWARD AIR CONTROLLER

face [feɪs] *verb* **(a)** to look towards; ***they occupied positions near the top of the hill, facing north*** **(b)** to be likely to have to undergo something; ***he faced a court-martial after his ship rammed the harbour wall***

face-veil [ˈfeɪs veɪl] *noun* garment made of net-like fabric, which is usually worn round the neck as a scarf but can also be used as a small camouflage net; ***we used a couple of face-veils to break up the outline of the machine-gun***

facilitate [fəˈsɪlɪˌteɪt] *verb* to make something easier (ie. to assist); ***our mission is to facilitate the extraction of the patrol***

faction [ˈfækʃən] *noun* small group which disagrees with the main body of an organization *or* population and makes trouble; ***the conflict between the different factions may lead to civil war***

factory [ˈfæktri] *noun* large building or complex where things are manufactured

FAE [ˌef eɪ ˈiː] *see* FUEL-AIR EXPLOSIVE

fag [fæg] *noun GB slang* cigarette; ***put that bloody fag out !***

fall [fɔːl] **1** *noun* **(a)** descent to the ground (usually out of control); ***he broke his leg in the fall***; **fall of shot** = place where a projectile strikes **(b)** *(of places)* capture; ***the fall of Singapore*** **2** *verb* **(a)** to descend to the ground (usually out of control); ***he fell out of the vehicle*** **(b)** to be killed in action; ***his grandfather fell at the Battle of the Somme*** **(c)** *(of places)* to be captured; ***the town finally fell to the rebels*** (NOTE: **falling - fell - have fallen**)

fall back [ˌfɔːl 'bæk] *verb* to withdraw (usually under strong pressure from the enemy *or* as a result of a strong enemy threat); ***3 Brigade has fallen back towards Soltau***; *see also* PULL BACK

fall in [ˌfɔːl 'ɪn] *verb* to take your place on a formal parade; ***the squad fell in in front of the sergeants' mess***

fall out [ˌfɔːl 'aʊt] *verb* to leave a formal parade

fallout ['fɔːlaʊt] *noun* radioactive dust produced by a nuclear explosion

fall upon ['fɔːl əˌpɒn] *verb* to attack; ***government soldiers fell upon the refugee column***

FA MAS [ˌef eɪ 'mʌs] *noun* French-designed 5.56mm assault weapon

Fantan ['fæntæn] *noun* NATO name for the Chinese-designed Q-5 fighter aircraft

farm [fɑːm] *noun* **(a)** area of land used for the cultivation of crops or rearing and fattening of livestock **(b)** complex of buildings, including the farmhouse, which form part of a farm

farmer ['fɑːmə] *noun* person who owns and manages a farm

farmhouse ['fɑːmhaʊs] *noun* home of a farmer

farrier ['færɪə(r)] *noun* person who fits horseshoes onto horses

fascine [fæ'siːn] *noun* bundle of logs or plastic pipes or other material used to fill a ditch so that vehicles can cross it

fatigue [fə'tiːg] *noun* **(a)** non-military task or duty (such as cleaning toilets, clearing up rubbish, peeling potatoes, etc.) **(b) fatigues** = clothing worn for carrying out a fatigue **(c)** being tired; ***they are all suffering from fatigue***

FAV [ˌef eɪ 'viː] *noun* = FAST ATTACK VEHICLE American-designed light all-terrain vehicle fitted with medium machine-guns or ATGM or other weapons systems

FC [ˌef 'siː] = FORCE COMMANDER

FDC [ˌef diː 'siː] *noun* = FIRE DIRECTION CENTRE command post which coordinates the fire of several batteries

FDO [ˌef diː 'əʊ] *noun GB* = FLIGHT DECK OFFICER person who controls the taking off and landing of aircraft on an aircraft carrier

feature ['fiːtʃə] *noun* **(a)** any natural or man-made thing which is visible on the ground; **man-made features** = buildings, bridges, canals, embankments, pylons, roads, etc.; **natural features** = hills, ridges, rivers, valleys, woods, etc. **(b)** distinctive piece of high ground (such as a hill, knoll, ridge, saddle, etc.); ***we must capture that feature***

FEBA ['fiːbə] = FORWARD EDGE OF THE BATTLE AREA

F Echelon ['ef ˌeʃlɒn] *noun* fighting elements of a tactical grouping

federal riot gun (FRG) [ˌfedrəl 'raɪət ˌgʌn] *noun* gun designed to fire baton rounds

feint [feɪnt] *noun* attack which is not followed through, but is intended merely to test the enemy's defences or to give him a false idea of your own intentions; ***the attack was just a feint***; *see also* DIVERSION

fence [fens] *noun* barrier made of vegetation, wire or wood, which encloses an area of ground and is designed to control or prevent access

fence out [ˌfens 'aʊt] *noun* to prepare a jet fighter for action (ie. switching on your weapons systems, RWR, HUD, etc)

Fencer ['fensə] *noun* NATO name for the Soviet-designed SU-24 fighter-bomber

Ferret ['ferɪt] *noun* small British-designed armoured car

ferry ['feri] **1** *noun* boat used to transport people or vehicles across a river or lake or narrow stretch of sea, as part of a regular service; ***the brigade will cross the Channel by ferry*** **2** *verb* to carry people, vehicles, etc., across a river, lake, narrow stretch of sea; ***small boats ferried the whole battalion across the river***

fertilizer ['fɜːtəlaɪzə(r)] *noun* chemical substance used by farmers to stimulate the growth of crops

> COMMENT: fertilizers with a high nitrogen content are often used by terrorists to produce home-made explosive

fever ['fiːvə(r)] *noun* sickness, where a person's body temperature is higher than normal; ***he has a slight fever***

FF = FRIGATE (with guns)

FFG = FRIGATE (with guided missiles)

FGA [ˌef dʒiː 'eɪ] = FIGHTER GROUND-ATTACK

fiasco [fɪ'æskəʊ] *noun* complete failure, resulting in chaos; ***the beach landing was a fiasco***

FIBUA ['fɪbjʊə] = FIGHTING IN BUILT-UP AREAS; ***I am going on a FIBUA course***

field [fiːld] *noun* **(a)** well-defined piece of agricultural land (usually enclosed by a fence); ***they bivouacked in a corner of a field*** **(b) field of fire** = area of ground in which there is sufficient visibility to shoot at targets; ***this position has excellent fields of fire*** **(c) the field** = area where a battle or other military operation takes place; ***he performs far better in the field than in barracks***; *see also* BATTLEFIELD; **field ambulance** = battalion-sized medical unit (usually attached to a brigade); **field dressing** = camouflaged dressing designed to treat serious wounds; **field engineering** = tasks carried out by engineers in support of ground forces (such as the construction, repair and demolition of bridges, construction of field fortifications, construction and removal of obstacles, etc.); **field fortification** = improvised fortification prepared on the battlefield (eg. anti-tank ditch, bunker, trench, etc.); **field gun** = artillery piece designed to be moved easily over all types of ground; **field hospital** = mobile hospital set up on or near to the battlefield, which is capable of providing surgery; **field officer** = army officer of any rank above captain and below general; **field rank** = any army rank above captain and below general; **field training exercise (FTX)** = training exercise in which military skills are practised in field conditions

fieldcraft ['fiːldkrɑːft] *noun* basic infantry skills of camouflage and concealment and tactical use of ground

field-glasses ['fiːld ˌglɑːsɪz] *noun* optical instrument with a lens for each eye, designed for looking at distant objects; *see also* BINOCULARS, TELESCOPE

field marshal (FM) ['fiːld ˌmɑːʃəl] *noun GB* most senior officer rank in the army

fife [faɪf] *noun* musical instrument, like a little metal pipe; **Fifes and Drums** = band of drummers and fife-players belonging to a battalion or regiment

fifth-columnist [ˌfɪfθ 'kɒləmɪst] *noun* agent or saboteur operating secretly within the territory of an enemy state

fight [faɪt] *verb* to use physical force against another person, army, etc., in order to defend yourself or to inflict injury; ***the battle of Waterloo was fought outside the town of Brussels***; ***they fought for possession of the hilltop*** (NOTE: **fighting - fought**)

fight back [ˌfaɪt 'bæk] *verb* to defend yourself, to resist attack; ***we were surprised that the garrison fought back so strongly***

fight dirty [ˌfaɪt 'dɜːtɪ] *verb (after a chemical agent has been used)* to remain in NBC suits and respirators until the engagement has finished completely; ***if they do use chemicals, we'll have to fight dirty from then on***

fighter ['faɪtə] *noun* **(a)** light fast-moving aircraft designed to attack other aircraft; **fighter-bomber** = aircraft which is designed to drop bombs on or fire missiles at targets on the ground, and is also capable of defending itself against enemy fighter aircraft; **fighter ground-attack (FGA)** = attack by fighter aircraft on a target on the ground, in support of ground forces; **fighter controller** = air traffic controller who works in permanent partnership with the pilot of a fighter aircraft, following his progress on radar and directing him to intercept enemy aircraft **(b)** fighting soldier; ***guerilla fighters attacked our troops from the cover of the woods***; ***the Gurkhas are some of the best fighters in the world***; *see also* ATTACK AIRCRAFT, STRIKE AIRCRAFT

> COMMENT: the word "fighter" is often used by the layman to describe any light fast-moving aircraft. Specialists (eg aircrew, air-defence, etc) would be

more specific, and would classify an aircraft by its primary role: eg. attack, interceptor, EW, etc

fighting ['faɪtɪŋ] *noun* action of warfare; ***fighting continued along the whole front line***; ***their troops are experienced in guerilla fighting***; **fighting in built-up areas (FIBUA)** = special skills relating to combat in towns and villages; ***I am going on a FIBUA course***; **hand-to-hand fighting** = close fighting, especially with the bayonet

Fighting Falcon [ˌfaɪtɪŋ 'fɔːlkən] *see* F-16

fighting patrol ['faɪtɪŋ pəˌtrəʊl] *noun* large well-armed patrol sent out on an offensive operation (eg. snatching a prisoner for interrogation)

fighting strength ['faɪtɪŋ ˌstreŋθ] *noun* number of men or vehicles available to a unit for the purposes of fighting

figure ['fɪgə *US* 'fɪgjə] *noun* **(a)** number; ***a six-figure grid reference*** **(b)** *(radio terminology)* **figures** = minutes; ***I will be with you in figures ten***

file [faɪl] **1** *noun* tactical formation where men or vehicles move one behind the other; **single file** = single line of men or vehicles moving one behind the other; **double file** = two parallel lines of men or vehicles moving one behind the other **2** *verb* to move in single file; ***the men filed along the jungle path***

fin [fɪn] *noun* thin, flat projection on an aircraft, missile or other projectile, which provides extra stability during flight

fir (tree) [fɜː] *noun* tree which does not lose its leaves in winter (such as pine, spruce, etc.); ***the enemy position is behind that line of firs***; *see also* CONIFER, EVERGREEN

fire ['faɪə] **1** *noun* **(a)** any substance which is burning; ***we saw a fire in the distance***; **to catch fire** = to start burning; **on fire** = in the act of burning; ***the tank was on fire*** **(b)** discharge of a gun or missile **(c)** effect of bullets or other projectiles hitting a target and its vicinity; ***the platoon came under heavy fire from the farm***; **fire and manoeuvre** = tactic of moving in bounds, where one soldier or vehicle or sub-unit moves, while another soldier or vehicle or sub-unit gives covering fire; ***the platoon made good use of fire and manoeuvre in the assault***; **fire control** = control exercised by a commander over the weapons under his command; ***good fire control will be essential in this attack***; **fire-mission** = specific artillery or mortar task; ***'Hullo 42C this is 2; fire mission, over'***; ***we only have enough rounds for one more fire mission***; **fire position =** any location from which a weapon is discharged; ***the sniper found himself a good fire position in the church tower***; **fire support** = additional fire provided by another unit or arm; ***a squadron from the Royal Hussars will be providing fire support for this attack***; **baptism of fire** = the first occasion of being shot at; ***he received his baptism of fire in Vietnam***; **covering fire** = fire designed to neutralize the enemy so that another person or unit can move; ***prepare to give covering fire***; **direct fire** = fire from weapons which are actually pointed at their targets (for example a rifle, anti-tank gun, guided missile); **effective enemy fire** = situation where the enemy's fire starts to cause casualties amongst your own troops; **field of fire** = area of ground in which there is sufficient visibility to shoot at targets; ***this position has excellent fields of fire***; **final protective fire (FPF)** = pre-determined artillery target, registered on or just in front of your own position, as a final defensive measure in the event of being overrun by the enemy; ***the company commander called for his FPF***; **friendly fire** = incident where friendly forces fire on their own troops or vehicles by mistake; ***he was killed by friendly fire***; **harassing fire** = random artillery fire-mission directed at a likely area of enemy activity, in order to disturb the enemy's rest, disrupt his movements and inflict casualties, thereby affecting his morale; **indirect fire** = artillery or mortar fire; **under fire** = situation of being shot at; ***'Hullo 2, this is 22, we are under fire from the village, wait out'***; **fire support coordination line (FSCL)** = real *or* imaginary line behind the forward line of enemy troops (FLET), beyond which friendly aircraft can attack targets without requiring the directions *or* permission from friendly forward air controllers (FAC) **2** *verb* to discharge a gun or missile or to detonate an explosive device; ***he fired at the leading tank***; ***the Claymore failed to fire***; **fire-and-forget weapon** = missile equipped with a guidance system which requires no further action from its operator, once it is locked onto its target; **fire discipline** = personal judgement preventing unnecessary wastage of ammunition; ***the section quickly ran out of ammunition because of their poor fire discipline***; *(of artillery or mortars)* **to fire for effect** = to fire rounds as quickly as possible; ***ten rounds, fire for effect!***; **free fire zone** = area of ground in which any person or vehicle should be considered hostile and may therefore be shot at; **fire! or open fire! =start shooting; cease fire!** = stop shooting; **hold your fire!** = don't shoot

firearm ['faɪəɑːm] *noun* hand-held gun (such as a pistol, rifle or assault weapon)

firebase ['faɪəbeɪs] *noun* **(a)** *US* fortified base location containing heavy weapons which can provide fire support to troops operating in the surrounding area; ***B Company is at Firebase Alfa*** **(b)** troops giving fire support during an attack; ***the firebase was ordered to move forward***

fire-bomb ['faɪəbɒm] **1** *noun* bomb designed to set buildings alight; *see also* INCENDIARY BOMB **2** *verb* to drop fire-bombs on; ***the centre of the town was fire-bombed***

fire-break ['faɪəbreɪk] *noun* open space between two areas of woodland, which is designed to prevent the spread of fire, but can also often be used by vehicles

fire brigade ['faɪə brɪˌgeɪd] *noun* officially organized body of men trained to fight fires

firefight ['faɪəfaɪt] *noun* exchange of fire between opposing forces; ***a fierce firefight developed on the edge of the village***

Firefly™ ['faɪəflaɪ] *noun* small pocket-sized strobe; ***we'll need a Firefly to signal to the chopper***

fireman ['faɪəmən] *noun* member of the fire brigade; ***several firemen were attacked during the riot***

fireplan ['faɪəplæn] *noun* **(a)** plan outlining the sequence in which specific or potential targets will be engaged by individual weapons or sub-units; ***the battalion mortars were included in the artillery fireplan*** **(b)** document showing the fireplan; ***all company commanders were given a copy of the artillery fireplan***

firepower ['faɪəˌpaʊə] *noun* destructive capacity of guns, missiles and other weapons; ***the firepower available to the brigade was enormous***

fire-storm ['faɪə stɔːm] *noun* extensive burning caused by fire-bombs, producing artificial winds which can suck heavy objects into the fires

fireteam *US* **fire team** ['faɪətiːm] *noun* **(a)** *GB* infantry grouping of 4 men (half of a section) **(b)** *US* infantry grouping of 4 men (one third of a squad)

fire-trench ['faɪə trenʃ] *noun* hole in the ground used by infantrymen as a fire position and as shelter from enemy fire (NOTE: American English is **foxhole**)

firing ['faɪərɪŋ] **1** *adjective* relating to the firing of weapons or the detonation of explosives; **firing party** = detachment of riflemen delegated to fire a salute over a soldier's grave; **firing pin** = little metal pin which hits the end of a round in the breech of a rifle to detonate it; **firing point** = location from which an engineer detonates an explosive device (as in bridge demolition); **firing port** = aperture in the side of a vehicle through which a soldier can fire his personal weapon; **firing post** = missile launcher; ***the anti-tank platoon had three firing posts on the forward edge of the village***; **firing squad** = detachment of soldiers delegated to execute a condemned prisoner **2** *noun* act of firing weapons; ***we heard firing away to the right***

first aid [ˌfɜːst 'eɪd] *noun* basic emergency treatment given to a casualty before proper medical treatment is available

first lieutenant [ˌfɜːst lef'tenənt *US* luː'tenənt] *noun US* junior officer in the army, marines or air force (equivalent of a lieutenant in the British Army)

first light [ˌfɜːst 'laɪt] *noun* time of day when daylight first appears; ***be ready to move at first light***

first-line [ˌfɜːst 'laɪn] *adjective* relating to resources (eg ammunition, fuel, rations) actually carried by the fighting troops, as opposed to those carried by the echelons or stored in dumps or depots.; ***all companies have drawn their first-line ammunition***

First Parade [ˌfɜːst pə'reɪd] **(1)** *noun* first daily task for any unit *or* sub-unit equipped with vehicles, where routine maintenance and daily checks are carried out on each vehicle; ***he was late for First Parade* 2**; *(first-parade) verb* to carry out the routine tasks of First Parade; ***your platoon has not been first-parading its vehicles properly***

first sergeant [ˌfɜːst 'sɑːdʒənt] *noun* *US* senior non-commissioned officer (SNCO) in the army or marines (normally responsible for administration and discipline within a sub-unit)

Fishbed ['fɪʃbed] *noun* NATO name for the Soviet-designed MiG-21 fighter aircraft

fitter ['fɪtə] *noun* vehicle mechanic

fix [fɪks] *verb* **(a)** to mend; ***he's trying to fix the radio*** **(b)** to engage *or* threaten an enemy force (usually from the front) in order to divert his attention, and hold him in his current positions, while your main forces manoeuvre to envelop him *or* mount a flanking attack

fixed-wing aircraft [ˌfɪkstwɪŋ 'eəkrɑːft] *noun* conventional aircraft, with wings fixed to the fuselage (as opposed to helicopters and VTOL aircraft)

flag [flæg] *noun* square or rectangular piece of fabric attached to a pole, displaying colours or insignia; **flag-captain** = captain of a flagship; **flag-lieutenant** = aide-de-camp (ADC) to an admiral; **flag-officer** = naval rank of admiral, vice admiral or rear admiral; **Union flag** = the national flag of Great Britain (NOTE: also known as the **Union Jack**); **flag of truce** *or* **white flag** = white flag displayed by soldiers wishing to surrender, or by a messenger indicating to the enemy that they should stop shooting

flagship ['flægʃɪp] *noun* warship used as a command vessel by the commander of a fleet or flotilla

flail [fleɪl] *noun* set of revolving chains attached to the front of a tank or armoured engineer vehicle, designed to clear a way through a minefield by detonating the mines in its path

flak [flæk] *noun* anti-aircraft fire; **flak jacket** = vest fitted with panels of synthetic material (eg. Kevlar) designed to protect a soldier from shrapnel and low-velocity bullets; *see also* BODY ARMOUR, BULLETPROOF VEST

flame [fleɪm] *noun* burning gas which forms the yellow part of a fire; **flame-thrower** = weapon which squirts a jet of burning liquid at a target; **in flames** = in the act of burning

flammable ['flæməbl] *adjective* easy to set on fire (NOTE: means the same as **inflammable**)

flank [flæŋk] **1** *noun* left- or right-hand side of a military force which is deployed in a defensive position or tactical formation; ***the army's right flank was exposed***; **flank in the air** = exposed *or* unprotected flank; ***the enemy's right flank is in the air*** **2** *verb* **(a)** to manoeuvre around the enemy's flank; ***the enemy tried to flank us on the right***; **to turn a flank** **(b)** to be positioned next to; ***6 Brigade is flanking us on the left***; ***the president stood at the saluting base, flanked by several officers***

flanker ['flæŋkə] *noun* **(a)** soldier, vehicle or sub-unit assigned to guard a formation's flank on the move; ***B Squadron will provide flankers for the advance*** **(b) Flanker** = NATO name for the Soviet-designed SU-27 fighter aircraft

flanking ['flæŋkɪŋ] *adjective* relating to movement on the enemy's flank; **flanking attack** = attack on the enemy's flank; **flanking movement** = manoevre around the enemy's flank; **left-flanking** = relating to an attack on the left flank of an enemy position (as you are looking at it); ***Hello 2, this is 22, am going left-flanking, over***; **right-flanking** = relating to an attack on the right flank of an enemy position

flannelette [ˌflænəl'et] *noun* strip of fabric used to clean the barrel of a weapon

flare [fleə] *noun* pyrotechnic which emits a bright light in order to improve visibility at night; ***flares were dropped over the target zone***; **trip-flare** = flare which is activated by a trip-wire

flash [flæʃ] **1** *noun* **(a)** **flash eliminator** = perforations at the muzzle of a machine-gun *or* assault weapon, designed to reduce the flashes produced when firing sudden emission of bright light; ***we saw a bright flash to the left*** **(b)** flame and heat given off by an explosion; ***several sailors were injured by flash***; **flash hood** = fire-resistant covering for the head and face, worn by sailors in battle to reduce the effects of flash **(c)** coloured patch of cloth worn on the uniform to distinguish a unit or grouping; ***he was wearing yellow flashes***; ***some Scottish regiments wear tartan flashes on their bonnets*** **2** *verb* **(a)** to produce a brief emission of light; ***he flashed his torch three times*** **(b)** to send a radio message with the highest priority; ***we flashed the information to HQ***; **flash message** *or* **signal** = high priority radio message

flashlight ['flæʃlaɪt] *noun US* hand-held battery-powered device for producing light; ***we used flashlights to attract attention***; *see also* TORCH

flat [flæt] **1** *adjective* **(a)** *(of surfaces)* completely level; ***the terrain is flat*** **(b)** *(of batteries)* without electrical charge; ***the battery is flat*** **2** *noun* dwelling, consisting of a set of rooms all on one storey of a building; ***we'll have to clear that block of flats***

flèchette [fleɪ'ʃet] *noun* anti-personnel projectile in the form of a tiny dart, designed to be released in large numbers by certain types of explosive projectile (especially canister rounds)

flee [fliː] *verb* to run away from danger; ***the civilian population fled into the hills*** (NOTE: **fleeing - fled**)

fleet [fliːt] *noun* large number of warships; ***the American Pacific Fleet was based at Pearl Harbor***; ***the enemy fleet could be seen on the horizon***; **Fleet Air Arm (FAA)** = air force forming part of the British Royal Navy; *see also* ADMIRAL

flesh wound ['fleʃ ˌwuːnd] *noun* wound which does not damage a bone or organ

flight [flaɪt] *noun* **(a)** act of flying; ***we were able to observe the flight of the missile***; ***the plane exploded in flight*** **(b)** journey by air; ***the flight to the target area only takes two minutes*** **(c)** **(i)** sub-unit of an air-force squadron; **(ii)** small tactical grouping of aircraft; **(iii)** administrative air-force grouping of approximately 30 men (equivalent to a platoon in the army) *GB* **flight lieutenant (Flt Lt)**= middle-level officer in the air force, above flying officer and below squadron leader *GB* **flight officer (Flt Off)** = female Royal Air Force rank corresponding to flight lieutenant *GB* **flight sergeant (Flt Sgt)** = senior non-commissioned officer (SNCO) in the Royal Air Force **(d)** act of running away from danger; ***with the flight of the civilian government, all resistance collapsed***

flight crew ['flaɪt kruː] *noun US* aircrew

flight deck ['flaɪt dek] *noun* **(a)** deck of an aircraft carrier used for the take-off and landing of aircraft **(b)** compartment for the pilot, navigator, etc., on a large aircraft

flight control ['flaɪt kənˌtrəʊl] *noun* **(a)** the direction of aircraft within a specific airspace **(b)** unit (usually located on the ground) which directs aircraft within a specific airspace

flight path ['flaɪt pɑːθ] *noun* **(a)** course of an aircraft or missile in flight; ***a flock of birds flew across our flight path*** **(b)** recognised air route; ***you will have to avoid all civil flight paths***

fling [flɪŋ] *verb* to throw; ***he flung a couple of grenades into the bunker*** (NOTE: **flinging - flung**)

float [fləʊt] *verb* to rest upon the surface of water

Flogger ['flɒgə] *noun* NATO name for the Soviet-designed MiG-23 and MiG-27 fighter aircraft

flood [flʌd] **1** *noun* overflow of water or other liquid beyond its normal limits; ***the region has been affected by floods*** **2** *verb* **(a)** *(of rivers or the sea)* to cover dry land with water; ***the sea has flooded most of the town***; **flood plain** = valley bottom which becomes covered by water when the river floods **(b)** to cause a flood; ***the enemy have flooded the valley*** **(c)** *(of motor engines)* to overfill the carburettor with fuel; ***the engine is flooded***

FLOT [flɒt] = FORWARD LINE OF OWN TROOPS

flotilla [flə'tɪlə] *noun* small group of warships or other vessels; ***he commanded a British flotilla in the Baltic***

flotsam ['flɒtsəm] *noun* debris or other objects found floating in water

Flt Lt = FLIGHT LIEUTENANT

Flt Off = FLIGHT OFFICER

Flt Sgt = FLIGHT SERGEANT

fluorescent [fluə'resent] *adjective; (of colours)* very bright and highly visible; ***the ground crew wore fluorescent orange jackets***

fly [flaɪ] *verb* **(a)** to move through the air; ***the aircraft were flying towards the coast*** **(b)** to travel by aircraft; ***we flew to Brussels*** **(c)** to transport men or objects by aircraft; ***reinforcements are being flown into the area*** (NOTE: **flying - flew - have flown**)

flying officer (FO) ['flaɪɪŋ ˌɒfɪsə] *noun* *GB* junior officer in the air force, below a flight lieutenant

flyover ['flaɪˌəʊvə(r)] *noun* embankment and bridge carrying one road over another

flypast ['flaɪpɑːst] *noun* flying of aircraft over a certain place as part of a ceremony; *compare* MARCH PAST

flysheet ['flaɪʃiːt] *noun* outer-covering of a tent, designed to give extra protection from the weather

FM = FIELD MARSHAL

fmn = FORMATION

FN [ˌef 'en] *noun* = FABRIQUE NATIONALE series of small arms manufactured the Fabrique Nationale in Belgium; **FN-FAL** = 7.62mm assault weapon; **FN-MAG** = 7.62mm general purpose machine-gun; **FN-Minimi** = 5.56mm light machine-gun

> COMMENT: the British-made GPMG (general purpose machine-gun) and SLR (self-loading rifle) were modified from the original FN designs. However, the FN-FAL has an automatic capability while the SLR is semi-automatic

FO = FLYING OFFICER

FOB [ˌef əʊ 'biː] *noun (= FORWARD OPERATING BASE)* supply dump (especially of ammunition and fuel) which is located in the battle area

fo'c'sle ['fəʊksl] *see* FORECASTLE

FOD [fɒd] = FOREIGN OBJECT DAMAGE

foe [fəʊ] *noun* outdated word for "enemy"

fog [fɒg] *noun* thick cloud of water vapour at ground level, which reduces visibility; **fog bank** = a mass of fog at sea; **fog-bound** = unable to travel because of fog; **fog of war** = state of confusion on the battlefield owing to smoke, noise and limited information

foggy ['fɒgi] *adjective* obscured by fog

foghorn ['fɒghɔːn] *noun* noise-making device designed to warn ships in foggy conditions

foliage ['fəʊliɪdʒ] *noun* mass of leaves or other vegetation

follow ['fɒləʊ] *verb* **(a)** to move behind someone or something else; ***follow that truck - it's going to the HQ*** **(b)** to pursue; ***we followed the enemy patrol back to their camp***; ***they were too exhausted to follow the enemy into the mountains*** **(c)** *(of an event)* to come after another event; **to follow up** = to take further action; ***we followed up the attack by shelling the enemy as they withdrew***; **to follow through** = to continue an action to its finish; ***the enemy were able to counter-attack because the assault was not followed through***

follow-on force attack [ˌfɒləʊ ɒn ˌfɔːs ə'tæk] *noun* strategy by which reinforcements are attacked by aircraft and missiles as they are moving towards the battle area

follow-on forces ['fɒləʊ ɒn ˌfɔːsɪz] *plural noun* subsequent waves of an advancing force, which are in a position to reinforce the leading elements *or* take over the lead when required; ***we will attack the enemy's follow-on forces with aircraft and missiles***

folly ['fɒli] *noun* **(a)** stupid action; ***to attack now would just be folly*** **(b)** ornamental building (in a park or garden); ***there was a sniper in the folly***

FOO [ˌef əʊ 'əʊ] = FORWARD OBSERVATION OFFICER

foot [fʊt] *noun* **(a)** part of the body on which a person or animal walks; **on foot** = not in a vehicle; **to go on foot** = to walk **(b)** unit of linear measurement corresponding to 12 inches or 30.48 centimetres (NOTE: in meaning (a) and (b), the plural of **foot** is **feet**) **(c)** *(historical)* **the foot** *or* **foot soldiers** = the infantry

footbridge ['fʊtbrɪdʒ] *noun* small bridge designed for pedestrians (ie not suitable for vehicles)

foothold ['fʊthəʊld] *noun* capture of a small area of enemy territory, which can be used as a base for mounting or supporting further attacks; ***we have gained a foothold on the enemy position***

foothold line ['fʊthəʊld ˌlaɪn] *noun* tactical monoeuvre carried out under fire, in which men *or* vehicles withdraw to form an extended line out of direct enemy fire, prior to breaking contact *or* redeploying; *compare* BASELINE

-FOR [fɔːr] *noun* suffix meaning FORCE, used in the titles of contingents engaged in international peacekeeping operations; **BRITFOR** = British Force**; UNPROFOR** = United Nations Protection Force

forage ['fɒrɪdʒ] *verb* to search for food; *see also* LIVE OFF THE LAND; **forage cap** = peaked military head-dress with a flat top (usually worn in barracks or on parade)

> COMMENT: foraging usually refers to the act of taking food from the civilian population without paying for it

foray ['fɒreɪ] *noun* operation into enemy territory (usually a raid or reconnaissance); ***he was wounded on a foray behind the enemy lines*** (NOTE: the verb form is to **make a foray**)

force [fɔːs] **1** *noun* **(a)** use of physical strength; ***we had to use force to remove the protesters*** **(b)** military power; ***if diplomacy fails, we will have to consider using force***; **economy of force** = not wasting your military assets and, if practicable, using fewer forces than normal on your initial tasks, so that you will have the maximum forces avaliable for your main effort; **force multiplier** = any activity or equipment which increases the combat effectiveness of a military grouping without actually increasing its firepower (such as engineer support, electronic warfare, deception, surprise); **force projection** = identifying and planning for possible future operations **(c)** unspecified military grouping; ***a large force of tanks is heading this way***; **residual force** = small security force which remains in a war zone after hostilities have ended and the main force has withdrawn; **task force** = large combined arms grouping formed for a specific operation or campaign; ***the government is sending a special task force to recapture the islands*** **(d) the forces** = general title for army, navy and air force; ***the forces are expected to receive a pay increase in the new year***; **ground** *or* **land forces** = military forces which operate on land; **naval forces** = military forces which operate at sea **(e) air force** = branch of the armed forces which operates in the air **2** *verb* **(a)** to make someone do something against his/her will; ***the enemy forced the captured soldiers to tell them where their commander was***; **forced landing** = emergency landing made by an aircraft on unprepared ground; ***we had to make a forced landing in a potato field*** **(b)** to achieve something with the use of strength or great effort; ***they forced their way through the crowd***; **forced march** = movement by infantry over a long distance on foot; ***a series of forced marches enabled us to reach the Danube in four days*** **(c)** to break a lock or fastening; ***the window has been forced***

◊ **in force** *phrase* **(a)** in large numbers; ***enemy armour is crossing the border in force near Landshut*** **(b)** valid; ***the regulations have been in force since Tuesday***

ford [fɔːd] **1** *noun* place on a river or stream where the water is shallow enough for men to wade across or for vehicles to drive through **2** *verb* to cross a river or stream by wading or driving through the water

forecastle ['fəʊksl] *noun* forward part of a ship (NOTE: also spelled **fo'c'sle**)

foreign ['fɒrən] *adjective* **(a)** belonging to or coming from another country; **foreign legion** = force of foreign volunteers serving in a state's army (such as the

French Foreign Legion or the Spanish Foreign Legion); **foreign national** = person who is a citizen of another country; *see also* ALIEN **(b)** coming from outside; ***he had a foreign object in his eye***; **foreign object damage (FOD)** = damage to a jet aircraft caused by an object being sucked into the air intakes

foreigner ['fɒrənə] *noun* person who comes from another country; *see also* ALIEN, FOREIGN NATIONAL

forest ['fɒrɪst] *noun* very large area of woodland

Forger ['fɔːdʒə] *noun* NATO name for the Soviet-designed YAK-38 fighter aircraft

fork [fɔːk] *noun* place where a single road divides into two; *see also* Y-JUNCTION

form [fɔːm] *verb* to make; ***form columns of four***

formation [fɔː'meɪʃn] *noun* **(a)** arrangement of aircraft, ships, troops or vehicles for movement or tactical purposes; ***the tanks were moving in formation*** **(b)** military grouping; ***several large enemy formations have crossed the river*** **(c)** grouping; ***he was posted to a different formation***; **higher formation** = grouping to which a smaller grouping or sub-unit belongs; ***your request for a transfer has been passed to the higher formation***

forming-up point (FUP) [fɔːmɪŋ 'ʌp ˌpɔɪnt] *noun* place where soldiers or vehicles of a grouping get into tactical formation before commencing the next phase of an operation

form up [ˌfɔːm 'ʌp] *verb* to get into tactical formation prior to commencing the next phase of an operation

fort [fɔːt] *noun* small fortified building; ***the rebels are based in hill forts to the north of the capital***

fortification [fɔːtɪfɪ'keɪʃn] *noun* **(a)** act of preparing buildings or ground for defence; ***he was put in charge of the fortification of the town*** **(b)** earthwork, structure or obstacle prepared for defensive purposes; ***there is a strong line of fortifications along the coast***; **field fortification** = improvised fortification prepared on the battlefield (eg. bunker, trench, etc.)

fortify ['fɔːtɪfaɪ] *verb* to prepare buildings or ground for defence; ***the enemy occupies strongly fortified positions to the north of the river***

fortress ['fɔːtrəs] *noun* fortified place or town; ***they retreated to a fortress overlooking the Danube***

forward ['fɔːwəd] **1** *adverb* **(a)** straight ahead; ***the tank moved forward slowly*** **(b)** beyond; ***the OP was deployed forward of the main position*** **(c)** to the front line; ***the battalion is moving forward tonight*** **2** *adjective* **(a) in front: forward slope** = side of a hill which is facing the enemy; ***the enemy are dug in on the forward slope***; *see also* REVERSE SLOPE **(b)** in the front line; ***the forward positions were under fire for several hours***; **forward air controller (FAC)** = air-force or artillery officer or NCO operating from an aircraft or attached to ground troops in order to direct close air support; **forward area** = area occupied by troops who are close to or in contact with the enemy; ***everyone had to wear body armour in the forward areas***; *see also* FRONT, FRONT LINE; **forward defence** = NATO doctrine of stopping an invasion as close to the border as possible; **forward edge of the battle area (FEBA)** *or* **forward line of own troops (FLOT)** = line formed by the positions of friendly forces which are closest to the enemy *GB* **forward observation officer (FOO)** *or* **forward observer** = artillery officer or NCO attached to an infantry or armoured unit, or operating from an aircraft in order to direct artillery fire; *see also* SPOTTER

432 [ˌfɔː θriː 'tuː] *short for AFV-432*

Fox [fɒks] *noun* British-designed wheeled armoured reconnaissance vehicle (CVR)

Foxbat ['fɒksbæt] *noun* NATO name for the Soviet-designed MiG-25 fighter aircraft

foxhole ['fɒkshəʊl] *noun US* hole in the ground used by infantrymen as a fire position and as shelter from enemy fire (NOTE: British English is **fire-trench**)

Foxhound ['fɒkshaʊnd] *noun* NATO name for the Soviet-designed MiG-31 interceptor aircraft

Foxtrot ['fɒkstrɒt] sixth letter of the phonetic alphabet (Ff)

FPF [ˌef piː 'ef] *noun* = FINAL PROTECTIVE FIRE pre-determined artillery target, registered on or just in front of your own position, as a final defensive measure in the event of being overrun by the enemy; ***the company commander called for his FPF***

frag [fræg] *verb (slang)* US to deliberately wound an unpopular or unreliable comrade, while giving the appearance that he was hit by enemy fire; ***the sergeant was fragged by his own men*** (NOTE: this word is derived from **fragmentation grenade,** a common means of 'fragging' in Vietnam)

fragment 1 ['frægmənt] *noun* piece which has broken off an object; ***he was hit by fragments from a shell*** **2** [fræg'ment] *verb* to break up into separate parts; ***the minefields will force the enemy formations to fragment***

fragmentation grenade [frægmən'teɪʃn grəˌneɪd] *noun* hand-grenade, which is designed to explode into fragments

FRAGO ['frægəʊ] *noun* US = FRAGMENT OF AN ORDER ammendment to part of a set of orders

FRAGPLAN ['frægplæn] *noun* US = FRAGMENTARY PLAN list of actions for dealing with different contingencies which might occur during the course of an operation

fraternization [ˌfrætənaɪ'zeɪʃən] *noun* act of fraternizing; ***fraternization with civilians is prohibited***

fraternize ['frætəˌnaɪz] *verb* to be friendly towards enemy troops *or* enemy civilians; ***we aren't allowed to fraternize with the local population***

fratricide ['frætrɪsaɪd] *noun* US casualties caused by friendly units firing on each other in error; *see also* BLUE ON BLUE, FRIENDLY FIRE

freddy ['fredɪ] *noun (slang)* fighter controller; ***my freddy is on leave***

free [friː] **1** *adjective* **(a)** not in custody or under another person's domination; ***the hostages are now free*** **(b)** unrestricted; **free fire zone** = area of ground in which any person or vehicle should be considered hostile and may therefore be shot at **(c)** not engaged in any other activity; ***I am free at the moment*** **2** *verb* to release someone from custody; ***the hostages have been freed***

freedom ['friːdəm] *noun* state of being free; **freedom-fighter** = person who uses armed force to resist foreign domination or the established government

free-fall [ˌfriː 'fɔːl] *noun* method of opening a parachute, where the parachutist descends some distance through the air before pulling a rip-cord; *compare* STATIC LINE; *see also* HALO

Freestyle ['friːstaɪl] *noun* NATO name for the Soviet-designed YAK-41 fighter aircraft

freight [freɪt] *noun* goods or supplies carried by an aircraft, vehicle or train; *compare* CARGO

frenzy ['frenzɪ] *adverb (forward air controller jargon)* the target has not been destroyed; ***Hello Cowboy this is G33, frenzy, frenzy, over***

frequency ['friːkwənsi] *noun* radio setting on which a signal is transmitted and received; **low frequency (LF)** = range of radio frequencies from 30 - 300 kilohertz (kHz); **high frequency (HF)** = range of radio frequencies from 3 - 30 megahertz (MHz); **ultra high frequency (UHF)** = range of radio frequencies from 300 - 3,000 megahertz (MHz); **very high frequency (VHF)** = range of radio frequencies from 30 - 300 megahertz (MHz)

FRG [ˌef ɑː 'dʒiː] *noun* = FEDERAL RIOT GUN gun designed to fire baton rounds

friction ['frɪkʃən] *noun* disruptive effect of unforeseen factors upon the execution of a plan

friendly ['frendli] *adjective* on the same side in a conflict or war (ie not enemy); **friendly fire** = incident where friendly forces fire on their own troops or vehicles by mistake; ***he was killed by friendly fire***; *see also* BLUE ON BLUE, FRATRICIDE; **friendly forces** = your own forces or the forces of your allies (ie not the enemy); *compare* ENEMY

frigate ['frɪgət] *noun* medium-sized warship used to escort other ships or to carry out missions on its own

COMMENT: in the British Navy, the frigate's primary mission is anti-submarine warfare (ASW)

FROG [frɒg] *noun* = FREE ROCKET OVER-GROUND NATO name for Soviet-designed ballistic tactical surface-to-surface missile

Frogfoot ['frɒgfʊt] *noun* NATO name for the Soviet-designed SU-25 ground-attack aircraft

frogman ['frɒgmən] *noun* person equipped with special clothing and breathing apparatus in order to operate underwater; ***naval frogmen attached mines to the ship's hull***; *see also* DIVER

front [frʌnt] *noun* **(a)** forward edge or forward part of something; ***he aimed at the front of the tank***; **front line** = forward positions of an army at the front; **front-line troops** = soldiers serving on the front line **(b)** zone occupied by military forces which are fighting or preparing to fight the enemy; ***we are moving up to the front tonight***; ***the Germans are now fighting on two fronts***; ***he was drafted into the army and immediately sent to the front***; ***we attacked the enemy along a wide front***

frontage ['frʌntɪdʒ] *noun* distance along the front of a tactical formation or defensive position; ***the brigade advanced on a wide frontage***

frontal ['frʌntl] *adjective* towards the front of something; **frontal attack** = attack on the front of an enemy position (as opposed to the flank)

frontier ['frʌntɪə] *noun* region on the border between two states

frost [frɒst] *noun* **(a)** freezing conditions, which cover the ground and other surfaces with ice; ***there's going to be a hard frost tonight*** **(b)** frozen grass; ***we could see tracks in the frost***

frostbite ['frɒstbaɪt] *noun* injury caused to fingers, toes, ears, nose *or* other parts of the body as a result of freezing conditions; ***he lost six toes through frostbite***

FSCL [ef ˌes iː 'el *or* 'fɪs(ə)l] *noun* = FIRE SUPPORT COORDINATION LINE real *or* imaginary line behind the forward line of enemy troops (FLET), beyond which friendly aircraft can attack targets without requiring the directions *or* permission from friendly forward air controllers (FAC)

COMMENT: during a rapid advance by friendly forces, it will be necessary to constantly readjust the FSCL, not only to prevent fratricide from friendly aircraft, but also to allow those aircraft the maximum freedom to engage enemy targets to the rear. One solution to this problem is to have several pre-planned FSCLs which can be activated as the advance proceeds

FTX [ˌef tiː 'eks] = FIELD TRAINING EXERCISE

fuel ['fjuːəl] *noun* substance which is burnt in order to provide heat or power (for example diesel, gas, petrol, wood)

fuel-air explosive (FAE) [ˌfjuːəl eə ɪk'spləʊsɪv] *noun* munition containing a highly inflammable substance (such as ethylene oxide) which is delivered over a target as a cloud of vapour and then detonated, producing a powerful blast, intense heat, and depriving persons in the target area of oxygen

Fulcrum ['fʊlkrəm] *noun* NATO name for the Soviet-designed MiG-29 fighter aircraft

Fuller's earth [ˌfʊləz 'ɜːθ] *noun* hydrous aluminium silicate powder, which is used to decontaminate things after a chemical attack

fumigate ['fjuːmɪgeɪt] *verb* to disinfect *or* kill insects with smoke *or* chemical vapour; ***the MO ordered us to fumigate the tents***

funnel ['fʌnl] *noun* chimney for a ship's engine

FUP [ˌef juː 'piː] = FORMING-UP POINT

furlough ['fɜːləʊ] *noun US* permission to be absent from your unit for a specific period (NOTE: British English is **leave**)

furze [fɜːz] *noun (in Ireland)* bush with spiked leaves and yellow flowers (NOTE: British English is **gorse**)

fuse *US also* **fuze** [fjuːz] *noun* **(a)** length of fast-burning cord which is lit from a safe distance in order to activate an explosive device **(b)** component designed to detonate a bomb, shell or other explosive device on impact or after a specific period;

variable-time fuse (VT) = fuse fitted to an artillery shell, which causes it to explode at a specified height above the ground **(c)** component designed to prevent serious damage or fire, by safely burning out when an excessive electrical charge passes through it

fuselage ['fju:zəlɑ:ʒ] *noun* main body of an aircraft

fusilier *US* **fusileer** [fju:zə'lɪə] *noun (historical)* **(a)** infantryman armed with a light musket (called a 'fusil') **(b)** title of a normal infantryman in a grenadier regiment

> COMMENT: many infantry regiments still retain their historical title of Fusiliers

fusillade [fju:zə'leɪd] *noun* prolonged period of firing of small guns

FV-432 *see* AFV-432

fwd = FORWARD

GOLF - Gg

G1 [ˌdʒiː ˈwʌn] *noun* department of a headquarters responsible for personnel

G2 [ˌdʒiː ˈtuː] *noun* department of a headquarters responsible for intelligence

G3 [ˌdʒiː ˈθriː] *noun* department of a headquarters responsible for operations and training

G4 [ˌdʒiː ˈfɔː] *noun* department of a headquarters responsible for logistics

GA [ˌdʒiː ˈeɪ] *noun* = DIMETHYLAMINOETHOXYCYANOPHOSPHINE OXIDE type of nerve agent (NOTE: also known as **Tabun**)

G-agent [ˈdʒiː ˌeɪdʒənt] *noun* non-persistent nerve agent

gain [geɪn] **1** *noun* achievement or result; ***we lost all our gains next day when the enemy counterattacked*** **2** *verb* to achieve something; ***we have gained most of our objectives***; **to gain ground =** to move forwards; ***we have gained a lot of ground since yesterday***; ***after the battle we found we had only gained 200m***

Gainful [ˈgeɪnfʊl] *noun* SA-6, Soviet-designed low to medium altitude surface-to-air missile (SAM)

gaiter [ˈgeɪtə(r)] *noun* garment of fabric *or* leather, which is worn over the ankle and lower leg in order to keep your trousers dry and to prevent small stones and other objects going into your boots; ***the soldiers wore white belts and gaiters***

gale [geɪl] *noun* very strong wind

gallantry [ˈgæləntri] *noun* bravery

galley [ˈgæli] *noun* cabin or compartment on an aircraft or ship where food is prepared

gallon [ˈgælən] *noun* unit of measurement for liquids

> COMMENT: in Britain one gallon (the imperial gallon) equals 4.546 litres; in the USA, a gallon equals 3.78 litres

Gammon [ˈgæmən] *noun* SA-5, Soviet-designed long-range surface-to-air missile (SAM)

Ganef [ˈgænef] *noun* SA-4, Soviet-designed medium to high altitude radar-guided surface-to-air missile (SAM)

gang [gæŋ] *noun* group of people who act together for some illegal purpose; ***gangs of youths have been looting the town centre***

gangrene [ˈgæŋgriːn] *noun* condition where tissues die and decay as a result of bacterial action, because the blood supply has been lost through injury or disease of an artery; **gas gangrene =** complication of severe wounds in which the bacterium *Clostridium welchii* breeds in the wound and then spreads to healthy tissue which is rapidly decomposed with the formation of gas

gap [gæp] *noun* interval or space; ***he went through a gap in the fence***; ***there are large gaps between our positions***

garden [ˈgɑːdən] *noun* area of ground (usually next to or surrounding a house) used for the growing of flowers and plants or fruit and vegetables; **market garden =** large area of ground used for the commercial cultivation of fruit and vegetables

garrison [ˈgærɪsən] **1** *noun* troops who occupy a fortress or town in order to defend it; ***the garrison held out for three weeks***; **garrison town =** town in which troops are permanently stationed **2** *verb* to occupy a fortress or town with troops in order to defend it; ***the general garrisoned the town with troops loyal to the president***; ***the***

troops garrisoned in the town complained about the lack of amenities

gas [gæs] **1** *noun* **(a)** substance which behaves like air by completely filling the space which it occupies; *see also* GANGRENE **(b)** chemical weapon in the form of gas, used to irritate the skin, to blind, to choke or to kill; ***the president launched gas attacks on the civilian population***; **Gas! Gas! Gas!** = verbal alarm given for a chemical attack; **gas mask** = protective face-covering containing an apparatus to filter air; *see also* RESPIRATOR; **CS** *or* **tear gas** = gas which irritates the eyes and causes choking; **mustard gas** = gas which causes blisters on exposed skin **(c)** gas used as a fuel (such as butane) **(d)** *US (informal)* gasoline or petrol; ***we are out of gas*** **2** *verb* to use poisonous gas as a weapon; ***he was gassed during the war***

Gaskin ['gæskɪn] *noun* SA-9, Soviet-designed low altitude surface-to-air missile (SAM), normally issued to motor rifle and tank regiments, and used in conjunction with the ZSU-23 anti-aircraft cannon

gasoline ['gæsəliːn] *US noun* liquid fuel made from petroleum, used by motor vehicles (NOTE: British English is **petrol;** in many other languages it is **benzine**)

gate [geɪt] *noun* barrier which can be opened and closed in order to allow access through a fence

Gatling gun ['gætlɪŋ ˌgʌn] *noun* machine-gun with a cluster of barrels, which revolve in order to fire

Gazelle [gə'zel] *noun* French-designed and British-made reconnaissance helicopter

GB [ˌdʒiː 'biː] *noun* = METHYLISOPROPOXYFLUOROPHOSPHINE OXIDE type of nerve agent (NOTE: also known as **Sarin**)

GBU-15 [ˌdʒiː biː ˌjuː fɪf'tiːn] *noun* American-designed glide bomb

GCI [ˌdʒ iː siː 'aɪ] *noun* = GROUND CONTROLLED INTERCEPTION ground-based radar; ***we destroyed a GCI site***

GCT [ˌdʒ iː siː 'tiː] *noun* French-designed 155mm self-propelled gun

GD [ˌdʒiː 'diː] *noun* = METHYPLINACOLYLOXYFLUOROPHOSPHINE OXIDE type of nerve agent (NOTE: also known as **Soman**)

Gds = GUARDS

Gdsm = GUARDSMAN

GDP [ˌdʒiː diː 'piː] = GENERAL DEPLOYMENT POSITION

Gecko ['gekəʊ] *noun* SA-8, Soviet-designed low altitude surface-to-air missile (SAM)

Geiger counter ['gaɪgə ˌkaʊntə] *noun* instrument for measuring levels of radiation

Gen = GENERAL

general ['dʒenrəl] **1** *adjective* **(a)** not restricted, not specialized; ***this is a general warning to all base personnel*** **(b)** common to everyone or everything; **general cease-fire** = cease-fire observed by all participants in a war or armed conflict; **general deployment position (GDP)** = pre-selected position that a unit or sub-unit will deploy to in the event of war; **general headquarters (GHQ)** = headquarters of an army commander; **general staff** = staff which has supreme control over a state's armed forces **2** *noun* **(a)** senior army commander (not necessarily holding the rank of general); ***Napoleon was one of the greatest generals in history;*** *US* **one-star general** = brigadier general; **two-star general** = major general; **three-star general** = lieutenant general; **four-star general** = general; *US* **commanding general (CG)** = commander of a large tactical grouping (eg division, corps, army) *GB* **general officer commanding (GOC)** = general in command of a large army grouping (usually a division) **(b)** senior rank in the British army or marines *US* senior rank in the army, marines or air force; **general of the army** = top rank in the US army (equivalent to a British field-marshal); **general of the air force** = top rank in the US air force (equivalent to the British marshal of the RAF) (NOTE: shortened to **Gen** in this meaning); *see also* BRIGADIER GENERAL, LIEUTENANT GENERAL, MAJOR GENERAL

general purpose (GP) ['dʒenrəl 'pɜːpəs] *adjective* suitable for a variety of different uses; **general purpose (GP) bomb** = bomb which is simply dropped onto a target by an aircraft; *see also* BALLISTIC BOMB, IRON BOMB

general purpose machine-gun (GPMG) [ˌdʒenrəl ˌpɜːpəs məˈʃiːn gʌn] *noun* **(a)** medium-sized machine-gun which can be used for a variety of roles (such as air defence, infantry weapon, sustained fire (SF), vehicle armament, etc.) **(b)** British-made 7.62mm machine-gun modified from the Belgian-designed FN-MAG

generator [ˈdʒenəreɪtə] *noun* machine designed for producing electricity

generic planning [dʒəˌnerɪk ˈplænɪŋ] *noun* making plans for future operations where various elements have still to be identified

Geneva Convention [dʒəˌniːvə kənˈvenʃn] *noun* international agreement concerning the conduct of military personnel in war, and dealing with subjects such as treatment of prisoners, care of the wounded, protection of civilian lives and property, etc.

genocide [ˈdʒenəʊsaɪd] *noun* large-scale killing of people of a specific nationality *or* ethnic *or* sectarian group

Gepard [ˈgepɑːd] *noun* German-designed self-propelled anti-aircraft gun

germ [dʒɜːm] *noun* bacterium *or* virus which causes a disease; **germ warfare** = biological warfare

ghetto [ˈgetəʊ] *noun* part of a city (usually with poor housing), which is prodominantly occupied by an ethnic *or* sectarian minority

ghillie *or* **gillie suit** [ˈgɪlɪ suːt] *noun* *GB* camouflaged suit worn by a sniper, consisting of a set of khaki overalls with large quantities of scrim sewn onto it, in order to break up the outline of his body

GHQ [ˌdʒiː eɪtʃ ˈkjuː] = GENERAL HEADQUARTERS

GI [ˌdʒiː ˈaɪ] *noun (informal)* American soldier (NOTE: plural is **GIs** [ˌdʒiː ˈaɪz])

Giant Viper [ˌdʒaɪənt ˈvaɪpə(r)] *noun* apparatus for clearing a lane through a minefield, consisting of a long length of flexible tube filled with explosive, which is fired into the minefield by means of a rocket, and then detonated; ***we require Giant Viper at grid 443659***; *see also* MICLIC

gillie suit *see* GHILLIE SUIT

Gimlet [ˈgɪmlət] *noun* SA-16, Soviet-designed hand-held surface-to-air missile (SAM)

gimpy [ˈdʒɪmpi] *noun* *GB (slang)* general purpose machine-gun (GPMG)

gipsy *or* **gypsy** [ˈdʒɪpsɪ] *noun* member of an ethnic group of people who traditionally lead a nomadic lifestyle throughout Europe and the Near East

> COMMENT: the word **gipsy** is considered by many people to be derogatory. A more acceptable alternative these days is **traveller**

give in [ˌgɪv ˈɪn] *verb* to admit that you have been beaten; ***groups of snipers are still refusing to give in***

give up [ˌgɪv ˈʌp] *verb* **(a)** to hand something over; ***the enemy gave up their weapons without a fight***; ***during our attack they gave up several kilometres of territory*** **(b)** to admit you cannot do something; ***after trying for six hours to capture the hill they gave up and retreated to base***

glacier [ˈglæsɪə(r)] *noun* large mass of ice in arctic *or* mountainous regions, which moves slowly downhill

glasshouse [ˈglɑːshaʊs] *noun* military prison

glen [glen] *noun (Scotland and Ireland)* valley in the mountains

glide bomb [ˈglaɪd bɒm] *noun* aerodynamic bomb which is released by an aircraft several kilometres from its target and which then makes a ballistic descent to the target controlled by a guidance system

glide path [ˈglaɪd pɑːθ] *noun* path which an aircraft follows as it comes down from its cruising altitude in to land

glider [ˈglaɪdə] *noun* aircraft without an engine, which is used to land troops and is normally towed to the landing zone by a powered aircraft

global [ˈgləʊbl] *adjective* relating to the whole world; **Global Positioning System (GPS)** = satellite navigation system

Globemaster ['gləʊb,mɑːstə] *see* C-17

glory ['glɔːri] *noun* fame and honour as a result of a great achievement (such as winning a battle)

GMT [,dʒiː em 'tiː] = GREENWICH MEAN TIME

Gnr = GUNNER

Goa ['gəʊə] *noun* SA-3, Soviet-designed short-range surface-to-air missile (SAM)

Goalkeeper ['gəʊl,kiːpə(r)] *noun* Dutch-designed radar-controlled 30mm naval anti-aircraft cannon (CIWS), which automatically detects, tracks and engages targets

GOC [,dʒiː əʊ 'siː] = GENERAL OFFICER COMMANDING

gofer ['gəʊfə(r)] *noun (slang)* person who goes and gets things for other people; ***get one of the gofers to fetch my kit***

go firm [,gəʊ 'fɜːm] *verb* to stop moving and take up a position of defence; ***the platoon went firm on the edge of the wood*** (NOTE: **going - went - have gone**)

goggles ['gɒglz] *plural noun* spectacles with plain glass lenses, for protecting the eyes from dust, rain, wind, etc.; ***he was wearing goggles***

going ['gəʊɪŋ] *noun* conditions for movement (such as the state of the ground, effects of enemy fire, amount of traffic, etc.); ***the going was extremely difficult***

Golf [gɒlf] seventh letter of the phonetic alphabet (Gg)

gong [gɒŋ] *noun (slang)* medal

goose-egg ['guːs eg] *noun* tactical map-marking symbol for a defended locality, consisting of a circle *or* oval, intersected by a symbol for the size of grouping which occupies the locality; ***you've marked B Company's goose-egg on the wrong hill***

goose-step ['guːs step] **1** *noun* style of ceremonial marching, with raised steps in which the leg is kept straight **2** *verb* to march with the legs kept straight; ***the guards goose-stepped past the President's tomb***

Gopher ['gəʊfə] *noun* SA-13, Soviet-designed surface-to-air missile, usually attached to armoured and mechanized groupings

gorge [gɔːdʒ] *noun* deep and narrow valley (usually with rocky sides)

gorse [gɔːs] *noun* bush with spiked leaves and yellow flowers (NOTE: Irish English is **furze**)

Gortex™ ['gɔːteks] *noun* type of waterproof material which allows condensation produced by a person's body to escape through it

government (Govt) ['gʌvənmənt] *noun* official body of people who control all the activities of the state; ***the government is sending a task force to the area***

Govt = GOVERNMENT

GP [,dʒiː 'piː] = GENERAL PURPOSE

GP BOMB [,dʒiː ,piː 'bɒm] = GENERAL PURPOSE BOMB

Gp Capt = GROUP CAPTAIN

GPMG [,dʒiː piː em 'dʒiː] = GENERAL PURPOSE MACHINE-GUN

GPO [,dʒiː piː 'əʊ] *noun* *GB* = GUN POSITION OFFICER officer in charge of the guns of an artillery battery when they are deployed on the gun line

> COMMENT: the battery commander (BC) normally accompanies the commander of the battle group which he is supporting

GPS [,dʒiː piː 'es] = GLOBAL POSITIONING SYSTEM

GR-7 [,dʒiː ɑː 'sevən] *see* HARRIER

grade [greɪd] *noun* level of proficiency, quality, rank, etc.

gradient ['greɪdiənt] *noun* steepness of a slope

> COMMENT: a gradient is usually measured as a ratio: such as 1:4 (say 'one in four')

Grail [greɪl] *noun* SA-7, Soviet-designed hand-held optically-tracked surface-to-air missile (SAM)

grain [greɪn] *noun* **(a)** edible seeds from cereal plants such as barley, maize, oats *or* wheat; ***this is a major grain-producing region*** **(b)** *(of the country)* predominant direction in which the high ground and rivers of a region run (thereby affecting the direction of routes); ***we'll have the disadvantage of advancing across the***

grain of the country; *see also* CROSS-GRAIN

grappling-hook ['græplɪŋ ˌhʊk] *noun* metal hook with three or more prongs, which is attached to a rope, and can be used as an aid to climbing *or* to catch and drag in an object (especially one floating in water); ***we used a grappling-hook to get over the wall***

graticule ['grætɪkjuːl] *noun* one of a number of fine lines visible in the lens of an optical instrument, as an aid to measuring distance or sighting objects

grave [greɪv] **1** *adjective* serious or threatening; ***the situation is extremely grave*** **2** *noun* hole in the ground, in which a dead body is buried; ***they discovered the bodies of women and children buried in shallow graves***; **Graves Registration Unit** = unit responsible for recording the location of temporary graves in wartime

> COMMENT: for reasons of hygiene and logistics, soldiers killed in action are usually buried in temporary graves on or close to the battlefield, until such a time as the bodies can be returned home to relatives or, alternatively, ried in a proper military cemetery

graveyard ['greɪvjɑːd] *noun* area of ground containing graves; *see also* CEMETERY, CHURCHYARD

green [griːn] *adjective* **(a)** colour of vegetation the vehiples were pæinted green **(b)** inexperienced; ***the troops were completely green***

Green Berets [ˌgriːn bə'reɪz] *noun* American special forces unit; British marines; US Army airborne special forces organization; *compare* BLUE BERETS, RED BERETS

Greenwich Mean Time (GMT) [ˌgrenɪtʃ 'miːn taɪm] *noun* local time on the meridian at Greenwich, London; used to calculate international time

> COMMENT: Greenwich Mean Time is used by NATO forces on operations and is referred to as **Zulu time**

greeny ['griːnɪ] *noun (informal)* strong plastic container designed to carry two mortar rounds and fitted with a sling for man-packing; ***greenies will be dumped at the company RV***

Gremlin ['gremlɪn] *noun* SA-14, Soviet-designed hand-held surface-to-air missile (SAM)

gremlin ['gremlɪn] *noun (informal)* imaginary goblin who is blamed for unexplained mechanical failures in aircraft; ***there must be a gremlin at work in this plane !***

grenade *or* **hand-grenade** [grə'neɪd] *noun* small bomb designed to be thrown by hand; **grenade launcher** = gun designed to fire small explosive projectiles; **fragmentation grenade** = anti-personnel grenade designed to explode into fragments; **rifle grenade** = grenade designed to be fired from the muzzle of a rifle; **smoke grenade** = grenade which releases smoke; **stun grenade** = blast grenade designed to stun its victim, but not inflict physical injury; **grenade necklace** = improvised booby trap, consisting of a series of grenades which are secured to trees *or* other firm objects, with trip-wires attached to the safety-pins

grenadier [ˌgrenə'dɪə] *noun (historical)* elite infantryman; **panzer grenadier** = German armoured infantryman

grid [grɪd] *noun* **(a)** system of numbered squares printed on a map in order to produce references to particular points; **grid bearing** = direction, in mils or degrees, from one location to another, obtained from the map with the aid of a protractor; **grid north** = north as shown on a map; *compare* MAGNETIC NORTH; **grid reference** = six or eight figure reference number, obtained from the coordinates of the grid, used to identify an exact location on the map; ***the grid reference for the church is 656364*** (NOTE: also called **map reference**); **grid square** = segment of a map grid formed by two eastings and two northings, normally showing an area of one square kilometre (NOTE: the horizontal lines of a map grid and their coordinates are known as **northings,** while the vertical lines and their coordinates are known as **eastings**); **military grid reference system (MGRS)** = world-wide series of maps which are compatible with the GPS satellite navigation system **(b)**

grid reference; ***the bridge is at grid 423019; 'hullo 2, this is 22, request recovery at grid 559321, over'*** **(c)** framework of spaced parallel bars designed to prevent entry; ***the entrance to the tunnel was protected by a metal grid***

Gripen ['grɪpən] *see* SAAB-39

groom [gru:m] *noun* person who looks after a horse

ground [graʊnd] **1** *noun* surface of the earth; **ground attack** = attack by aircraft on a target on the ground; **ground crew** = air-force personnel who maintain an aircraft, but do not fly in it; **ground forces** = military forces which operate on the ground (such as armour, artillery, engineers, infantry, etc.); *see also* LAND FORCES; **ground of tactical importance** = area of ground which, if captured by the enemy, could seriously affect a unit or sub-unit's ability to fulfill its mission; **ground personnel** *or* **ground staff** = non-flying personnel of an air-force unit; **ground zero** = point on the ground directly under the explosion of a nuclear weapon; **dead ground** = area of ground which provides cover from view (eg. the reverse slope of a hill); **to gain ground** = to move forwards; **to give ground** = to withdraw; ***the enemy was forced to give ground***; **killing ground** = area of ground selected as a place to destroy an enemy force; **vital ground** = area of ground which, if captured by the enemy, will make it impossible for a unit or sub-unit to fulfill its mission **2** *verb* **(a)** to stop an aircraft from flying; ***the squadron was grounded by fog*** **(b)** to stop a pilot or member of an aircrew from flying; ***he was grounded until the investigation was completed***

> COMMENT: the **vital ground** of a sub-unit (such as a platoon) will often constitute the **ground of tactical importance** of its higher formation (ie the company). If a unit's vital ground is captured, then that unit has effectively lost its part of the battle

groundsheet ['graʊndʃi:t] *noun* waterproof sheet which can be spread on the ground or used to construct an improvised shelter

> COMMENT: in many armies, the groundsheet can also be used as a waterproof cape or poncho

group [gru:p] **1** *noun* **(a)** number of people or things which are close together; ***a group of trees*** **(b)** number of people who work together; **gun group** = infantry machine-gunner and his loader **(c)** division of the air force *GB* **group captain (Gp Capt)** = senior officer in the air force, above a wing commander **2** *verb* to organize people or things into groups; ***the recruits were grouped into squads***

grouping ['gru:pɪŋ] *noun* **(a)** number of people, vehicles or sub-units organized together for a specific role (such as a brigade, company, regiment, squadron, etc.) **(b)** group of bullet-holes in a target, made by several shots which have been fired in order to test the accuracy of the weapon; ***your grouping is two inches to the right of the aiming mark***

Grumble ['grʌmbl] *noun* SA-10, Soviet-designed medium-range surface-to-air missile (SAM)

grunt [grʌnt] *noun* *US* *(slang)* infantryman

guard [gɑ:d] **1** *noun* **(a)** person who protects other people or things; ***there are four guards at the front gate***; **guard dog** = dog trained to attack intruders **(b)** person who keeps control of prisoners; ***the prisoners managed to kill their guards*** **(c)** military force assigned to protect other people or things; **guard force** = force assigned to cover a likely enemy approach; **guard of honour** *or* **honour guard** = detachment of servicemen assigned to salute a dignitary on a parade or other formal occasion; **the Changing of the Guard** = ceremonial parade in which a detachment of soldiers who are guarding a royal or presidential palace hands over to another detachment **(d) the Guards** = elite troops, who traditionally guard a monarch or head of state (such as a president); **Brigade of Guards** = the five guards infantry regiments of the British Army (Grenadier Guards, Coldstream Guards, Scots Guards, Irish Guards and Welsh Guards); *US* **the National Guard** = volunteer force of part-time soldiers, which can be used for home defence or the maintenance of public order, but is not deployed outside the USA; **Presidential Guard** = elite troops whose duty is to protect the president; **Republican Guard** = elite troops in some armies; French ceremonial troops **3** *verb* **(a)** to protect other people or things **(b)** to keep control of prisoners

guardroom ['gɑːdrʊm] *noun* secure location in a barracks or base, providing a command post and accommodation for the guard and also secure accommodation for prisoners

guardsman (Gdsm) ['gɑːdzmən] *noun* **(a)** member of an elite guards regiment **(b)** *GB* private soldier in the Brigade of Guards

guerilla *or* **guerrilla** [gə'rɪlə] *noun* irregular soldier fighting against regular troops; *see also* PARTISAN

guidance ['gaɪdns] *noun* **(a)** process of directing someone or something to a destination; **guidance system** = component which directs a missile towards its target **(b)** advice or supervision; ***he is able to work without guidance***

guide [gaɪd] **1** *noun* **(a)** person who shows another person the way to a destination; ***local guides led the soldiers over the mountain pass*** **(b)** written instructions; ***you will find the safety instructions set out in the guide*** **2** *verb* **(a)** to show someone the way to a destination; ***they tried to find some local men to guide them across the desert*** **(b)** to advise someone how to carry out a task; ***we must be guided by our instructions from HQ***

guided ['gaɪdɪd] *adjective* equipped with a guidance system or by remote control; **guided missile** = missile which is directed to its target by a guidance system or by remote control; **guided weapon** = weapon whose projectile is directed to its target by a guidance system or by remote control

Guideline ['gaɪdlaɪn] *noun* SA-2, Soviet-designed medium-range surface-to-air missile (SAM)

Guild [gɪld] *noun* SA-1, Soviet-designed medium-range surface-to-air missile (SAM)

gulf [gʌlf] *noun* **(a)** very large area of sea partly enclosed by land; ***oil exploration in the Gulf of Mexico*** **(b) the Gulf** = **(i)** the Persian Gulf; **(ii)** the Gulf War of 1991, following the invasion of Kuwait by Iraq; ***I was in the Gulf*** *or* ***I served in the Gulf***; *see also* DESERT STORM

gully ['gʌli] *noun* small re-entrant

gun [gʌn] *noun* **(a)** any type of firearm; **gun cotton** = type of explosive **(b)** artillery piece; **field gun** = artillery piece designed to be moved easily over all types of ground; **gun crew** = soldiers who operate an artillery piece; **gun layer** = person who aims an artillery piece; **gun line** = fire position of a battery

gunboat ['gʌnbəʊt] *noun* small vessel with heavy guns, designed to operate in shallow waters; **gunboat diplomacy** = political negotiation supported by the threat of military action

gunfire ['gʌnfaɪə] *noun* firing of guns; ***we could hear gunfire in the distance***

gung-ho [ˌgʌŋ 'həʊ] *adjective; (informal)* very eager to take military action; ***the general is a very gung-ho type***

gunman ['gʌnmən] *noun* person equipped with a firearm for criminal or terrorist purposes

gunner ['gʌnə] *noun* **(a)** artillery soldier; **the Gunners** = the Royal Artillery **(b)** *GB* private in the artillery (NOTE: shortened to **Gnr** in this meaning: **Gnr Jones)** **(c)** warrant officer or senior non-commissioned officer in charge of a battery on a warship

gunnery ['gʌnəri] *noun* training and operational use of artillery or large calibre guns; **gunnery officer** = officer responsible for gunnery within a unit or warship; *US* **gunnery sergeant** = senior non-commissioned officer (SNCO) in the marines who acts as operations and training coordinator for a company

gunny ['gʌni] *US* = GUNNERY SERGEANT

gunpowder ['gʌnˌpaʊdə] *noun* obselete explosive substance, used as propellant for muzzle-loading firearms

gunrunner ['gʌnˌrʌnə(r)] *noun* person who imports weapons illegally; ***gunrunners are supplying weapons to the guerrillas***

gunship ['gʌnʃɪp] *noun* another name

for certain types of ground-attack aircraft (both fixed-wing and rotary) eg. AC-130 , Mi-24, etc; **(helicopter) gunship** = heavily armed attack helicopter

gunsight ['gʌnsaɪt] *noun* device on a weapon, which is used by the firer to aim at a target

gunsmith ['gʌnsmɪθ] *noun* person who makes firearms; ***local gunsmiths are producing good copies of the AK-47 assault weapon***

gunwale ['gʌn(ə)l] *noun* top edge of the side of a boat *or* ship; ***our only air defence was a couple of machine-guns clamped to the gunwale***

Gurkha ['gɜːkə] *noun* inhabitant of the mountains in Nepal, serving in a Gurkha regiment of the British or Indian Army

gut [gʌt] *noun* intestine; **to have guts** = to be brave

guy [gaɪ] *noun (slang)* man

gypsy ['dʒɪpsɪ] *see* GIPSY

HOTEL - Hh

H [eɪtʃ] *noun* = 2.2-DICHLORO-DIETHYL SULPHIDE type of blister agent (NOTE: also known as **Levinstein Mustard**)

hack [hæk] *verb* ***1* (a)** to chop or cut with a sharp tool or weapon; ***they hacked their way through the jungle*** **(b) to hack into a programme =** to gain unauthorised access to a computer programme **2** *adverb* (forward air controller jargon) minutes before an aircraft fires its weapons system at a target; ***hello G33 this is Cowboy, hack three, over***

hackle ['hækl] *noun* cluster of feathers worn in a military head-dress

haemorrhage *US* **hemorrhage** ['hemərɪdʒ] **1** *noun* loss of blood from a damaged blood-vessel **2** *verb* to suffer from a haemorrhage

hail [heɪl] *noun* small round pellets of frozen rain falling to the ground; **hail of bullets** *or* **shrapnel =** heavy small arms or artillery fire; ***we advanced into a hail of bullets***

half-colonel ['hɑːf ˌkɜːnl] *noun GB* lieutenant-colonel

half-track ['hɑːf træk] *noun* amoured military vehicle with two wheels in front and tracks at the rear

HALO ['heɪləʊ] *noun* = HIGH ALTITUDE LOW OPENING covert method of deploying troops by parachute: the aircraft flies at a very high altitude in order to avoid detection, and the parachutists then descend a considerable distance through the air before opening their parachutes

HALO Trust ['heɪləʊ trʌst] *noun* non-governmental organization (NGO) dedicated to clearing mines after the cessation of an armed conflict

halt [hɒlt] **1** *noun* temporary stop during a journey; ***after six hours marching through jungle we decided to call a halt*** **2** *verb* to stop moving (especially as a command); ***Halt, or I fire!***; **Halt! Who goes there? =** traditional challenge given by sentries

hamlet ['hæmlət] *noun* very small village

hammock ['hæmək] *noun* bed, made from fabric or netting, which is suspended by cords from trees or walls (formerly traditionally used by seamen)

hand [hænd] *noun* any member of a ship's crew; ***all hands to action stations !***

handcuff ['hændkʌf] **1** *noun (handcuffs)* pair of lockable metal bracelets joined by a short chain, which are used to bind the hands of a prisoner; ***the policeman produced a pair of handcuffs*** **2** *verb* to secure a prisoner with handcuffs; ***he was handcuffed and placed in the vehicle***; *see also* PLASTICUFF

hand-grenade *or* **grenade** ['hænd grəˌneɪd] *noun* small bomb designed to be thrown by hand; *see also* GRENADE

handguard ['hængɑːd] *noun* part of an apparatus or machine, which prevents the operator from harming his hands while using it; ***the handguard is made of wood***

handgun ['hængʌn] *noun* small gun designed to be held in one hand and fired without holding it to the shoulder; *see also* PISTOL, REVOLVER

handle ['hændl] **1** *noun* part of an object, which enables it to be carried by hand; ***the machine-gun has a carrying handle*** **2** *verb* to operate with the hands; ***you should wear protective clothing when handling gas canisters***

handset ['hændset] *noun* radio or telephone apparatus containing an earpiece and mouth piece, which is designed to be held in the hand

hand-to-hand fighting [ˌhænd tə ˌhænd 'faɪtɪŋ] *noun* close fighting, especially with the bayonet

hangar ['hæŋə] *noun* large building for the housing of aircraft

harass ['hærəs *US* hə'ræs] *verb* to attack again and again in order to disrupt the enemy's activities

harassing fire ['hærəsɪŋ *US* hə'ræsɪŋ ˌfaɪə] *noun* random artillery fire directed at a likely area of enemy activity, in order to disturb the enemy's rest, disrupt his movements and inflict casualties, and so affect his morale

harbour *US* **harbor** ['hɑːbə] **1** *noun* **(a)** natural or man-made place where ships can shelter from the weather; ***the ship was unable to enter the harbour*** **(b)** secure area in the field, where units can rest and reorganize before starting the next phase of an operation; ***we set up a harbour in the woods*** **2** *verb* to provide shelter (especially for a deserter or escaped prisoner); ***he was arrested for harbouring an escaped prisoner***

hard-target [ˌhɑːd 'tɑːgət] *verb* to move across ground in such a way, so as not to present an easy target to the enemy; ***we had to hard-target across the square***

harm [hɑːm] **1** *noun* injury or damage; ***he is safe from harm***; ***was any harm done to the radio?*** **2** *verb* to cause injury or damage; ***I won't let you harm the prisoners***; ***it is almost impossible to harm this device***

HARM [hɑːm] *noun* = HIGH-SPEED ANTI-RADIATION MISSILE American-designed air-to-ground anti-radar missile (ARM)

harmful ['hɑːmfʊl] *adjective* able to cause harm

harmless ['hɑːmləs] *adjective* not able to cause harm

harmonize ['hɑːmənaɪz] *verb* **(a)** to make things similar, to standardize operations **(b)** to make sure that all guns on an aircraft are aimed at the same target

harness ['hɑːnəs] *noun* straps which fasten a piece of equipment to a person's body; ***his parachute harness was not attached properly***

Harpoon [hɑː'puːn] *noun* American-designed anti-ship missile

Harrier ['hærɪə] *noun* British-designed fighter aircraft with a vertical take-off capability; **Harrier AV-8** = American ground-attack variant, designed to operate from aircraft carriers and certain other ships; **Harrier GR-7** = British multirole variant, which is specially useful for ground attack; **Sea Harrier** = multirole fighter, designed to operate from aircraft carriers

HAS [ˌeɪtʃ eɪ 'es] *noun* = HARDENED AIRCRAFT SHELTER shelter designed to protect an aircraft from artillery, bomb or missile attack

hatch [hætʃ] *noun* opening in an aircraft, ship or vehicle, which is fitted with a cover

haven ['heɪvən] *noun* **(a)** old-fashioned word for harbour **(b)** place of safety; **safe haven** = peacekeeping term for a secure area in which members of an ethnic *or* sectarian minority can seek safety from the hostility of the main population

haversack ['hævəsæk] *noun* canvas or webbing bag, carried slung over the shoulder

havoc ['hævək] *noun* **(a)** great confusion; **to cause havoc in** *or* **to wreak havoc on** = to throw something into confusion; ***atmospherics wrought havoc on our communications*** **(b) Havoc** = NATO name for the Soviet-designed Mi-28 attack helicopter

Hawkeye ['hɔːkaɪ] *see* E-2

hawser ['hɔːzə] *noun* thick metal wire or rope used to moor or tow a ship (NOTE: also called a **cable**)

haybox ['heɪbɒks] *noun* insulated container, designed to keep food hot for several hours; ***the food was brought up to the position in hayboxes***

hazard ['hæzəd] *noun* danger; **hazard beacon** = warning beacon indicating that there is some danger to aircraft

haze [heɪz] *noun* reduction in visibility caused by dust or hot air

HCN [ˌeɪtʃ siː 'en] = HYDROGEN CYANIDE type of blood agent (NOTE: also known as **AC**)

HD [ˌeɪtʃ 'diː] *noun* = 2.2-DICHLORO-DIETHYL SULPHIDE type of blister agent (NOTE: also known as Distilled Mustard)

> COMMENT: HD is simply a purer form of H (Levinstein Mustard)

HE [ˌeɪtʃ 'iː] *noun* **(a)** = HIGH EXPLOSIVE **(b)** projectile or other explosive ordnance containing high explosive; ***load with HE!***

headcount ['hedkaʊnt] *noun* act of counting people; ***the sergeant did a quick headcount***

head-dress ['heddres] *noun* cap or hat worn as part of a uniform

heading ['hedɪŋ] *noun* **(a)** direction; ***we moved on a heading of 3.340 mils*** **(b)** title at the head of a page or part of a document; ***he prepared his report under a number of headings***

heads [hedz] *noun* toilets on a ship (or in a naval or marine establishment)

headquarters (HQ) [hed'kwɔːtəz] *noun* **(a)** administrative and command centre of a tactical grouping; **main headquarters =** primary resourcing and planning headquarters for a large tactical grouping (normally located to the rear of the forward troops); **rear headquarters =** primary logistical headquarters for a large tactical grouping (normally located well to the rear of the front line); **tactical headquarters (TAC) =** small mobile headquarters, used by a commander when he is moving around the battlefield **(b)** staff of a headquarters; ***brigade headquarters are being accommodated in the school***

headset ['hedset] *noun* part of a radio or other audio-equipment, consisting of a set of earphones, sometimes with a microphone attached for speaking, which is worn on the head and is used to listen to transmissions or signals

heads-up [hedz 'ʌp] *noun US* notification by a commander to his subordinate commanders, informing them what he is thinking of doing

> COMMENT: a *heads-up* should not be confused with a **warning order**, which is issued when the commander has definitely decided what he is going to do

hearts and minds [ˌhɑːts ənd 'maɪnz] *noun* ***to win the hearts and minds of the population*** = philosophy of trying to win the support of the civilian population in your area of operations (eg. by ensuring good behaviour of troops, providing free medical care, assistance to local authorities, entertainment, etc)

heat [hiːt] *noun* state of being hot; *see also* HEAT-SEEKING MISSILE

HEAT [hiːt] *noun* = HIGH EXPLOSIVE ANTI-TANK anti-tank projectile with a shaped-charge warhead; ***load with HEAT!***; ***the tank was destroyed by a HEAT round***

heat exhaustion ['hiːt ɪgˌzɔːstʃ(ə)n] *noun* physical collapse, which is caused by carrying out intense physical activity (eg. running) in hot weather, and is potentially fatal; ***we've got three cases of heat exhaustion***

heat-seeking missile [ˌhiːt siːkɪŋ 'mɪsaɪl] *noun* missile equipped with a guidance system which homes in on a source of heat (such as the jet pipes of an aircraft engine)

heath [hiːθ] *noun* uncultivated area of dry sandy soil, covered with bracken, heather and small bushes

heather ['heðə] *noun* low-growing plant with wooden stems and purple flowers, which covers wide areas of ground on heathland and moorland

heathland ['hiːθlænd] *noun* terrain consisting mainly of heath

heave to [hiːv 'tuː] *verb* to bring a ship to a halt; ***heave to immediately !***; ***we were ordered to heave to***

heavy ['hevi] *adjective* **(a)** which weighs a lot; ***the troops moved through the jungle carrying heavy packs*** **(b)** difficult to lift or support; ***the firing-post is too heavy for one man to carry*** **(c)** intense; ***we came under heavy fire*** **(d)** *(of weapons)* having a large calibre; ***they moved up heavy artillery***

Heckler & Koch [ˌheklə ənd 'kɒk] *noun* German-designed 7.62mm assault weapon

hedge *or* **hedgerow** [hedʒ *or* 'hedʒrəʊ] *noun* fence made of living plants

height [haɪt] *noun* vertical distance from the ground

hel = HELICOPTER

HELARM ['helɑːm] *noun* use of attack helicopters in support of ground forces; ***HELARM is available on request***

helicopter ['helɪkɒptə] *noun* aircraft without wings, which obtains its upward lift by means of horizontally rotating blades (known as rotors), which are fitted to the top of the airframe; **attack helicopter** = helicopter equipped with weapons to attack other helicopters or targets on the ground; **utility** *or* **transport helicopter** = helicopter designed to transport men or equipment; *see also* CHOPPER, GUNSHIP, ROTARY-WING AIRCRAFT

helipad ['helɪpæd] *noun* prepared landing and take-off area for helicopters

Hellfire ['helfaɪə] *noun* American-designed laser-guided air-to-ground missile

helm [helm] *noun* wheel or bar with which a ship or boat is steered; **to take the helm** = to start steering a ship

helmet ['helmət] *noun* protective head covering

heliograph ['hiːlɪəˌgrɑːf] *noun* signalling device which uses a mirror to produce flashes of sunlight; ***I used my shaving-mirror as a heliograph***

helo ['hiːləʊ] *noun US (informal)* helicopter

hemisphere ['hemɪsfɪə] *noun* half of the earth's surface; **Northern Hemisphere** = area of the earth's surface north of the Equator; **Southern Hemisphere** = area of the earth's surface south of the Equator

Herc [hɜːk] *noun* informal name for the Hercules C-130 transport aircraft

Hercules ['hɜːkjuːliːz] *see* C-130

hero ['hɪərəʊ] *noun* man acknowledged by other people to have carried out an act of bravery; ***he was the hero of the battle***

heroine ['herəʊɪn] *noun* woman acknowledged by other people to have carried out an act of bravery; ***the media are calling her a heroine***

heroism ['herəʊɪzm] *noun* bravery in the face of danger

HESH [heʃ] *noun* = HIGH EXPLOSIVE SQUASH-HEAD anti-armour warhead which flattens on impact before exploding, thereby creating a shock wave which causes a part of the armour's interior surface to break off and ricochet around the inside of the vehicle; ***load with HESH!***; ***we used HESH to destroy the bunker***; *see also* SPALL

> COMMENT: HESH can also be used as a general purpose high explosive round by tanks and other large calibre direct-fire guns

hessian ['hesɪən] *noun* coarse fabric used as camouflage or to make sandbags (NOTE: American English is **burlap**)

HET [ˌeɪtʃ iː'tiː *or* het] *noun US* = HEAVY EQUIPMENT TRANSPORTER large wheeled vehicle designed to carry a tank *or* other armoured vehicle over long distances by road; *see also* TANK TRANSPORTER

HF [ˌeɪtʃ 'ef] = HIGH FREQUENCY

H-Hour ['eɪtʃ ˌaʊə] *noun* time at which an operation is due to begin; **H-minus-ten** = ten minutes before H-Hour; **H-plus-ninety** = ninety minutes after H-Hour; **H-minus-three hours** = three hours before H-hour; **H-plus-five hours** = five hours after H-hour

hide [haɪd] **1** *noun* **(a)** concealed location where a unit or sub-unit can rest or wait in reserve; ***B Company is in a hide at grid 221434*** **(b)** hiding place used by guerrillas or terrorists to conceal weapons or explosives; ***there is a weapons hide in the wood*** **2** *verb* **(a)** to conceal something; ***he hid the gun in the attic*** **(b)** to conceal oneself; ***they were hiding in the cellar*** (NOTE: **hiding - hid - has hidden**)

high altitude low opening *see* HALO

high explosive (HE) [ˌhaɪ ɪk'spləʊsɪv] *noun* **(a)** powerful explosive substance used in bombs, grenades, shells, etc. **(b)** bomb or projectile containing high explosive; *see also* HEAT, HESH

> COMMENT: ordnance containing high explosive is normally painted dark green, with yellow lettering and markings

high frequency (HF) [ˌhaɪ ˈfriːkwənsi] *noun* range of radio frequencies from 3 - 30 megahertz (Mhz); **ultra high frequency (UHF)** = range of radio frequencies from 300 - 3,000 megahertz (MHz); **very high frequency (VHF)** = range of radio frequencies from 30 - 300 megahertz (MHz)

highlander [ˈhaɪləndə] *noun* member of a British infantry regiment which traditionally recruits its soldiers from the mountainous regions of Scotland; ***the Argyll and Sutherland Highlanders***

high-mobility multipurpose wheeled vehicle *see* HMMW-V

high-tech *or* **hi-tech** [haɪ ˈtek] *adjective* = HIGH TECHNOLOGY using very sophisticated technology (especially computers); ***the enemy has very little hi-tech surveillance equipment***

high velocity [ˌhaɪ vəˈlɒsəti] *adjective; (of projectiles)* designed to travel faster than the speed of sound; ***he was hit by a high velocity bullet***; *compare* LOW VELOCITY

highway [ˈhaɪweɪ] *noun GB* public road *US* main road

hijack [ˈhaɪdʒæk] *verb* to seize control of an aircraft or vehicle; ***the guerillas hijacked a truck and killed the driver***

hijacker [ˈhaɪdʒækə] *noun* person who seizes control of an aircraft or vehicle; ***the hijackers threatened to blow up the plane if their demands were not met***

hill [hɪl] *noun* area of high ground

hillfort [ˈhɪlfɔːt] *noun* ancient fortification (usually an earthwork)

hillock [ˈhɪlək] *noun* small hill

Hind [haɪnd] *noun* NATO name for the Soviet-designed Mi-24 and Mi-35 attack helicopters and Mi-17h

Hip [hɪp] *noun* NATO name for the Soviet-designed Mi-8 and Mi-17 attack helicopters

hit [hɪt] **1** *noun* shot which strikes the target at which it is aimed; ***we scored a direct hit on the fuel storage depot*** **2** *verb* **(a)** to shoot at a target succesfully **(b)** to strike a person or thing (NOTE: **hitting - hit**); ***They couldn't hit an elephant at this distance - last words of Major General John Sedgewick at the Battle of Spotsylvania 1864***

HL [ˌeɪtʃ ˈel] *noun* type of blister agent

> COMMENT: HL is a combination of the agents HD and L

HLS [ˌeɪtʃ el ˈes] = HELICOPTER LANDING SITE

HMS [ˌeɪtʃ em ˈes] *abbreviation* = HER MAJESTY'S SHIP *or* HIS MAJESTY'S SHIP prefix given to all ships of the Royal Navy and also to some naval establishments; ***I served on board HMS Sheffield***

HMMW-V [ˈhʌmviː] *noun* = HIGH-MOBILITY MULTIPURPOSE WHEELED VEHICLE American-designed all-terrain vehicle (similar to a jeep) (NOTE: **known informally as Hummer**)

HN [ˌeɪtʃ ˈen] *noun* = NITROGEN MUSTARD type of blister agent; **HN-1** = 2.2' dichloro-triethylamine; **HN-2** = 2.2' dichloro-diethyl methylamine; **HN-3** = 2.2' 2-trichloro-triethylamine

HNS = HOST NATION SUPPORT

hoax [həʊks] *noun* false alarm intended to waste time and resources; ***the bomb warning turned out to be a hoax***

hog's back [ˈhɒgz bæk] *noun* ridge of high ground

Hokum [ˈhəʊkəm] *noun* NATO name for a Soviet-designed Ka-50 attack helicopter

hold [həʊld] **1** *noun* storage area in an aircraft or ship; ***we found three men hiding in the hold*** **2** *verb* **(a)** to have something in your hand; ***he was holding a pistol*** **(b)** to have possession of; ***the enemy are still holding the bridge*** **(c)** to prevent the enemy from capturing; ***we must hold this position until last light*** **(d)** to keep someone in custody; ***he is being held by the police***

holding attack [ˈhəʊldɪŋ əˌtæk] *noun* attack mounted to halt the advance of an enemy and keep him occupied, while other friendly forces conduct operations elsewhere; ***5 Brigade will mount a holding***

attack around Fallingbostel to enable the rest of the division to withdraw

hold on [ˌhəʊld ˈɒn] *verb* to wait for a short period; ***hold on, I haven't finished yet***

hold out [ˌhəʊld ˈaʊt] *verb* to continue to defend or resist; ***small units of the enemy are holding out in the mountains***; ***we can't hold out much longer***

hollow [ˈhɒləʊ] *noun* depression in the ground

holster [ˈhəʊlstə] *noun* carrying case for a pistol, which is worn on a belt or harness

home defence [ˌhəʊm dɪˈfens] *noun* defence of a state's own territory in the event of war (as opposed to territory belonging to another state); ***the division will be used for home defence***; *compare* CIVIL DEFENCE

home in [ˌhəʊm ˈɪn] *verb (of guidance systems and locating equipment)* **to home in on something** = to be guided towards something (such as an emission of radiation, heat source, radio signal, etc.)

homogeneous [hɒməˈdʒiːniəs] *see* ROLLED HOMOGENEOUS ARMOUR

honor *see* HONOUR

honorable *see* HONOURABLE

honour *US* **honor** [ˈɒnə] **1** *noun* **(a)** moral code of conduct; ***never surrendering to the enemy is a matter of honour*** **(b)** official or public recognition of a person's achievement; **battle honour** = official recognition of a unit's achievements or conduct during a battle, which gives that unit the right to carry the name of the battle on its colours; **honour guard** *or* **guard of honour** = detachment of servicemen assigned to salute a dignitary on a parade or other formal occasion **2** *verb* to acknowledge a person's achievement

honourable *US* **honorable** [ˈɒnrəbl] *adjective* **(a)** behaving in a good and moral way **(b)** worthy of honour

hooch [huːtʃ] *noun US informal* improvised shelter, usually with some protection from enemy fire; ***make sure that your hooches are well camouflaged***

Hook [hʊk] *noun* NATO name for the Soviet-designed Mi-6 transport helicopter

horizon [həˈraɪzn] *noun* line in the far distance, on which the sky and ground appear to meet; ***two ships appeared on the horizon***

horn [hɔːn] *noun* instrument fitted to a vehicle, which makes a noise as a warning signal; ***one of the signals for a chemical attack is sounding the horn of your vehicle***

Hornet [ˈhɔːnɪt] *see* FA-18

horse [hɔːs] *noun* large four-legged animal which can be ridden *or* used as a transport animal

> COMMENT: horses still play a prominent role in military life, especially in the British Army. Horses are ridden on ceremonial occasions and the army encourages participation in equestrian sports such as polo, steeplechasing, show jumping and fox-hunting. Although they are now equipped with tanks, British and American cavalry regiments still retain many of the traditions and expressions from the days when their troopers fought on horseback

hose [həʊz] *noun* **a** long flexible tube made of rubber *or* waterproof fabric, which is designed to convey liquid over a short distance and then pour it into a container *or* over an object; ***all vehicles should carry a hose for siphoning fuel***; ***the firemens' hoses weren't long enough***; **hose down** = to spray an object with liquid from a hose; ***they hosed down the vehicles*** **b**; *(no plural form)* long sock, reaching to just below the knee, which is worn with shorts or a kilt; ***the pipers were wearing kilts, tartan hose and spats***

hospital [ˈhɒspɪtl] *noun* establishment which provides surgery, medical treatment and nursing to ill and injured people; **field hospital** = mobile hospital set up on or near to the battlefield, which is capable of providing surgery

host [həʊst] *noun* person who invites other people as guests; **host nation** = nation which receives NATO forces on its territory; **host nation support (HNS)** = support given by a host nation to help NATO's efforts

hostage [ˈhɒstɪdʒ] *noun* person who is seized and held, in order to force other people to do something (for example paying a sum of money) or to deter them from doing something (for example attacking soldiers); ***the aim of the operation is to secure the release of the hostages*** (NOTE: the verb form is **to take someone hostage**)

hostile ['hɒstaɪl *US* 'hɒstl] *adjective* **(a)** enemy; ***all aircraft should be considered hostile*** **(b)** unfriendly; ***he was extremely hostile towards me***

> COMMENT: **hostile** is normally used to refer to the forces of a state which is aggressive and threatening towards your own country, but not officially at war. Whereas **enemy** is used when war has actually been declared

hostility [hɒ'stɪləti] *noun* **(a)** aggressive or threatening behaviour directed towards another person or state **(b) hostilities =** military action; **state of hostilites =** armed conflict

> COMMENT: **state of hostilities** is normally used to describe a situation where fighting occurs between the armed forces of two states, but they are not officially at war. Thus, the Falklands conflict of 1982 was described as a state of hostilities rather than a war, because Great Britain never actually declared war on Argentina

HOT [hɒt] *noun* = HIGH SUBSONIC, OPTICALLY GUIDED, TUBE FIRE European-produced wire-guided anti-tank missile (ATGW)

Hotel [həʊ'tel] eighth letter of the phonetic alphabet (Hh)

hour ['aʊə] *noun* **(a)** unit of time, corresponding to sixty minutes; *see also* H-HOUR **(b)** *(used in the plural after a 4-digit number, to show the time)* ***H-Hour is at 0600hrs***

> COMMENT: military timings are always given using the **twenty-four hour clock,** usually followed by the word **hours** which when written, is abbreviated to **hrs.** Thus, 8.15am is 0815hrs, 1pm is 1300hrs, 6.30pm is 1830hrs, etc. (say 'zero eight fifteen hours', 'thirteen hundred hours', 'eighteen thirty hours')

Household ['haʊshəʊld] *adjective GB* relating to the elite troops who traditionally guard the monarch; **Household Cavalry =** the Life Guards and the Blues and Royals; **Household Division =** Household Cavalry and the Brigade of Guards; **Household Troops =** Household Division and the King's Troop of the Royal Horse Artillery

housewife ['haʊswaɪf or 'hʌzɪf] *noun GB* small mending kit, consisting of needles, thread and spare buttons

hovercraft ['hɒvəkrɑːft] *noun* amphibious vehicle which travels over land or water supported on a cushion of air

howitzer ['haʊɪtsə] *noun* short-barrelled artillery piece designed to fire shells at high trajectories; **self-propelled howitzer (SPH) =** howitzer in the form of an armoured fighting vehicle (AFV)

HQ [ˌeɪtʃ 'kjuː] = HEADQUARTERS

hrs = HOURS

HUD [hʌd] *noun* = HEAD-UP DISPLAY display of instrument readings *or* other data, which is projected onto the windscreen of an aircraft, so that the pilot doesn't have to look down at inconvenient moments

Huey ['hjuːi] *noun* American-designed utility/transport helicopter (UH-1); **Huey Cobra =** American-designed AH-1 attack helicopter

hull [hʌl] *noun* **(a)** outer covering of a ship or boat; ***her hull almost touched bottom as she was entering the harbour*** **(b)** lower part of an armoured fighting vehicle; ***the explosion penetrated the ship's hull***

hulk [hʌlk] *noun* ***a*** body of an old ship, used as a target *or* as a floating storehouse **b** old tank *or* other armoured vehicle, used as a target on a range

hull-down [ˌhʌl 'daʊn] *adjective* positioned so that only the top parts, such as the turret of a tank) are visible and exposed; ***we could see the enemy fleet hull-down on the skyline***; ***the tank was hull-down behind a wall***

humanitarian [ˌhjuːmænɪ'teərɪən] *adjective* intended to prevent *or* reduce human suffering and hardship; ***the battalion is being sent to the disaster area to give humanitarian assistance***

HUMINT ['hjuːmɪnt *or* 'hʌmɪnt] *noun* = HUMAN INTELLIGENCE information about the enemy obtained from people (eg. friendly forces, agents, civilians, POW)

Hummer ['hʌmə(r)] *noun* ***US*** informal nickname for the HMMW-V all-terrain vehicle

hunter-killer [ˌhʌntə 'kɪlə] *noun* submarine designed to locate and destroy other submarines

hussar [hu'zɑː] *noun (historical)* light cavalryman

> COMMENT: some armoured regiments still retain their historical title of Hussars

hut [hʌt] *noun* simple wooden shelter

Hydra ['haɪdrə] *noun* American-designed unguided rocket, designed to be fired by an aircraft at a ground target

hydrogen cyanide (HCN) [ˌhaɪdrədʒən 'saɪənaɪd] *noun* type of blood agent (NOTE: also called **AC**)

hygiene ['haɪdʒiːn] *noun* practice of keeping oneself and your surroundings clean, in order to prevent disease; *see also* SANITATION

hypothermia [ˌhaɪpəʊ'θɜːmɪə] *noun* abnormally low body temperature, usually caused by exposure to wind, rain *or* extreme cold, which is potentially fatal; ***he was suffering from hypothermia***

INDIA - Ii

IAAG ['aɪæg] *noun* = IMPROVISED ANTI-ARMOUR GRENADE IRA-designed home-made hand-thrown grenade, containing a shaped-charge warhead which is designed to explode when it hits the side of a vehicle; ***there have been several IAAG attacks over the past month***

IA drill [aɪ 'eɪ ˌdrɪl] *noun* = IMMEDIATE ACTION DRILL standard procedure to be carried out in the event of something going wrong (such as an ambush, weapon misfire, equipment malfunction, etc.)

IC *or* **i/c** [ˌaɪ 'siː] *abbreviation* **(a)** in command; **2IC** = second-in-command **(b)** in command of or in charge of; ***he is i/c rations***

ICBM [ˌaɪ siː biː 'em] = INTERCONTINENTAL BALLISTIC MISSILE

ICM [ˌaɪ siː 'em] *noun* = IMPROVED CONVENTIONAL MUNITION artillery shell filled with a quantity of anti-personnel or anti-tank bomblets, which is designed to explode in the air and scatter the bomblets onto the target area below

ice [aɪs] *noun* frozen water

ice up [ˌaɪs 'ʌp] *verb* to become covered with ice; ***the aircraft crashed because the cockpit canopy had iced up***

ICP [ˌaɪ siː 'piː] *noun* = INCIDENT CONTROL POINT location from which the follow-up action to an incident is controlled; ***"Hello 2, this is 22. ICP at grid 434621. Over."***

ICRC [ˌaɪ siː ɑː'siː] = INTERNATIONAL COMMITTEE OF THE RED CROSS

ID [ˌaɪ 'diː] *noun (informal)* **(a)** identity; **ID card** = card issued by a government or organization as a means of identification; **ID disc** *or* **disk** = metal or plastic disc bearing a soldier's personal details, which is worn round the neck **(b)** proof of identity (such as an ID card, driving licence, passport, etc.); ***do you have any ID?*** **(c)** identification; ***he made a positive ID on the gunman***

identification [aɪˌdentɪfɪ'keɪʃn] *noun* act of identifying someone or something; **identification beacon** = radio beacon which sends out a signal by which a reference point can be identified; *see also* IFF

identify [aɪ'dentɪfaɪ] *verb* **(a)** to establish the identity of a person or thing; ***we haven't identified the dead man*** **(b)** to recognize a person or thing; ***he identified the gunman***

identity [aɪ'dentɪti] *noun* who a person is (ie name, date of birth, nationality, etc.); **identity card (ID card)** = card issued by a government or organization as a means of identification; **identity disc** *or* **disk (ID disc** *or* **disk)** = metal or plastic disc *or* disk bearing a soldier's personal details, which is worn round the neck (NOTE: also called a **dog-tag**)

ideology [ˌaɪdi'ɒlədʒi] *noun* system of ideas and principles (especially political)

IED [ˌaɪ iː 'diː] = IMPROVISED EXPLOSIVE DEVICE) home-made bomb or mine

IFF [ˌaɪ ef 'ef] = IDENTIFICATION FRIEND OR FOE technology carried in an aircraft which utilizes coded radio signals to identify other friendly aircraft

IFV [ˌaɪ ef 'viː] = INFANTRY FIGHTING VEHICLE

II [ˌaɪ 'aɪ] = IMAGE INTENSIFICATION; IMAGE INTENSIFIER

IL-76 [ˌaɪ el sevənti 'sɪks] *noun* Soviet-designed transport aircraft (NOTE: known to NATO as **Candid**)

Illum [ɪ'luːm] = ILLUMINATION ROUND

illuminate [ɪ'luːmɪneɪt] *verb* to light up (with artificial light)

illumination [ɪˌluːmɪ'neɪʃn] *noun* artificial light; **illumination round** = artillery or mortar projectile designed to produce light (NOTE: also called **illum** for short)

> COMMENT: illumination rounds are usually painted white, with black lettering and markings

image ['ɪmɪdʒ] *noun* appearance of an object as viewed through an optical instrument or other equipment (such as a thermal imager, radar, etc.); **image intensification (II)** = passive night-viewing technology which utilizes natural light (such as ambient light, moonlight, starlight); **image intensifier (II)** = night-viewing device which uses image intensification; **thermal image (TI)** = image produced by equipment which can identify the varying levels of heat given off by different objects

imager ['ɪmɪdʒə] *noun* **thermal imager (TI)** = optical instrument which produces a thermal image

immediate action drill [ɪˌmiːdiət 'ækʃn ˌdrɪl] *see* IA

immersion foot [ɪ'mɜːʃən fʊt] *noun* severe fungal infection of the feet, caused by wearing wet boots over a long period; *also known as* TRENCH FOOT

immigrant ['ɪmɪgrənt] *noun* person who enters a foreign country in order to live there; **illegal immigrant** = immigrant who enters a country without official permission to do so; ***our main role is to catch and arrest illegal immigrants***

immobilize [ɪ'məʊbɪlaɪz] *verb* **(a)** to do something to a vehicle, so that it cannot be driven **(b)** to do something to a machine or weapon, so that it does not work; *compare* MOBILIZE

impact ['ɪmpækt] *noun* act of one object hitting another object; **impact area** = part of a live firing range where projectiles strike or come to rest

implementation [ˌɪmpləmən'teɪʃn] *noun* putting something into effect; **implementation plans** = plans which allow NATO commanders to put into action operations which have been agreed between NATO and local forces

impregnable [ɪm'pregnəbl] *adjective; (of fortifications)* impossible to take by force; ***the guerillas are based in impregnable hilltop camps***

improved conventional munition [ɪmˌpruːvd kənˌvenʃənəl mjuː'nɪʃn] *see* ICM

improvise ['ɪmprəvaɪz] *verb* to do or make something without any proper planning; ***we improvised a shelter out of branches***; **improvised explosive device (IED)** = home-made bomb, booby-trap or mine

IMR [ˌaɪ em 'ɑː(r)] *noun* Soviet-designed armoured engineer vehicle

inaccessible [ˌɪnək'sesəbl] *adjective* impossible to get to; ***the village was inaccessible after the bridge was destroyed***

incendiary [ɪn'sendiəri] **1** *adjective* designed to set things on fire; ***incendiary bomb***; ***incendiary grenade***; **incendiary bullet** *or* **incendiary round** = bullet which is designed to ignite after firing and burn in flight, so that the fall of shot can be observed (NOTE: also called a **tracer bullet**) **2** *noun* person who sets buildings, vehicles or other objects on fire as an act of sabotage; ***several incendiaries have been arrested***

inch [ɪntʃ] *noun* unit of linear measure corresponding to 2.54 centimetres (NOTE: inch is sometimes represented as **in (8in)**, or as **" (8")**)

incident ['ɪnsɪdənt] *noun* significant event (such as an accident, explosion, terrorist attack, etc.); ***three people were injured in the incident***; ***the observers reported no further incidents during the night***

incite [ɪn'saɪt] *verb* to encourage other people to do something (normally acts of disorder); ***he was inciting the crowd to attack the police***

incline [ɪn'klaɪn] *noun* slope

inclusive [ɪn'kluːsɪv] *adjective* including; ***our area is inclusive of the main road***; *compare* EXCLUSIVE

incoming ['ɪnkʌmɪŋ] *noun* artillery fire which is landing or about to land on your position

incompetent [ɪn'kɒmpɪtənt] *adjective* unable to perform your role satisfactorily (due to lack of knowledge or motivation)

incursion [ɪn'kɜːʃn] *noun* act of entering the territory of another state, without the authorization or permission of that state (usually for offensive purposes)

indecent assault [ɪnˌdiːsənt ə'sɔːlt] *noun* sexual assault on a person which does not go as far as actual rape; ***an allegation of indecent assault has been made against you*** (NOTE: the verb form is **to indecently assault**; ***many of the female prisoners had been indecently assaulted***)

indecisive [ɪndɪ'saɪsɪv] *adjective* **(a)** unable to make decisions easily; ***he is very indecisive*** **(b)** having no clear result; ***the battle was indecisive***

indent 1 ['ɪndent] *noun* official request for ammunition, equipment, rations, etc.; ***all indents are to be submitted by 2200hrs*** **2** [ɪn'dent] *verb* to submit an official request for ammunition, equipment, rations, etc.; ***he indented for winter clothing***

India ['ɪndiə] ninth letter of the phonetic alphabet (Ii)

indicate ['ɪndɪkeɪt] *verb* **(a)** to draw someone's attention to something; ***he indicated a tree on the skyline*** **(b)** *(of vehicles)* to signal the intention to turn left or right; ***the lorry indicated left***

indication [ɪndɪ'keɪʃn] *noun* act of indicating something; **target indication** = sequence of verbal instructions for informing your comrades of the exact location of a target

indicator ['ɪndɪkeɪtə] *noun* instrument which shows something; **engine temperature indicator** = instrument on a dashboard or in a cockpit which shows the temperature of the engine

indirect fire [ˌɪndaɪrekt 'faɪə] *noun* fire of weapons which are not pointed directly at the target (ie artillery or mortar fire)

indirect weapon [ˌɪndaɪrekt 'wepən] *noun* weapon which is not pointed directly at its target (for example an artillery piece or mortar)

indoctrinate [ɪn'dɒktrɪneɪt] *verb* to teach political *or* religious *or* nationalist ideas in such a way that a person accepts them without question

inf = INFANTRY

infantier [ˌɪnfən'tɪə(r)] *noun* exponent of infantry tactics; ***of course, the infantiers among us may disagree with me on this point.***

infantry ['ɪnfəntri] *noun* soldiers who fight on foot; ***the infantry will advance at daybreak***; ***the infantry attack was beaten back***; **air-assault infantry** = infantry equipped with their own transport helicopters and supporting attack helicopters; **air-portable infantry** = infantry who are not equipped with armoured fighting vehicles (AFV) and can therefore be deployed to an area of operations by transport aircraft; **infantry fighting vehicle (IFV)** = armoured personnel carrier, fitted with a gun or cannon, which is designed to transport a section of infantry around the battlefield and provide them with fire support once they are fighting on foot; **armoured infantry** = infantry equipped with infantry fighting vehicles (IFV); **mechanized infantry** = infantry equipped with armoured personnel carriers (APC) or infantry fighting vehicles (IFVs) (NOTE: no plural; the word is followed by a plural verb); ***the infantry are crossing the bridge***; *see also* MARINE

> COMMENT: although most modern infantry are now equipped with some form of vehicle to move them around the battlefield, their basic role remains to get out of the vehicles and fight on foot with the rifle and bayonet; *"Infantry is the queen of battles." Napier*

infantryman ['ɪnfəntrimən] *noun* infantry soldier; ***British infantrymen are famous for their discipline***

infect [ɪn'fekt] *verb* to transmit a disease or illness; ***the whole arm soon became infected***

infectious [ɪn'fekʃəs] *adjective; (of diseases)* capable of being transmitted to other people; ***this strain of flu is highly infectious***

inferior [ɪn'fɪəriə] **1** *adjective* **(a)** of a lower rank than another person; ***a corporal is inferior to a sergeant*** **(b)** smaller or weaker than something else; ***the enemy artillery had an inferior rate of fire to our guns*** **(c)** of worse quality than something else; ***our boots are inferior to those of the Germans*** **2** *noun* person who holds a lower rank than another person; ***he always treats his inferiors with respect***; *compare* SUPERIOR

infestation team [ˌɪnfes'teɪʃən ˌtiːm] *noun* small special forces grouping, which operates behind enemy lines, directing artillery fire and air strikes

infiltrate ['ɪnfɪltreɪt] *verb* to move into enemy territory in small groups by different routes, in order to avoid detection, and then to join up in order to attack an objective in force; *compare* EXFILTRATE

infiltration [ˌɪnfɪl'treɪʃn] *noun* act of infiltrating; *compare* EXFILTRATION

inflammable [ɪn'flæməbl] *adjective* easy to set on fire; ***many sailors were badly burnt because their clothing was made of inflammable material***

inflict [ɪn'flɪkt] *verb* to do something unpleasant to another person; ***the bombing inflicted heavy casualties on the civilian population***; ***we inflicted the worst defeat of the war on the enemy*** (NOTE: this verb is usually followed by **on** *or* **upon**)

inform [ɪn'fɔːm] *verb* **(a)** to tell someone something; ***he informed me of the change in the timings*** **(b) to inform on someone** = to tell the authorities about another person's illegal activities; ***he informed on his neighbour***

information [ˌɪnfə'meɪʃn] *noun* facts (whether accurate or not) which are passed on from one person to another; **information warfare** = hacking into an enemy's computer network in order to disrupt it (eg. by sowing a virus) or to obtain information or to insert false information

informer [ɪn'fɔːmə] *noun* someone who tells the authorities about another person's illegal activities

infrared (IR) [ˌɪnfrə'red] *adjective* relating to a form of red light which is used in some night-viewing devices because it is invisible to the naked eye

> COMMENT: night-viewing devices which utilize infrared light are described as **active,** while those which do not are described as **passive**

infrastructure ['ɪnfrəˌstrʌktʃə] *noun* basic amenities and facilities upon which a modern society relies in order to function properly (such as electricity, roads and railways, telecommunications, water, etc.)

inhabitant [ɪn'hæbɪtənt] *noun* person who lives in a place; ***the original inhabitants were removed from their villages***

initiate [ɪ'nɪʃɪeɪt] *verb* **(a)** to start or introduce something; ***the commanding officer was forced to initiate legal proceedings*** **(b)** to activate an explosive or pyrotechnic device; ***the Claymores were initiated electronically***

initiative [ɪ'nɪʃətɪv] *noun* **(a)** ability to assess a situation and take action, without asking for guidance; ***he lacks initiative*** **(b)** ability to make the enemy conform to your own movements; ***we must not allow the enemy to regain the initiative***; **to lose the initiative** = to stop being able to force the enemy to react to your actions

> COMMENT: the initiative is usually held by whichever side is attacking. As soon as the attackers are stopped and forced to defend themselves, they are considered to have lost the initiative

inject [ɪn'dʒekt] *verb* to put a liquid drug *or* vaccine into a person's body, using a syringe *or* syrette; ***I injected him with morphine ten minutes ago***

injection [ɪn'dʒekʃ(ə)n] *noun* act of injecting; ***he was given an injection of morphine***

> COMMENT: there are three types of injection: **subcutaneous (SC)** = under the skin; **intramuscular (IM)** = into a muscle; **intravenous (IV)** = into a vein

injure ['ɪndʒə] *verb* to do physical harm to someone

injury ['ɪndʒəri] *noun* physical harm to a person

inland ['ɪnlænd] *adverb* away from the sea; ***strong enemy forces are moving inland***

inlet ['ɪnlet] *noun* place where a lake or the sea cuts into the land

INMARSAT ['ɪnmɑːsæt] *noun* = INTERNATIONAL MARITIME SATELLITE insecure satellite telephone system

inoculate [ɪ'nɒkjʊleɪt] *verb* to treat someone with a vaccine or serum, in order to prevent them contracting a disease

inoculation [ɪˌnɒkjʊ'leɪʃn] *noun* act of inoculating someone; **battle inoculation** = preparing soldiers for battle by the use of live rounds and simulated battle effects

inoperable [ɪ'nɒpərəbl] *adjective* impossible to carry out; ***the plan is now inoperable***

inoperative [ɪ'nɒpərətɪv] *adjective* not working properly; ***all our communications equipment was made inoperative by the nuclear explosion***

insect repellant ['ɪnsekt rɪˌpelənt] *noun* liquid designed to stop insects biting a person's skin

insecure [ɪnsɪ'kjʊə(r)] *adjective* **(a)** *(of objects)* not properly attached *or* closed; ***your magazine is insecure*** **(b)** *(of locations)* undefended; ***the bridge is insecure*** **(c)** *(of communications)* not coded *or* scrambled; ***the enemy is equipped with insecure radios***

internal security [ɪnˌtɜːnəl sɪ'kjʊərəti] *noun* State's use of its armed forces (usually in support of its civil police) to maintain *or* restore law and order within its own territory; ***several battalions are engaged in internal security duties***

insert ['ɪnsɜːt] *verb* **(a)** to put one thing into another; ***he inserted a new battery into his torch*** **(b)** to move into an area of operations; ***we will insert by helicopter***; *compare* EXTRACT

insertion [ɪn'sɜːʃn] *noun* act of inserting; *compare* EXTRACTION

in-service [ɪn 'sɜːvɪs] *adjective* which is currently being operated; **in-service support** = support which is organized when in service

inshore ['ɪnʃɔː] **1** *adjective* designed for use close to the shore; ***an inshore fishing boat*** **2** *adverb* **(a)** at sea but close to the shore; ***the boat was anchored inshore*** **(b)** towards the shore; ***the landing craft moved inshore***

insignia [ɪn'sɪgniə] *noun* decorative symbol (used to denote the identity of a unit, specialist qualification, rank, etc.)

inspect [ɪn'spekt] *verb* **(a)** *(of people)* to examine a person in order to ensure that he is correctly dressed or that he has all his equipment; ***the general inspected his troops*** **(b)** to examine something in order to ensure that it is in working order; ***he inspected his platoon's weapons*** **(c)** *(in the event of an accident or malfunction)* to examine something in order to look for defects or signs of damage; ***the vehicle was inspected for defects*** **(d)** to examine a weapon in order to ensure that it is clean or unloaded; ***upon inspecting the man's rifle he found a live round in the chamber***

inspection [ɪn'spekʃn] *noun* act of inspecting a person or thing; ***the soldiers stood to attention ready for inspection***

installation [ˌɪnstə'leɪʃn] *noun* building, complex or other permanent structure, which contains some form of technical equipment (such as communications equipment, radar, weapons system, etc.); ***our target was the radar installations along the north coast***

instruct [ɪn'strʌkt] *verb* **(a)** to teach; ***he instructs soldiers in the use of the bayonet*** **(b)** to tell someone to do something; ***I instructed him to clean his rifle***

instruction [ɪn'strʌkʃn] *noun* **(a)** act of teaching something; ***we have two periods of map-reading instruction today*** **(b)** act of telling someone to do something; ***he did not carry out my instructions*** **(c)** written directions telling someone how to do something; ***the instructions are printed on the side of the container***

instructor [ɪn'strʌktə] *noun* teacher; ***Cpl Smith is our bayonet instructor***

insubordinate [ˌɪnsə'bɔːdɪnət] *adjective* **(a)** unwilling to carry out orders **(b)** unwilling to show respect to your superiors; ***he was quite insubordinate and had to be restrained***

insubordination [ɪnsəˌbɔːdɪ'neɪʃn] *noun* **(a)** failure or refusal to carry out an order **(b)** lack of respect to a superior

insurgency [ɪn'sɜːdʒənsi] *noun* armed resistance to the established government or foreign domination; **counter-insurgency (COIN)** = action taken to destroy an insurgency; *see also* INSURRECTION, REBELLION, REVOLT, REVOLUTION

insurgent [ɪn'sɜːdʒənt] *noun* person who uses force to resist the established government or foreign domination; *see also* FREEDOM FIGHTER, REBEL, REVOLUTIONARY

insurrection [ˌɪnsə'rekʃn] *noun* armed resistance to the established government or foreign domination; *see also* INSURGENCY, REBELLION, REVOLT, REVOLUTION

int [ɪnt] *GB* = INTELLIGENCE

intake ['ɪnteɪk] *noun* place where air, fuel or water in drawn into a machine; **air intake** = part of an engine which draws air in from outside

integrate ['ɪntɪgreɪt] to link up several things to form a whole; **integrated logistic support** = support which is integrated into all parts of the project as it is being developed

intel ['ɪntel] *US* = INTELLIGENCE

intelligence [ɪn'telɪdʒəns] *noun* **(a)** any information which may be useful (especially information about the enemy); ***we have received some fresh intelligence on the enemy artillery***; **electronic intelligence (ELINT)** = information obtained through monitoring the enemy's electronic transmissions (usually by specially equipped aircraft); **human intelligence (HUMINT)** = information obtained from people (eg. friendly forces, agents, civilians, POW); **signals intelligence (SIGINT)** = information obtained by listening to the enemy's radio transmissions **(b)** people and equipment involved in the gathering, analysis and dissemination of intelligence; ***we are feeding false information to the enemy's intelligence***; **intelligence officer (IO)** = officer responsible for intelligence; **intelligence summary (intsum)** = simple report on the enemy's locations, strength, organization, intentions, etc., during a specific period of time

> COMMENT: the department responsible for intelligence in a headquarters is known as **G2**

intend [ɪn'tend] *verb* to decide to do something

intent ['ɪntent] *noun US* document issued to subordinate commanders, explaining a commander's idea of how a future operation is likely to proceed

intention [ɪn'tenʃn] *noun* what a person has decided to do; ***we do not know the enemy's intentions***

intercede [ˌɪntə'siːd] *verb* to enter a dispute in order to support another person or group; ***the priest tried to intercede on behalf of the young men***

intercept [ˌɪntə'sept] **1** *verb* **(a)** to meet up with or catch persons or vehicles as they move from one place to another; ***we intend to intercept them at the river***; ***two squadrons of fighters were sent to intercept the bombers*** **(b)** to locate and listen to another person's radio transmissions; ***the message was intercepted*** **2** *noun* enemy message which has been intercepted; ***here is a transcript of our last intercept***

interception [ˌɪntə'sepʃn] *noun* **(a)** the act of intercepting a person or vehicle **(b)** the act of locating and listening to another person's radio transmission; the people who locate and listen to another person's radio transmissions; ***these countermeasures are designed to confuse the enemy interception***

interceptor [ˌɪntə'septə] *noun* fighter aircraft which is capable of flying great distances in order to intercept and engage enemy aircraft

interchangeability [ɪntəˌtʃeɪnʒə'bɪləti] *noun* being able to be exchanged one for another

interchangeable [ˌɪntə'tʃeɪnʒəbl] *adjective* which can be exchanged one for another; ***the two parts are interchangeable***

intercom ['ɪntəkɒm] *noun* internal telecommunication system within an aircraft, ship or vehicle allowing crew members to speak to each other

COMMENT: the intercom is normally incorporated into the radio system

intercontinental ballistic missile (ICBM) [ɪntəkɒntɪˌnentl bəˌlɪstɪk 'mɪsaɪl] *noun* guided missile which flies from one continent to another and ends its flight in a ballistic descent

interdict ['ɪntədɪkt] *verb* to deny the enemy the use of something; ***the airstrikes were intended to interdict the enemy's supply lines***

interdiction [ˌɪntə'dɪkʃn] *noun* the act of denying the enemy the use of something (typically, air attacks on enemy reinforcements and supply columns as they move towards the battle area); ***our principal role is the interdiction of the enemy railway network***; **battlefield interdiction** = interdiction against targets close to the battle area.d; **deep interdiction** = interdiction against targets in the enemy's rear areas *or* home territory

interfere [ˌɪntə'fɪə] *verb* **(a) to interfere in** = to involve oneself in another person's activity without being asked; ***he is always interfering in the running of my platoon*** **(b) to interfere with** = to obstruct or hinder; ***these air strikes are designed to interfere with the enemy's supply system***

interference [ˌɪntə'fɪərəns] *noun* **(a)** act of interfering **(b)** obstruction of a radio signal by other radio waves

interior lines [ɪnˌtɪəriə 'laɪnz] *noun* routes available in territory which is protected by the forward and flank positions of an army or other large tactical grouping; ***the general failed to take advantage of his interior lines***

COMMENT: good interior lines enable a commander to move troops directly from one part of his line to another over comparatively short distances, while enemy forces may have to travel a considerable distance around his perimeter in order to redeploy against him. Of course, on the modern battlefield, he would need air and electronic superiority to enjoy this advantage

interlocking arcs of fire [ˌɪntəlɒkɪŋ ˌɑːks əv 'faɪə] *noun* situation where a weapon's arc of fire overlaps with that of its neighbour

intern [ɪn'tɜːn] *verb* to confine a person in custody (usually for reasons of national security rather than because they have committed an offence); ***all enemy aliens will be interned for the duration of the war***

COMMENT: internment applies to civilians; military personnel would be treated as prisoners of war

internment [ɪn'tɜːnmənt] *noun* act of interning people; ***the government has ordered the internment of all enemy aliens***; **internment camp** = secure location where people are interned

interoperability [ɪntəˌɒpərə'bɪləti] *noun* being able to operate in the place of something else

interoperable [ɪntə'ɒpərəbl] *adjective* which can operate in place of something else

interpreter [ɪn'tɜːprətə] *noun* person who acts as a translator in a conversation between two people who do not understand each other's language

interrogate [ɪn'terəgeɪt] *verb* to ask a series of questions in a systematic way, in order to obtain information

COMMENT: this word normally implies a confrontational situation (as when an intelligence officer is interrogating a prisoner) You interrogate an enemy, but you would debrief someone from your own side

interrogation [ɪnˌterə'geɪʃn] *noun* act of interrogating; ***under interrogation he revealed the location of the camp***

intersection [ˌɪntə'sekʃən] *noun* US junction; ***turn left at the next intersection***

interval ['ɪntəvəl] *noun* space between two persons *or* groups *or* vehicles; ***I want intervals of twenty-five metres between vehicles***; ***vehicles will set off at five minute intervals***

intervasion [ˌɪntə'veɪʒən] *noun* = INTERVENTION & INVASION invading a State's sovereign territory with the reluctant approval of that State's government in order deal with hostile elements within the population

intervene [ˌɪntə'viːn] *verb* to enter a dispute between two other parties; ***we were forced to intervene when they attacked the refugees***

intervention [ˌɪntə'venʃn] *noun* act of intervening; ***there is a strong possibility of foreign intervention***

intimate support tank [ˌɪntɪmət sə'pɔːt ˌtæŋk] *noun* tank which is attached to an infantry platoon for a specific phase of an attack and which fights alongside that platoon throughout the action

intimidate [ɪn'tɪmɪdeɪt] *verb* to use threats in order to make someone do *or* not do something; ***the guerillas are trying to intimidate the local people***

intimidation [ɪnˌtɪmɪ'deɪʃən] *noun* use of threats in order to make someone do *or* not do something; ***the guerillas use intimidation to gain the support of the local people***

intraregional [ˌɪntrə'riːdʒənəl] *adjective* inside a region; **intraregional mobility** = ability of forces to move rapidly inside a given region

intruder [ɪn'truːdə] *noun* **(a)** person who enters an area or building without the authority or right to do so **(b)** unidentified person, vehicle or aircraft which enters your area of responsibility; **intruder alarm** = device designed to detect movement **(c)** **Intruder** = American-designed A-6 ground-attack aircraft

INTSUM *or* **intsum** ['ɪntsʌm] = INTELLIGENCE SUMMARY

invade [ɪn'veɪd] *verb* to enter another state's territory with military forces, in order to conquer it; ***troops are massed on the border ready to invade***; ***the country was invaded by rebel armies from the south***

invasion [ɪn'veɪʒn] *noun* act of invading; ***the invasion of Britain by the Normans in 1066***; ***they planned the invasion of the island***

invest [ɪn'vest] *verb* to surround an area or town occupied by an enemy force, in order to prevent its being reinforced and resupplied and also to prevent any withdrawal, usually with the ultimate intention of capturing the place or of compelling the enemy force inside it to surrender; ***our main objective is to invest Port Stanley***; *see also* BESIEGE

investment [ɪn'vestmənt] *noun* act of investing a town or fortress; *see also* SIEGE

IO [ˌaɪ 'əʊ] = INTELLIGENCE OFFICER

IR [ˌaɪ 'ɑː] = INFRARED

Irish Defence Force [ˌaɪrɪʃ dɪ'fens 'fɔːs] *noun* army of the Republic of Ireland (Eire) (NOTE: the Irish Defence Force should not be confused with the Irish Republican Army (IRA), which is a terrorist organization seeking to unite the British-controlled province of Northern Ireland with the Republic of Ireland (Eire))

iron bomb ['aɪən bɒm] *noun* bomb which is simply dropped onto a target by an aircraft; *see also* BALLISTIC BOMB, GENERAL PURPOSE (GP) BOMB

iron ration [ˌaɪən 'ræʃn] *noun* small pack of food carried by a soldier in case of emergency

irregular [ɪ'regjʊlə] **1** *adjective* not part of the regular army; ***the refugees were attacked by irregular troops*** **2** *noun* member of a unit which is not part of the regular army (ie guerillas, mercenaries, partisans, etc.); ***he commands a small unit of irregulars***

irreparable [ɪ'repərəbl] *adjective* which cannot be made good; ***the bombing has caused irreparable damage to the airfield***

irreplaceable [ˌɪrɪ'pleɪsəbl] *adjective* which cannot be replaced; ***he is irreplaceable because he speaks five different languages***

irresponsible [ˌɪrɪ'spɒnsəbl] *adjective* acting without considering the possible consequences of that action; ***your irresponsible behaviour could have caused an accident***

irresponsibility [ɪrɪˌspɒnsə'bɪlɪti] *noun* irresponsible behaviour; ***the operation has been compromised by your irresponsibility***

irretrievable [ˌɪrɪ'triːvəbl] *adjective* which cannot be retrieved; ***that data is irretrievable***

irrevocable [ɪ'revəkəbl] *adjective* which cannot be altered or cancelled; ***the order is irrevocable***

irritant ['ɪrɪtənt] *noun* something which causes irritation; ***the shells are releasing some sort of eye irritant***

irritate ['ɪrɪteɪt] *verb* **(a)** to anger or annoy; ***he was irritated by the soldier's behaviour*** **(b)** to cause discomfort or pain; ***this chemical agent irritates the eyes and respiratory system***

irritation [ˌɪrɪ'teɪʃn] *noun* **(a)** anger or annoyance; ***he looked at me with irritation*** **(b)** discomfort or pain; ***this substance causes irritation to the skin***

island ['aɪlənd] *noun* piece of land surrounded by water

ISB [ˌaɪ es 'biː] *noun* = INTERMEDIATE STAGING BASE administration area located en route to a war zone *or* in the rear area, where a deploying force can offload and assemble its equipment before moving forward to the front line

ISO container ['aɪsəʊ kənˌteɪnə(r)] *noun* huge metal container, which is designed to be fitted to a lorry or loaded onto a ship, in order to transport goods overseas.; ***we lived in ISO containers until the camp had been built***

isolate ['aɪsəleɪt] *verb* **(a)** to prevent movement to or from a location; ***the village has been isolated by snow*** **(b)** to surround a unit so that it can neither retreat, nor be reinforced or supported; ***the enemy battalion was isolated and then destroyed***; *see also* CUT OFF

isolated ['aɪsəleɪtɪd] *adjective; (of buildings, villages, etc.)* far from anyone or anything else; ***there are a few isolated villages in the hills***; ***the platoon was left isolated when the rest of the company withdrew***

issue ['ɪʃuː] **1** *noun* act of supplying servicemen with equipment, supplies, etc.; ***there will be an issue of rations at 1600hrs***; **on issue** = provided by the logistical system; ***these boots are no longer on issue*** **2** *verb* to supply equipment, supplies, etc.; ***ammunition will be issued at 1600hrs***

isthmus ['ɪsməs] *noun* narrow strip of land connecting two larger pieces of land

item ['aɪtəm] *noun* object or thing (especially one of several different things)

ITG [ˌaɪ tiː 'dʒiː] *noun US* = INITIAL TERMINAL GUIDANCE task carried out by pathfinders (ie. advance reconnaissance, security and marking of a DZ *or* LZ *or* beach-landing site)

IVIS ['aɪvɪs] *noun* = INTERVEHICULAR INFORMATION SYSTEM American-designed computer system fitted to armoured fighting vehicles (eg. Abrams M-1 tank), which shows the current locations of all other vehicles in the sub-unit and their ammunition and fuel states and which also has an e-mail facility for the transmission of orders

JULIET - Jj

J1 [ˌdʒeɪ 'wʌn] *noun* department of a joint headquarters (JHQ) responsible for personnel

J2 [ˌdʒeɪ 'tuː] *noun* department of a joint headquarters (JHQ) responsible for intelligence

J3 [ˌdʒeɪ 'θriː] *noun* department of a joint headquarters (JHQ) responsible for operations and training

J4 [ˌdʒeɪ 'fɔː(r)] *noun* department of a joint headquarters (JHQ) responsible for materiel

J5 [ˌdʒeɪ 'faɪv] *noun* department of a joint headquarters (JHQ) responsible for civil/military relations

JAAT [dʒæt] = JOINT AIR ATTACK TEAM

JAG [dʒæg] A *US* = JUDGE ADJUTANT GENERAL legal department for the US armed forces

Jaguar ['dʒægjʊə] *noun* British/French-designed attack aircraft

jam [dʒæm] *verb* **(a)** to block the enemy's radio transmissions by causing interference; ***we are being jammed*** **(b)** *(of automatic weapons)* to stop firing because of a mechanical failure; ***the machine-gun jammed***

jamming ['dʒæmɪŋ] *noun* act of blocking the enemy's radio transmissions by causing interference

jankers ['dʒæŋkəz] *noun GB slang* extra fatigue duty given as a punishment (especially unpleasant tasks: eg. cleaning out latrines)

Javelin ['dʒævlɪn] *noun* **(a)** British-designed hand-held optically-tracked surface-to-air missile (SAM) **(b)** American-designed precision-guided anti-tank missile

JDAM ['dʒeɪdæm] *noun* = JOINT DIRECT ATTACK MUNITION aircraft bomb which guides itself onto a target using the GPS system

jeep [dʒiːp] *noun* any type of light general purpose all-terrain military vehicle

> COMMENT: the original jeep was an American-designed vehicle used during the Second World War. The name is derived from GP (= general purpose)

jeopardize ['dʒepədaɪz] *verb* to place in a dangerous situation; ***your actions have jeopardized the entire mission***

jerrycan ['dʒerɪkæn] *noun* rectangular container for transporting fuel or water

jet [dʒet] *noun* **(a)** stream of fire, gas or water sent out under high pressure; **jet engine** = engine which uses jet propulsion; **jet propulsion** = forward movement caused by the backward ejection of a jet of gas at high speed **(b)** aircraft powered by jet propulsion; ***enemy jets bombed our positions***

jettison ['dʒetɪsən] *verb* to throw objects out of an aircraft, ship or vehicle because they are no longer needed or dangerous or in order to lighten the load; ***the aircraft was forced to jettison its bombs***

JHQ [ˌdʒeɪ eɪtʃ 'kjuː] = JOINT HEADQUARTERS

jigger ['dʒɪgə(r)] *see* CHIGGER

jihad [dʒɪ'hʌd] *noun Arabic* holy war, waged by muslims to defend the Islamic faith; *see also* MUJAHIDEEN

Jock [dʒɒk] *noun (informal)* Scottish soldier

join [dʒɔɪn] *verb* **(a)** to become a part of something; ***she joined the army*** **(b)** to come together; ***they joined us in the briefing room*** **(c)** to connect two objects to each other; ***he joined the wires together***

joint [dʒɔɪnt] *adjective* with two or more services working together, sharing a common purpose; **joint air attack team (JAAT)** = American doctrine for a coordinated attack by aircraft (both from the air force and army aviation assets), artillery and naval gunfire; **joint operation** = operation involving two or more different branches of the armed forces; **joint task force (JTF)** = large combined arms grouping involving different branches of the armed forces formed for a specific operation or campaign (NOTE: if all services are involved in an operation, then they are not specified individually; if only two services are involved jointly, then they are specified: **joint Navy-Air Force manoeuvres**); **joint headquarters (JHQ)** = headquarters of a joint force, with staff officers and other personnel from all three arms (navy, army and air force)

join up [ˌdʒɔɪn 'ʌp] *verb* to join the armed forces; ***he joined up when he left school***

Jolly Green Giant [ˌdʒɒli ˌgriːn 'dʒaɪənt] *noun* unofficial nickname for the American-designed CH-53 heavy transport helicopter

joystick ['dʒɔɪstɪk] *noun* stick used to steer an aircraft

JSOTF [ˌdʒeɪ es ˌəʊ tiː 'ef] *noun US* = JOINT SPECIAL OPERATIONS TASK FORCE elite unit of special forces personnel designed to deploy (usually by helicopter) at short notice and carry out specialized military tasks

JSTARS ['dʒeɪ stɑːz] = JOINT SURVEILLANCE AND TARGETING ATTACK RADAR SYSTEM) American-designed technology used by special aircraft to detect moving ground targets at long range

JTF [ˌdʒeɪ tiː 'ef] = JOINT TASK FORCE

Juliet *US* **Juliett** [ˌdʒuːli'et] tenth letter of the phonetic alphabet (Jj)

jump [dʒʌmp] *noun* parachute drop; ***he was injured on the last jump***

jump-jet ['dʒʌmp dʒet] *noun* fighter aircraft with vertical take-off capability (eg. Harrier, YAK-38, YAK-41)

jump-leads ['dʒʌmp liːdz] *plural noun* two lengths of electrical cable, which are used to convey electrical charge from one fully-charged vehicle battery to another which has lost its charge

jumpmaster ['dʒʌmpmɑːstər] *noun* aircrew member who controls paratroopers as they jump out of an aircraft

jump-start ['dʒʌmp stɑːt] *verb* to start a vehicle, whose battery has lost its charge, by pushing, towing *or* using jump-leads

junction ['dʒʌŋkʃn] *noun* place where two or more roads or railway lines meet each other; **T-junction** = place where two roads meet at right angles to each other; **Y-junction** = place where a single road divides into two; *see also* FORK, INTERSECTION

jungle ['dʒʌŋgl] *noun* area (in the tropics) which is covered by dense vegetation; **primary jungle** = jungle where the trees have grown to a considerable height, and there is little ground vegetation; **secondary jungle** = jungle where the original trees have been cleared (by man or by fire) and replaced by a dense growth of bushes and young trees

junior ['dʒuːniə] *adjective* of low rank; **junior non-commissioned officer** = corporal or lance corporal; **junior ranks** = junior non-commissioned officers and privates of a unit; *GB* **the Junior Service** = the Royal Air Force; *GB* **junior technician** = non-commissioned rank in the air-force (equivalent to an experienced or well-qualified private soldier in the army); *compare* SENIOR

junk [dʒʌŋk] *noun* traditional Chinese sailing ship

jurisdiction [ˌdʒʊərɪs'dɪkʃn] *noun* legal or other authority; ***we have jurisdiction over this area***

KILO - Kk

K [keɪ] *noun (used in speech)* kilometre; ***the village is 3 Ks beyond the river***

k = KNOT(S)

Ka-50 [ˌkeɪ eɪ 'fɪfti] *noun* Soviet-designed attack helicopter (NOTE: known to NATO as **Hokum**)

Kalashnikov (AK-47) [kə'læʃnɪkɒf] *noun* Soviet-designed 7.62mm assault weapon

karst [kɑːst] *noun* hilly terrain with crags and outcrops of limestone

Katyusha [kə'tjuːʃə] *noun* Soviet-designed multiple rocket launcher, mounted on a truck

KE [ˌkeɪ 'iː] = KINETIC ENERGY

keel [kiːl] *noun* structure forming the base of a boat or ship

keg [keg] *noun* small metal barrel, designed to hold beer; ***the device consisted of several beer kegs filled with homemade explosives***

kelp [kelp] *noun* seaweed with large brown leaves which grows in a dense mass close to the shore

kennel ['kenl] *noun* accommodation for dogs

kerosene ['kerəsiːn] *noun* fuel distilled from petroleum, which is suitable for aviation fuel and also for heating and lamps; *also known as* PARAFFIN OIL

Kevlar ['kevlɑː] *noun* synthetic material used in the manufacture of body armour, helmets, and some types of vehicle armour

key [kiː] **1** *adjective* of vital importance; **key point (KP)** = location or installation which is of strategic importance (such as docks, government or other administrative building, power installation, etc.); ***commandos will be used to seize the key points***; **key terrain** = ground which you must occupy or control in order to achieve your mission **2** *noun* **(a)** instrument used to operate a lock or to start an engine or motor; ***I don't have a key for this door*** **(b)** explanatory list of symbols which are shown on a map; ***the key is written in English and German*** **(c)** system for interpreting a code; ***we obtained a key to the enemy's codes*** **(d)** solution to a problem; ***the village of Ladna is the key to the enemy's defences***

kg = KILOGRAM

khaki ['kɑːki] *noun* brownish-green colour used for army uniforms and vehicle camouflage

kHz = KILOHERTZ

KIA [ˌkeɪ aɪ 'eɪ] = KILLED IN ACTION

kill [kɪl] **1** *noun* killing of an enemy soldier or destruction of an enemy aircraft, vehicle or ship, when viewed as a result; ***the patrol reported three kills*** **2** *verb* to deprive a person or animal of life; ***two of the hostages were killed in the gun battle***; ***their aim is to kill as many enemy soldiers as possible***

killing ['kɪlɪŋ] *noun* act of killing a person or animal; **killing area** *or* **killing ground** *or* **killing zone** = area of ground selected as a place to destroy an enemy force

Kilo ['kiːləʊ] eleventh letter of the phonetic alphabet (Kk)

kilometre *US* **kilometer (km)** [kɪ'lɒmɪtə] *noun* unit of linear measure, corresponding to 1,000 metres *or* 0.6214 mile

kiloton ['kɪlətʌn] *noun* unit of explosive power equivalent to 1,000 tons of TNT

kilt [kɪlt] *noun* garment, similar to a woman's skirt, traditionally worn by some Scottish and Irish regiments

Kim's Game ['kɪmz geɪm] *noun* activity designed to develop observation skills; a selection of different objects are placed on the ground and covered with a blanket. The blanket is removed for a few seconds and then replaced again. The participants then have to remember exactly what they saw.

> COMMENT: Kim's Game is taken from the adventure story "Kim" by Rudyard Kipling, in which the above exercise was used in the training of spies. It can be made as simple *or* as complex as you like, and is extremely worthwhile, especially if practised on a regular basis.

kinetic energy (KE) [kaɪˌnetɪk 'enədʒi] *noun* energy produced by an object moving at high speed; **kinetic energy (KE) round =** anti-tank projectile, made of a heavy metal (such as depleted uranium or tungsten carbide), which is fired at a very high velocity, in order to punch its way through armour; *see also* LONG-ROD PENETRATOR

King's Commission *see* QUEEN'S COMMISSION

King's Regulations *see* QUEEN'S REGULATIONS

kit [kɪt] *noun* **(a)** equipment; ***all the platoon's kit was left behind on the truck***; ***make sure that all your kit is secure*** **(b)** set of tools or other articles used for a specific purpose (for example cleaning kit, decontamination kit, first-aid kit, etc.)

kitbag ['kɪtbæg] *noun* long cylindrical canvas bag, for storing spare clothing and other personal effects

> COMMENT: kitbags are normally used for any additional clothing which is not carried in the rucksack

Kiwi ['kiːwiː] *noun (informal)* soldier from New Zealand

klaxon ['klæksən] *noun* noise-making warning device, similar to vehicle's horn

klick [klɪk] *noun (informal)* kilometre; ***it's three and a half klicks to the RV***; *see also* CLICK

km = KILOMETRE(S)

knife ['naɪf] **1** *noun* instrument used for cutting or stabbing, with a sharp metal blade fixed in a handle; ***he silenced the sentry with a knife*** **2** *verb* to kill or wound with a knife; ***he was knifed while he slept***

knock out ['nɒk 'aʊt] *verb* **(a)** to destroy a vehicle; ***we knocked out three tanks*** **(b)** to make someone unconscious; ***he was knocked out by the blast from a shell***

knoll [nɒl] *noun* very small hill

knot (k) [nɒt] *noun* unit of speed for a ship or aircraft, corresponding to one nautical mile per hour; ***the patrol boat was travelling at 15 knots*** *or* ***was doing 15 knots***

Kormoran ['kɔːmərʌn] *noun* German-designed anti-ship missile

KP [ˌkeɪ 'piː] = KEY POINT

KPH [ˌkeɪ piː 'eɪtʃ] = KILOMETRES PER HOUR

kukri ['kʊkri] *noun* fighting-knife carried by Gurkha soldiers

LIMA - LI

L [el] *noun* = DICHLORO (2-CHLOROVYNYL) ARSINE type of blister agent (NOTE: also known as **Lewisite**)

laager *or* **lager** ['lɑːgə] **1** *noun* encampment of armoured vehicles; ***the laager was attacked during the night*** **2** *verb* to form a laager; ***we will laager at grid 417339***; *see also* LEAGUER

LAC = LEADING AIRCRAFTMAN

LAD [ˌel eɪ 'diː] *noun GB* = LIGHT AID DETACHMENT detachment of vehicle mechanics from the Royal Electrical and Mechanical Engineers (REME), which is attached to an infantry battalion or armoured regiment on a permanent basis

lager ['lɑːgə] *see* LAAGER

lake [leɪk] *noun* large area of water surrounded by land

lamp [læmp] *noun* device which produces light

lance [lɑːns] *noun* **(a)** long spear used by cavalry soldiers **(b) Lance** = American-designed tactical surface-to-air missile

lance-bombardier (L/Bdr) [ˌlɑːns bɒmbə'dɪə] *noun GB* lance-corporal in the artillery

lance corporal (L/Cpl) [ˌlɑːns 'kɔːprəl] *noun US* junior non-commissioned officer in the marines

lance-corporal (L/Cpl) [ˌlɑːns 'kɔːprəl] *noun GB* lowest non-commissioned officer rank in the army or marines (usually second in command (2IC) of a section or equivalent-sized grouping); **lance-corporal of horse (L/CoH)** = corporal in the Household Cavalry

lancer ['lɑːnsə] *noun* **(a) Lancer** = American-designed B-1 long-range strategic bomber aircraft **(b)** *(historical)* cavalry soldier armed with a long spear or lance

> COMMENT: certain armoured regiments still retain their historical title of Lancers

lance rank ['lɑːns ˌræŋk] *noun* army rank between private and corporal

lance-sergeant (L/Sgt) ['lɑːns ˌsɑːdʒənt] *noun GB* corporal in the Brigade of Guards

land [lænd] **1** *noun* solid part of the earth's surface (ie not the sea); **land forces** = military forces which operate on land; ***allied land forces drove the enemy back to the coast*** **2** *verb* **(a)** to bring a flying aircraft back onto the ground; ***the squadron has just landed*** **(b)** to leave a ship and go back onto dry land; ***we will start landing tomorrow morning*** **(c)** to deploy troops from aircraft or ships; ***the invasion force landed near Bremen***; ***enemy paratroopers are landing to the north of Arnhem***

landing ['lændɪŋ] *noun* **(a)** act of landing an aircraft; ***landing on the jungle airstrip was difficult***; **short take-off and landing (STOL)** = technology which enables a fixed-wing aircraft to take off and land over considerably shorter distances than those required by conventional fixed-wing aircraft; **vertical take-off and landing** = technology which enables a fixed-wing aircraft to take off and land from a stationary position (ie without the need for a runway) **(b)** act of leaving a ship and going back onto dry land; ***the landing took several hours*** **(c)** deployment of troops from aircraft or ships; **beach landing =** act of disembarking troops and vehicles onto a beach; **landing craft =** small flat-bottomed boat designed to move troops and vehicles from a transport ship to a beach; *see also* LCAC, LCM, LACU, LCVP; **landing ship**

= transport ship designed to transport and launch landing craft and helicopters during amphibious operations; **landing-ship logistics;** *see* LSL

landing zone (LZ) ['lændɪŋ zəʊn] *noun* **(a)** area of ground selected for the landing or pick-up of troops by helicopter **(b)** *US* area of ground selected for the landing of troops by parachute

land-line ['lænd laɪn] *noun* telephone system which operates through electrical cable placed on, under or above the ground; ***I will call you on land-line***

landmark ['lændmɑːk] *noun* distinctive natural *or* man-made feature, which can be used as a reference point *or* as an aid to navigation; ***it's easy to get lost when there are so few landmarks***

landmine ['lændmaɪn] *noun* explosive device which is buried in or placed on the surface of the ground, and is designed to detonate when a person steps on it or a vehicle drives over it (often simply called a **mine**)

Landrover ['lændrəʊvə] *noun* British-designed all-terrain vehicle (similar to a jeep)

lane [leɪn] *noun* **(a)** narrow rural road; ***these lanes are not wide enough for tanks*** **(b)** cleared route through a minefield or other obstacle; ***the lane is marked with mine tape*** **(c)** one of several parallel routes; ***the brigade advanced along three parallel lanes*** **(d)** recognized route for aircraft or ships; ***we must avoid the main shipping lanes***

lanyard ['lænjəd] *noun* **(a)** cord used to attach a weapon or piece of equipment to a person's body **(b)** coloured cord, worn around the shoulder in order to denote the wearer's rank, role or unit **(c)** cord which is pulled in order to activate the firing mechanism of a gun

laser ['leɪzə] *noun* device which projects a beam of intense light; **laser-guided bomb** *or* **missile (LGB** *or* **LGM)** = bomb or missile designed to home in on a target which has been illuminated by a laser beam; **laser range-finder** = device which utilizes a laser beam in order to calculate the exact distance to an object; **laser target designator** = device which projects a laser beam onto a target in order to illuminate it for a laser-guided bomb or missile

lashings ['læʃɪŋz] *noun* ropes used to tie a cargo down

last light [ˌlɑːst 'laɪt] *noun* time of day when it becomes completely dark; ***no move before last light***

Last Post [ˌlɑːst 'pəʊst] *noun* bugle-call blown in barracks and bases at bedtime (usually around 2200hrs) and also at military funerals

latrine [lə'triːn] *noun* toilet (especially one constructed in the field)

launch [lɔːntʃ] **1** *noun* **(a)** act of launching a missile or rocket; ***the launch was delayed by 15 minutes*** **(b)** ceremony of placing a newly-built ship into the water for the first time; ***are you attending the launch next week?*** **(c)** small boat (especially one which is carried on a ship and is used to transport people or things to the shore or to other ships) **2** *verb* **(a)** to put a boat into the water (especially for the first time); ***the new destroyer was launched by the Queen*** **(b)** to fire a missile or rocket into the air; ***they launched rockets at the enemy positions*** **(c)** to set an attack or other offensive operation in motion; ***the enemy launched a furious attack on the castle***

launcher ['lɔːntʃə] *noun* device or vehicle used to launch a missile or rocket; **grenade launcher** = gun designed to fire small explosive projectiles; **rocket launcher** = apparatus or vehicle from which a rocket is fired

LAV [læv *or* ˌel eɪ 'viː] *noun* LIGHT ARMORED VEHICLE = American-designed multi-purpose air portable amphibious armoured fighting vehicle (AFV); **LAV-25** = armoured personnel carrier (APC); **LAV-AD** = air defence variant, fitted with Stinger missiles and a Gatling gun; **LAV-AT** = anti-tank variant, fitted with TOW missiles; **LAC-C2** = command vehicle; **LAV-L** = load-carrying variant, designed to transport supplies; **LAV-M** = mortar variant, fitted with an 81mm mortar; **LAV-R** = recovery variant

LAW [ˌel eɪ 'dʌb(ə)ljuː] *noun* = LIGHT ANTI-TANK WEAPON hand-held anti-tank rocket used by infantry

lay [leɪ] *verb* to place or position an object; ***the engineers are laying mines***; **to lay a gun** = to aim an artillery piece at a target by adjusting the direction and elevation of the barrel (NOTE: **laying - laid**)

laydown bombing ['leɪdaʊn ˌbɒmɪŋ] *noun* low altitude attack in which the aircraft passes very low over its target and releases bombs fitted with parachutes *or* other devices to slow down the descent, so that the aircraft can get clear before the bombs explode

L/Bdr = LANCE-BOMBARDIER

LBE [ˌel biː 'iː] = LOAD-BEARING EQUIPMENT

LCAC [ˌel siː eɪ 'siː] *noun* = LANDING CRAFT AIR CUSHIONED American-designed hovercraft which is used as a landing craft

LCM [ˌel siː 'em] *noun* = LANDING CRAFT MEDIUM landing craft of simple design which is capable of carrying a company of infantry or most types of equipment, except large armoured vehicles (it is smaller than the LCU)

L/Cpl = LANCE-CORPORAL

L/CoH = LANCE-CORPORAL OF HORSE

LCU [ˌel siː 'juː] *noun* = LANDING CRAFT UTILITY large landing craft which is capable of transporting main battle tanks (MBT) and is able to operate over long distances in all types of sea conditions

LCVP [ˌel siː viː 'piː] *noun* = LANDING CRAFT VEHICLE AND PERSONNEL small landing craft which is capable of carrying a platoon of infantry or a small vehicle

LD [ˌel 'diː] = LINE OF DEPARTURE

lead [liːd] **1** *adjective* **(a)** which is moving in front of others; ***the lead tank was destroyed by a mine*** **(b)** which is in charge; **lead nation** = nation which takes the responsibility for organizing a multinational force, and coordinates the forces of other countries **2** *noun* **(a)** act of moving in front of others; ***B Company will take the lead***; ***B Company was in the lead*** **(b)** distance which one must aim in front of a moving target in order to hit it; ***he did not give the tank enough lead*** **3** *verb* **(a)** to move in front of others; ***B Company was leading*** **(b)** to command men by inspiring them and setting a good example; ***he leads from the front*** **(c)** to aim in front of a moving target; ***you don't lead your targets enough*** (NOTE: **leading - led**)

leader ['liːdə] *noun* **(a)** person who leads; ***he is a natural leader*** **(b)** commander *US* **squad leader** = commander of an infantry squad

leadership ['liːdəʃɪp] *noun* ability to make other people carry out your orders effectively and willingly

leading ['liːdɪŋ] *adjective* **(a)** moving in front of others; ***we engaged the leading tank*** **(b)** senior; *GB* **leading aircraftman (LAC)** = junior non-commissioned rank in the air force (equivalent to an experienced private soldier in the army); *GB* **leading rating** *or* **leading seaman** = junior non-commissioned officer (NCO) in the navy (equivalent to a corporal in the army, marines or air force)

leaflet ['liːflət] *noun* piece of paper containing information or a message; ***leaflets were dropped over the enemy lines***

leaguer ['liːgə] **1** *noun* encampment of armoured vehicles **2** *verb* to form a leaguer; *see also* LAAGER, LAGER

leapfrog ['liːpfrɒg] *verb* to move in alternate bounds, with one person, vehicle or sub-unit stationary and giving or prepared to give covering fire, while the other moves past to occupy a fire position beyond (NOTE: **leapfrogging - leapfrogged**)

leave [liːv] **1** *noun* **(a)** permission; ***may I have your leave to carry on, Sir?*** **(b)** holiday or vacation; ***he is on leave***; **compassionate leave** = leave granted when a serviceman has problems at home (such as the death of a relative) **2** *verb* **(a)** to go away from a place; ***the CO has already left the barracks*** **(b)** to go away without something; ***he left his rifle in my room*** (NOTE: **leaving - left**)

Leclerc [lə'klɜːk] *noun* French-designed 1990s-era main battle tank (MBT)

leech [liːtʃ] *noun* worm-like creature which lives in water and attaches itself to an animal's skin in order to suck its blood

leg [leg] *noun* **(a)** lower limb of the human body; ***his leg had to be amputated***;

he was hit in the leg by shrapnel **(b)** one section of a journey; ***the first leg is 520 metres, on a bearing of 3214 mils***

legend ['ledʒənd] *noun* explanatory notes on the symbols shown on a map

legion ['liːdʒən] *noun* **(a)** *(historical)* division of the Roman army with about 5,000 men **(b)** grouping of soldiers or ex-servicemen; **American Legion** = group which protects the interests of American veterans; **Royal British Legion** = group which protects the interests of British ex-servicemen; **Foreign Legion** = force of foreign volunteers serving in a state's army (such as the French Foreign Legion or the Spanish Foreign Legion)

legionary ['liːdʒənri] *noun (historical)* member of a Roman legion

legionnaire [ˌliːdʒə'neə] *noun* **(a)** member of the French Foreign Legion **(b)** member of the American Legion

Leopard ['lepəd] *noun* German-designed main battle tank (MBT)

lethal ['liːθəl] *adjective* able to cause death; ***he received a lethal dose of radiation***

level-crossing [ˌlevəl 'krɒsɪŋ] *noun* place where a railway line crosses a road, and instead of using a bridge, the tracks are actually embedded into the tarmac of the road

Levinstein Mustard [ˌlevɪnstaɪn 'mʌstəd] *see* H

levy ['levi] *noun* soldiers who are forced to join the army; ***the army relies on levies from the provinces***

Lewisite ['luːɪsaɪt] *see* L

LF [ˌel 'ef] = LOW FREQUENCY

LGB [ˌel dʒiː 'biː] = LASER-GUIDED BOMB

LGM [ˌel dʒiː 'em] = LASER-GUIDED MISSILE

LGOP [ˌel dʒiː əʊ 'piː] *noun US* = LITTLE GROUPS OF PARATROOPERS small groups acting on their own initative during the confusion of a hot LZ; ***once the LGOPs had established a perimeter, we started to reorganize***

LHA [ˌel eɪtʃ 'eɪ] *noun* = LANDING HELICOPTER ASSAULT ship which is designed to transport and launch landing craft and helicopters during amphibious operations

LHD [ˌel eɪtʃ 'diː] *noun* = LANDING HELICOPTER DOCK ship which is designed to transport and launch landing craft and helicopters during amphibious operations (an updated version of the LHA)

LI [ˌel 'aɪ] = LIGHT INFANTRY

liaison officer (LO) [lɪ'eɪzən ˌɒfɪsə] *noun* officer who acts as a link between one tactical grouping and another or between a headquarters and its subordinate groupings

liberate ['lɪbəreɪt] *verb* **(a)** to set a person free from captivity; ***the prisoners were liberated by the Americans*** **(b)** to set a country, region or town free from occupation by a foreign power; ***Paris was liberated in 1944*** **(c)** *(slang)* to steal; ***we've liberated a case of beer***

liberation [ˌlɪbə'reɪʃn] *noun* act of liberating a person or place; ***the liberation of Europe began with the Normandy landings***

liberty ['lɪbəti] *noun* **(a)** freedom from captivity, oppression or foreign domination **(b)** *(in the navy)* free time; **liberty boat** = small boat taking naval ratings ashore for time off

lice [laɪs] *see* LOUSE

lie [laɪ] **1** *verb* **(a)** to be in or adopt a horizontal position; ***she lay on her bed***; ***he lay down on the floor*** (NOTE: **lie - lying - lay - have lain**) **(b)** to deliberately say something which is untrue; ***he lied about the troop movements*** (NOTE: **lie - lying - lied**) **2** *noun* statement which is untrue; ***that's a lie!***

lie up [ˌlaɪ 'ʌp] *verb* to rest or wait in a concealed position before continuing a patrol or other covert operation; **lie-up position (LUP)** = concealed position where a patrol can rest, wait or observe, before carrying out the next phase of a covert operation

lieutenant (Lt) [lef'tenənt *US* luː'tenənt] *noun* **(a)** *GB* junior officer in the army or marines (equivalent of a first lieutenant in the US Army; usually in command of a platoon or equivalent-sized grouping) **(b)** *GB* junior officer in the navy, below lieutenant-commander **(c)** *US* officer in the navy *US* **lieutenant junior**

grade = junior officer in the navy (equivalent of sub-lieutenant in British Royal Navy); *US* **first lieutenant (Lt)** = junior officer in the army, marines or air force (equivalent of a lieutenant in the British Army); *see also* SECOND LIEUTENANT, SUB-LIEUTENANT

lieutenant-colonel (Lt-Col) [lef,tenənt 'kɜːnl] *noun GB* officer in the army or marines (usually in command of a battalion or equivalent-sized grouping) sometimes referred to as half-colonel

> COMMENT: in some regiments of the British Army, the lieutenant-colonel commanding a battalion or its equivalent is addressed as 'Colonel' and referred to as **the Colonel.** In others, he is addressed as 'Sir' and referred to as **the Commanding Officer** or **the CO**

lieutenant colonel (Lt Col) [luː,tenənt 'kɜːnl] *noun US* officer in the army, marines or air force (usually in command of a battalion or equivalent-sized grouping)

lieutenant-commander (Lt-Cmdr) [lef,tenənt kə'mɑːndə] *noun GB* officer in the navy above lieutenant and below commander

lieutenant commander (Lt Cmdr) [luː,tenənt kə'mɑːndə] *noun US* officer in the navy

lieutenant-general (Lt-Gen) [lef,tenənt 'dʒenrəl] *noun GB* senior officer in the army or marines (junior to general and senior to major-general)

lieutenant general (Lt Gen) [luː,tenənt 'dʒenrəl] *noun US* senior officer in the army or marines (junior to general and senior to major general)

lifebelt ['laɪfbelt] *noun* plastic or wooden ring designed to keep a person floating in water

lifeboat ['laɪfbəʊt] *noun* small boat carried by a ship, in case the ship must be abandoned at sea

lifevest ['laɪfvest] *noun* jacket designed to keep a person floating in water

lift [lɪft] **1** *noun* **(a)** act of transporting equipment, men or supplies by air; ***we lost three planes during the last lift*** **(b)** *GB* machine which takes people up or down from one floor to another in a building; ***the lift was damaged by the explosion*** **2** *verb* **(a)** to raise an object to a higher position; ***the bomb exploded as they were lifting it*** **(b)** to transport equipment, men or supplies by air; ***we will start lifting supplies tomorrow*** **(c)** to remove a restriction; ***radio silence has been lifted***; *see also* AIRLIFT

light [laɪt] **1** *adjective* **(a)** bright enough to see **(b)** low in weight; **light aid detachment;** *see* LAD; **light anti-tank weapon;** *see* LAW; **Light Gun** = British-designed 105mm artillery piece, made of light alloy for ease of transportation and high mobility; *(historical)* **light infantry (LI)** = lightly equipped and highly mobile infantry, who specialized in reconnaissance and skirmishing; **light machine-gun (LMG)** = light man-portable machine-gun designed to be carried by infantry sections or squads; **light support weapon** = *see* LSW **2** *noun* **(a)** natural or artificial brightness which makes it possible to see; **first light** = time of day when daylight first appears; **last light** = time of day when it becomes completely dark **(b)** any source of artificial light; ***we saw a light in the distance*** **(c)** any pyrotechnical device or projectile which lights up an area of ground (for example, an illuminating round, shermuly, star shell, etc.); ***the platoon commander called for light*** **3** *verb* to apply fire to something; ***he lit his cigarette*** (NOTE: **lighting - lit - has lit**)

lightstick ['laɪtstɪk] *noun* simple light-producing device, consisting of a slim plastic cylinder containing two chemicals in liquid form. When the cylinder is bent, the two liquids mix together and cause a chemical reaction which produces a weak form of light, lasting for several hours

> COMMENT: lightsticks are avaliable in several different colours and are ideal for marking routes at night

light up [,laɪt 'ʌp] *verb* to project light onto an object or over an area; ***flares were dropped to light up the area***; *see also* ILLUMINATE

Lima ['liːmə] twelfth letter of the phonetic alphabet (Ll)

limber ['lɪmbə(r)] *noun* vehicle *or* trailer carrying artillery ammunition, which accompanies the guns

limit ['lɪmɪt] **1** *noun* level or point beyond which something cannot go; **limit of exploitation** = point on the ground beyond which the exploitation of a successful attack should not continue; ***our limit of exploitation is the rear edge of the wood***; **off limits** = prohibited (to the persons specified); ***this pub is off limits to officers and NCOs*** **2** *verb* to impose a limit on something; ***water has been limited to one litre per man***

line [laɪn] **1** *noun* **(a)** long thin feature which connects or appears to connect two or more points; **lines of communication** = main roads, air routes and sea routes which connect a military force to its operational base, along which supplies are moved and along which its supply depots and reserve forces are located; **line of defence** = line formed by a series of defensive positions and defended localities; **line of departure (LD)** = real or imaginary line, the crossing of which marks the start of an advance, attack or other offensive operation; ***our line of departure is formed by the main road***; *also known as* START LINE; **line of fire** = path of a bullet or other projectile from the weapon to the target; ***some of C Company wandered into our line of fire***; **line of march** = route taken by troops or vehicles from one location to another; **line of sight** = line from a gun's position to the target **(b)** *(often used in the plural)* **lines** = line or boundary formed by the positions of an army; **enemy lines** = the forward positions of the enemy; ***we were operating behind enemy lines***; **interior lines** = routes available in territory which is protected by the forward and flank positions of an army or other large tactical grouping; **passage of lines** = process whereby a unit or grouping moves through the established line of another unit or grouping; *see also* FRONT LINE **(c)** tactical formation where troops or vehicles move side by side in a single extended line; ***the platoon advanced in line formation*** **(d) in line** = **(i)** side by side forming a straight line; **(ii)** one behind the other; ***the battleships passed in line astern*** **(e)** *GB* **the Line** = armoured and infantry regiments not forming part of the Household Division **(f) the Line** = the Equator; ***we will be crossing the Line tomorrow*** **(g)** electrical cable used to connect field telephones to each other; ***we need to lay more line back to HQ*** **(h)** track of a railway; ***the line was blown up by partisans*** **(i)** length of rope or cord; ***he threw a line out to the lifeboat*** **2** *verb* to position or place in such a way as to form a line; ***the road was lined with tall trees***

line up [ˌlaɪn 'ʌp] *verb* to form a line; ***they lined up outside the armoury***

link [lɪŋk] **1** *noun* **(a)** something which connects **(b)** metal clip used to fasten rounds of machine-gun ammunition together, in order to form belts **(c)** machine-gun ammunition (which is fastened together by links); ***we need 10,000 rounds of 7.62mm link*** (NOTE: no plural in this meaning) **(d)** contact or means of communication; ***a liaison officer acts as a link between two different units***; ***we need another radio link with the Germans*** **2** *verb* to connect or join; ***we were linked to Brigade HQ***; ***the media have linked his name with the nationalist movement***

link up [ˌlɪŋk 'ʌp] *verb* to come together; ***the battalion linked up with the Royal Hussars on the far side of the river***

listening post ['lɪsnɪŋ ˌpəʊst] *noun* **(a)** small patrol, sent out in front of a defensive position at night, in order to listen for the approach of the enemy **(b)** small radio station where radio operators listen to enemy communications

littoral ['lɪtərəl] *noun* coast

live 1 [laɪv] *adjective* relating to real ammunition, which is designed to kill (as opposed to blank ammunition, which is designed to simulate the firing of a weapon); **live-firing exercise** = training exercise where live ammunition is used; **live round** = piece of real ammunition (as opposed to a blank round); *compare* BLANK, BLANK ROUND **2** [lɪv] *verb* ***to live off the land*** = to obtain food from the local area (as opposed to using your own supplies)

LMG [ˌel em 'dʒiː] = LIGHT MACHINE-GUN

LO [ˌel 'əʊ] = LIAISON OFFICER

load [ləʊd] **1** *noun* **(a)** something that is carried by an aircraft, person or vehicle; ***this helicopter is capable of carrying heavy loads*** **(b)** amount or weight of what is carried by an aircraft, person or vehicle; ***in the Falklands conflict, soldiers were carrying loads in excess of 100 pounds;*** *US* **load-bearing equipment (LBE)** = set of equipment pouches attached to a belt or harness **2** *verb* **(a)** to put a load or cargo onto an aircraft, vehicle or ship; ***they were attacked as they were loading the ship*** **(b)** to put ammunition into a weapon; ***have you loaded your weapon yet?; load with HE!*** **(c)** to put ammunition into a magazine; ***they are still loading magazines*** **(d)** to put a loaded magazine onto a weapon; ***with a magazine of 30 rounds, load!***

> COMMENT: a loaded weapon can be in one of two states: **made safe** means that a loaded magazine is fitted, but the weapon is not cocked and there is no round in the breech, whereas **made ready** means that a loaded magazine is fitted, the weapon is cocked and there is a round in the breech

loader ['ləʊdə] *noun* crew member responsible for loading an artillery piece, gun or other weapons system

loam [ləʊm] *noun US* soil; ***it was easy digging into the soft loam***

loan service ['ləʊn ˌsɜːvɪs] *noun* temporary secondment of servicemen to the armed forces of a friendly foreign state, usually to provide military expertise *or* training

local superiority [ˌləʊk(ə)l ˌsuːpɪəri'ɒrəti] *noun* having more troops than the enemy on one part of the battlefield, even though the enemy force as a whole may be equal in strength *or* even superior to your own

local time ['ləʊkl ˌtaɪm] *noun* time of the country in which one is operating; ***the general will be arriving at 1430hrs local time***

locate [ləʊ'keɪt] *verb* **(a)** to discover the exact location of something; ***we have located the enemy battery*** **(b)** to place or position something; ***the dressing station is located in the brigade administration area***

location [ləʊ'keɪʃn] *noun* **(a)** place where something is; ***he is not at this location*** **(b)** act of locating something; ***he is responsible for the location of the supply dumps; the location of the enemy positions is taking longer than expected***

locator [ləʊ'keɪtə] *noun* device or equipment designed to locate something

lock [lɒk] **1** *noun* **(a)** mechanism for securing one object to another (which usually requires a key to open it); ***we had to smash the lock in order to open the door*** **(b)** enclosed stretch of a canal *or* river, in which the water level can be raised *or* lowered by the use of gates; ***5 Platoon is dug in around the lock*** **2** *verb* **(a)** to secure with a lock; ***this door is to be locked at all times*** **(b)** to secure behind locked doors; ***the weapons will be locked in the guardroom overnight***

lock and load [ˌlɒk ənd 'ləʊd] *verb US (informal)* to operate the cocking lever of a weapon so that a round is placed in the chamber and the weapon is cocked and ready to fire (NOTE: British English is **make ready**)

locker ['lɒkə(r)] *noun* compartment *or* cupboard *or* wardrobe which can be locked; ***locker inspection at 1800hrs***

lock-on ['lɒk ɒn] *noun* moment when the operator of a guided weapon has the target in his sights and the guidance system is activated; *see also* ACQUISITION, SOLUTION

L of C = LINE OF COMMUNICATION

log [lɒg] = LOGISTICS

log [lɒg] **1** *noun* **(a)** official diary of a ship; ***the captain entered the ship's position in the log*** **(b)** chronological record of events; ***all radio operators must keep a log*** **(c)** record of journeys, maintenance, repairs, etc., for an aircraft, piece of equipment or vehicle; ***he inspected the vehicle's log*** **2** *verb* to make a chronological record; ***you are required to log all messages***

loggie ['lɒgɪ] *noun (informal)* person involved in logistics

logistic *or* **logistical** [lə'dʒɪstɪkl] *adjective* relating to logistics

logistics [lə'dʒɪstɪks] *noun* coordination of the supplying and resupplying of military units with the resources which they need in order to carry out their operational tasks (such as ammunition, equipment, food and water, fuel, medical facilities, replacement men and equipment, spare parts, transport, etc.); **consumer logistics** *or* **operational logistics** = logistics concerned with the receiving and storing of supplies

> COMMENT: logistics covers the design, development, acquisition, movement and storage of material; the movement of personnel; the construction and maintenance of buildings and other facilities; the provision of services such as medical services or food; the departments responsible for logistics in a headquarters are known as **G1 (personnel)** and **G4 (materiel)**

long-range ['lɒŋ ˌreɪndʒ] *adjective* used over a long distance; **long-range reconnaissance and patrolling (LRRP)** = special skills relating to covert patrolling far into enemy-held territory; ***he is going on a LRRP course***

long-rod penetrator [ˌlɒŋ rɒd 'penətreɪtə] *noun* anti-tank projectile, consisting of a simple metal dart (made out of a high density metal, such as tungsten carbide or depleted uranium and usually fitted with fins in order to provide extra stability in flight) which is fired at a very high velocity and uses kinetic energy to punch its way through armour; *see also* ARMOUR-PIERCING, KINETIC ENERGY (KE) ROUND

> COMMENT: long-rod penetrators are the only means of defeating many modern types of armour. Because a long-rod penetrator is considerably smaller than the diameter of the gun barrel through which it is fired, it is usually fitted with a metal collar or sleeve known as a **sabot,** which falls away once the projectile has left the barrel

loot [lu:t] **1** *noun* **(a)** any private property belonging to the enemy, which is taken for your own personal use or gain; ***we found plenty of loot on the enemy position*** **(b)** anything which is stolen in wartime or during a period of civil disorder; ***several valuable paintings were discovered amongst the loot*** **2** *verb* to steal during a period of disorder; ***soldiers are looting the town***

looter ['lu:tə] *noun* person who loots; ***the army was ordered to shoot into the air to discourage looters***

looting ['lu:tɪŋ] *noun* action of removing property during a period of civil disorder; ***there have been reports of widespread looting by the enemy***

lorry ['lɒri] *noun GB* large wheeled vehicle designed to transport men, equipment or supplies; *see also* TRUCK

lose [lu:z] *verb* **(a)** to be unable to find something; ***he has lost his rifle***; **to lose your way** = to be unaware of your exact location **(b)** not to have something any more, because it has been destroyed; not to have a person any more, especially one who has been killed; ***we lost three men yesterday***; ***B Squadron lost four tanks***; **to lose your nerve** = to be unable to control your fear **(c)** to be defeated; ***we have lost the battle*** (NOTE: **losing - lost**)

loss [lɒs] *noun* **(a)** act of losing something; ***he did not report the loss of his rifle*** **(b) losses** = casualties; ***we have suffered heavy losses***

lost [lɒst] *adverb* unaware of your exact location; ***we are lost***; **lost at sea** = missing, believed drowned

louse [laʊs] *noun* tiny parasitic insect, which lives on a person's body and in his clothes, breeding in great numbers and being easily transmitted to other people, thereby causing great discomfort and often transmitting disease; ***he was covered in louse bites*** (NOTE: the plural form is **lice** [laɪs] and is normally used in preference to the singular form; ***the prisoners were covered in lice***)

lower ['ləʊə] *adjective* below something else; ***we moved into the lower part of the town***; **lower case** = small letters written as a, b, c, etc. (NOTE: the opposite, capital letters, is **upper case**)

low frequency (LF) [ˌləʊ 'fri:kwənsi] *noun* range of radio frequencies from 30 - 300 kilohertz (kHz)

low-observables [ˌləʊ əb'zɜːvəblz] *plural noun* stealth technology; ***this aircraft design incorporates all the latest low-observables***

low velocity [ˌləʊ və'lɒsəti] *adjective; (of projectiles)* slower than the speed of sound; ***this helmet will resist a low velocity bullet***; *compare* HIGH VELOCITY

LPD [ˌel piː 'diː] *noun* = LANDING PLATFORM DOCK ship which is designed to transport and launch landing craft and helicopters during amphibious operations

LPH [ˌel piː 'eɪtʃ] *noun* = LANDING PLATFORM HELICOPTER ship designed to transport and launch helicopters and air-assault infantry during amphibious oprations

LRRP [lɜːp] = LONG-RANGE RECONNAISSANCE AND PATROLLING

L/Sgt = LANCE-SERGEANT

LS [el 'es] = LANDING SITE

LSD [ˌel es 'diː] *noun* = LANDING SHIP DOCK ship which is designed to transport and launch landing craft during amphibious operations

LSD [ˌel es 'diː] *noun* = LYSERGIC ACID DIETHYLAMIDE drug which causes intense hallucinations and can have serious long-term effects

> COMMENT: although LSD is usually associated with drug abuse, it could be delivered as a chemical agent, and would be particularly effective in causing disruption in rear areas

LSL [ˌel es 'el] *noun GB* (= LANDING SHIP LOGISTICS) ocean-going troop or supply ship

LSRV [elˌes ɑː 'viː] = LANDING SIGHT RENDEZVOUS

LST [ˌel es 'tiː] *noun* = LANDING SHIP TANK ship designed to transport and land armoured vehicles

LSW [ˌel es 'dʌb(ə)ljuː] *noun* = LIGHT SUPPORT WEAPON British-designed 5.56mm light machine-gun (LMG) based on the SA80 assault weapon

Lt = LIEUTENANT; LIGHT

2Lt = SECOND LIEUTENANT

Lt-Col = LIEUTENANT-COLONEL

Lt-Cmdr = LIEUTENANT-COMMANDER

Lt-Gen = LIEUTENANT-GENERAL

lubricant ['luːbrɪkənt] *noun* substance, such as oil, which is applied to machinery in order to make it run smoothly

lull [lʌl] *noun* temporary period of inactivity or quiet; ***he moved back during a lull in the battle***

luminous ['luːmɪnəs] *adjective; (of a substance, especially paint)* producing light (without electricity); ***my watch has a luminous face***

Lumocolor™ ['luːmɪkʌlə(r)] *noun* felt-tipped pen, which is suitable for writing on plastic (eg. map-cases, overlays, etc); ***the enemy positions are marked in red Lumocolor***

> COMMENT: Lumocolors come in two types: **water-soluble** will wash off on contact with any liquid (eg. rain-water, saliva, sweat, etc), while **permanent** can only be removed by some sort of cleaning fluid

LUP [ˌel juː 'piː] = LIE-UP POSITION

lurk [lɜːk] **1** *verb* to wait in a concealed position in the hope that a target might present itself; ***I think there's a tank lurking in that wood*** **2** *noun* patrol which waits in a likely area of enemy activity, in order to react to any incident which might occur there; ***tonight we're going to do a lurk by the railway bridge***

> COMMENT: a *lurk* is really a counter-insurgency term and differs from an *ambush* in that it is not set with the primary intention of killing anyone; in fact an arrest would be the more probable result

LVTP-7A1 [ˌel viː tiː piː ˌsevən eɪ 'wʌn] *noun* = LANDING VEHICLE TRACKED PERSONNEL American-designed amphibious tractor (NOTE: also known as **Amphibious Assault Vehicle Seven (AAV-7A1)**)

Lynx [lɪŋks] *noun* British-made multirole helicopter; **Navy or Sea Lynx** = Lynx helicopter modified for operating from a ship

LZ [ˌel 'zed *US* ˌel 'ziː] *noun* = LANDING ZONE **(a)** area of ground selected for the landing or pick-up of troops by helicopter; ***the LZ is at grid 941623***; ***B Company will secure the LZ*** **(b)** *US* area of ground selected for the landing of troops by parachute; **hot LZ** = landing zone which is under enemy fire (NOTE: the British Army uses the phrase **drop zone (DZ)** for parachute landings)

MIKE - Mm

M-1 [ˌem 'wʌn] *see* ABRAMS

M-2 [ˌem 'tuː] *see* BRADLEY

M-3 [ˌem 'θriː] *see* BRADLEY

M-16 [ˌem sɪks'tiːn] *see* ARMALITE

M-60 [ˌem 'sɪksti] *noun* **(a)** American-designed 1960s-era main battle tank (MBT) **(b)** American-designed 7.62mm general purpose machine-gun (GPMG) (NOTE: plural is **M-60s** [ˌem 'sɪkstiz])

M-61A1 [ˌem sɪksti wʌn 'eɪ wʌn] *noun* American-designed 20mm multi-barrelled anti-aircraft cannon (NOTE: also known as the **Vulcan**); *see also* M-163, PHALANX

M-82 [em ˌeɪti 'tuː] *see* BARRETT

M-109 [ˌem wʌn əʊ 'naɪn] *noun* American-designed 155mm self-propelled howitzer (SPH) (NOTE: plural is **M-109s** [ˌem wʌn əʊ 'naɪnz])

M-110 [ˌem wʌn wʌn 'əʊ] *noun* American-designed 203mm self-propelled gun (SPG) (NOTE: plural is **M-110s** [ˌem wʌn wʌn 'əʊz])

M-113 [ˌem wʌn wʌn 'θriː] *noun* American-designed 1960s-era armoured personnel carrier (APC) (NOTE: plural is **M-113s** [ˌem wʌn wʌn 'θriːz])

M-163 [em ˌwʌn sɪksti 'θriː] *noun* American-designed M-113 armoured personnel carrier fitted with the M-61A1 Vulcan 20mm multi-barrelled anti-aircraft cannon

M-198 [em ˌwʌn naɪnti 'eɪt] *noun* American-designed 155mm artillery piece (NOTE: plural is **M-198s** [em ˌwʌn naɪnti 'eɪts])

M-247 [ˌem tuː fɔːti 'sevən] *noun* American-designed self-propelled anti-aircraft gun (SPAAG) (NOTE: also known as the **SERGEANT YORK**)

M-249 [ˌem tuː fɔː 'naɪn] *noun* American-designed 5.56mm light machine-gun (LMG) (NOTE: known in the US Army as the **Squad Automatic Weapon (SAW)**)

M-551 [ˌem faɪv faɪv 'wʌn] *see* SHERIDAN

M-1973 [ˌem wʌn naɪn sevən 'θriː] *noun* Soviet-designed 152mm self-propelled gun (SPG) (NOTE: plural is **M-1973s** [ˌem wʌn ˌnaɪn sevən 'θriːz])

M-1974 [ˌem wʌn naɪn sevən 'fɔː] *noun* Soviet-designed 122mm self-propelled howitzer (SPH) (NOTE: plural is **M-1974s** [ˌem wʌn naɪn sevən 'fɔːz])

machete [mə'ʃeti] *noun Central America* long broad-bladed knife designed for clearing and often used as a weapon; *see also* PANGA, PARANG

machine-gun (MG) [mə'ʃiːn gʌn] **1** *noun* automatic firearm, which will continue to fire and reload for as long as its trigger is depressed; **coaxial machine-gun (COAX)** = machine-gun which shares the same sighting systems as the main gun of an armoured fighting vehicle (AFV); **general purpose machine-gun (GPMG)** = medium-sized machine-gun which can be used for a variety of roles (such as air defence, infantry weapon, sustained fire (SF), vehicle armament, etc.); **light machine-gun (LMG)** = light man-portable machine-gun designed to be carried by infantry sections or squads; **sub-machine-gun (SMG)** = small hand-held machine-gun, which is carried as a personal weapon **2** *verb* to shoot someone with a machine-gun; ***enemy gunships have been machine-gunning refugee columns***

machine-gunner [mə'ʃiːn ˌgʌnə] *noun* person who operates a machine-gun

Mach number ['mæk ˌnʌmbə] *noun* speed of an aircraft or missile in relation to the local speed of sound

> COMMENT: the Mach number will vary at different altitudes. An aircraft which travels faster than Mach 1 is said to be **supersonic**

made ready [ˌmeɪd 'redi] *adjective* state of a gun when a loaded magazine is fitted, the weapon is cocked and there is a round in the breech

made safe [ˌmeɪd 'seɪf] *adjective* state of a gun when a loaded magazine is fitted, but the weapon is not cocked and there is no round in the breech

maggot ['mægət] *noun* **(a)** larva of a fly; ***his wound was crawling with maggots*** **(b)** *slang* sleeping-bag; ***he's still in his maggot***

MAOT [ˌem eɪ əʊ 'tiː] *noun GB* = MOBILE AIR OPERATIONS TEAM small group of air traffic contollers for a temporary helicopter landing site

mag [mæg] *noun (informal)* magazine

magazine [ˌmægə'ziːn] *noun* **(a)** metal or plastic ammunition container, which is fitted to a gun and is designed to feed the rounds directly into the breech; ***every man is to carry five extra magazines*** (NOTE informally called **mag** in this meaning) **(b)** building used for storing ammunition and explosives; ***we have capture an enemy magazine*** **(c)** compartment in a ship, used for storing ammunition; ***the shell pierced the ship's armour and exploded in the magazine*** **(d)** building or compound, used for storing military supplies (such as ammunition, clothing, food, fuel, weapons, etc.); ***the enemy is resupplied by a network of magazines located in his rear areas***

magnetic [mæg'netɪk] *adjective* **(a)** having the property of attracting or repelling iron; **magnetic north** = direction in which the needle of a compass points; *compare* GRID NORTH **(b)** relating to magnetic north; **magnetic bearing** = bearing which is obtained using a compass; **magnetic variation** = difference between magnetic north and grid or true north (as shown on a map)

> COMMENT: the magnetic variation is normally shown on the key of a map. It is used to convert magnetic bearings to grid bearings and vice-versa. To convert a magnetic bearing to a grid bearing, subtract the magnetic variation. To convert a grid bearing to a magnetic bearing, add the magnetic variation

main [meɪn] **1** *adjective* **(a)** most important; **main battle area (MBA)** = part of the battlefield or operational area in which most of the activity is taking place; **main battle tank (MBT)** = heavily armoured tank, fitted with a large-calibre gun, which is primarily designed to destroy enemy tanks **(b)** largest or strongest; **main defence forces** = forces assigned to the major NATO commanders; **main headquarters** = primary resourcing and planning headquarters for a large tactical grouping (normally located to the rear of the forward troops) **2** *noun* ***Main*** = main headquarters; ***Main is located at grid 675784***

Mainstay ['meɪnsteɪ] *noun* NATO name for the Soviet-designed A-40 airborne early warning and control (AEW & C) aircraft

maintain [meɪn'teɪn] *verb* **(a)** to keep an activity going; ***we need to maintain the momentum of the attack*** **(b)** to look after equipment, so that it continues to function properly; ***you are responsible for maintaining our vehicles***

maintenance ['meɪntənəns] *noun* act of maintaining; ***this section is responsible for vehicle maintenance***; **maintenance check** = checking that a vehicle or weapon is in good functioning condition

> COMMENT: maintenance covers the inspection and repair of equipment and materiel to make sure it is kept in working order, the repair and upkeep of buildings and other facilities, and the continued supply of materiel to forces in the field

Maj-Gen = MAJOR-GENERAL

major ['meɪdʒə] **1** *adjective* very important; ***that road is a major line of communication for the enemy*** **2** *noun* **(a)** *GB* officer in the army or marines, below lieutenant-colonel and above captain (normally in command of a company or equivalent-sized grouping or employed as a staff officer); **brigade major** = chief of staff of a brigade **(b)** *US* officer in the army, marines or air force, below lieutenant colonel and above captain; *see also* SERGEANT MAJOR

COMMENT: in the Irish army, the equivalent of major is commandant

major-general (Maj-Gen) [ˌmeɪdʒə 'dʒenərl] *noun* *GB* senior officer in the army or marines (junior to lieutenant-general and senior to brigadier, usually in command of a division)

major general (Maj Gen) [ˌmeɪdʒə 'dʒenərl] *noun* *US* senior officer in the army, marines or air force (junior to lieutenant general and senior to brigadier general, usually in command of a division or equivalent-sized grouping)

Major NATO Command (MNC) [ˌmeɪdʒə ˌneɪtəʊ kə'mɑːnd] *noun* one of two areas of command (Allied Command Atlantic (ACLANT) and Allied Command Europe (ACE)) within NATO; Allied Command Europe is divided into three major subordinate commands (MSCs) which are south, central and north-west

Major NATO Commander (MNC) [ˌmeɪdʒə ˌneɪtəʊ kə'mɑːndə] *noun* one of two commanders of NATO forces: the Supreme Allied Commander Atlantic (SACLANT) and the Supreme Allied Commander Europe (SACEUR)

make [meɪk] **1** *verb* to construct or produce something; ***the soldiers made improvised shelters in the woods***; ***he you made your plan yet?***; ***he is making tea*** **2** *phrasal verb* ***to make contact*** = to see the enemy for the first time; *GB* **to make ready** = to operate the cocking mechanism of a loaded weapon, so that a round is fed into the breech and the weapon is cocked and ready to fire (NOTE: American English is **lock and load**); *GB* **to make safe** = to fully unload a cocked weapon and then replace the loaded magazine back onto the weapon; *see also* MADE READY, MADE SAFE (NOTE: **making - made**)

malaria [mə'leəriə] *noun* fever caused by the parasite *Plasmodium,* which is transmitted by the bite of a mosquito in tropical regions

malfunction [mæl'fʌŋkʃn] *(of equipment, instruments, machinery, etc.)* **1** *noun* failure to work properly; ***the accident was due to a malfunction in the steering system*** **2** *verb* to fail to work properly; ***the guidance system has malfunctioned***

man [mæn] **1** *noun* member of the armed forces; ***he sent six men to reconnoitre the road***; **man-management** = getting the best out of your soldiers, by treating them with respect and looking after their welfare; his man-management is very poor; *see also* ENLISTED MAN **2** *verb* to provide personnel to make something work; ***the battery is manned by six gunners***; *see also* UNMANNED

mandate ['mændeɪt] *noun* instruction *or* directive from an official organization (eg. EU, UN, etc); ***our mandate is to see that these people do not starve***

maneuver *see* MANOEUVRE

maneuverability *see* MANOEUVRABILITY

maneuverable *see* MANOEUVRABLE

manhole ['mænhəʊl] *noun* covered hole providing access to a sewer; ***we threw a grenade down every manhole***

manifest ['mænɪfest] *noun* list of passengers *or* cargo carried by an aircraft; ***his name isn't on the manifest***

manoeuvrability *US* **maneuverability** [məˌnuːvrə'bɪlɪti] *noun* ability to move easily over all types of terrain; ***the main advantage of this vehicle is its manoeuvrability***

manoeuvrable *US* **maneuverable** [mə'nuːvrəbl] *adjective* capable of moving easily over all types of terrain; ***the new tank is highly manoeuvrable***

manoeuvre *US* **maneuver** [mə'nuːvə] **1** *noun* **(a)** the art of moving troops and vehicles in order to achieve a military objective; ***the new tactical doctrine places great emphasis on manoeuvre*** **(b)** a planned movement by troops or vehicles designed to achieve a specific objective; ***the manoeuvre was supposed to cut off the enemy's line of retreat*** **(c) manoeuvres** = military training exercises; ***the manoeuvres will take place***

in April **2** *verb* **(a)** to perform a manoeuvre; ***the brigade manoeuvred against the enemy's flank*** **(b)** to perform a complicated movement with a vehicle; ***we manoeuvred the tank into the farmyard***

manoeuvre warfare [mə'nuːvə ˌwɔːfeə] *noun* military doctrine which seeks to break an enemy's will to fight by using mobility and constant aggression to shatter his cohesion and deprive him of any opportunity to reorganize; *compare* ATTRITIONAL WARFARE, POSITIONAL WARFARE

> COMMENT: manoeuvre warfare relies on the use of directive command for its success, since subordinate commanders must be free to use their own initiative whenever necessary

man-pack ['mænpæk] *verb* to transport something using people (as opposed to animals *or* vehicles *or* aircraft *or* boats); ***the ammunition will have to be man-packed up to the gun line***

MANPADS ['mænpædz] *noun* = MAN-PORTABLE AIR DEFENCE SYSTEM any hand-held surface-to-air missile (eg. Blowpipe, Grail, Stinger, etc)

man-portable ['mæn ˌpɔːtəbl] *adjective* designed to be carried by one or more persons; ***this missile launcher is man-portable***; ***the enemy is equipped with man-portable boats***

manual ['mænjʊəl] *noun* book of instructions; ***there is a detailed diagram in the manual***; ***remember to take a copy of the 'Manual of Military law' to the court-martial***

map [mæp] *noun* scale drawing of an area of ground, with symbols representing natural and man-made features; **map grid** = system of numbered squares printed on a map in order to produce map or grid references; **map reference** *or* **grid reference** = six- or eight-figure reference, obtained from the coordinates of a map grid, used to identify an exact location on the map

MAPEX ['mæpeks] *noun* = MAP EXERCISE exercise involving command elements of a grouping, in which a tactical scenario is played out on a map

march [mɑːtʃ] **1** *noun* **(a)** movement on foot; ***it was a long march to the assembly area***; **order of march (OOM)** = sequence in which the sub-units of a grouping move (either on foot or by vehicle) from one location to another; ***order of march: B Company will lead, followed by A Company, then D Company, then C Company*** **(b)** piece of music, traditionally played when a regiment is marching on a parade; ***our regimental march is 'The British Grenadiers'*** **2** *verb* **(a)** to move from one location to another on foot; ***we had to march to the concentration area*** **(b)** to walk in a smart military manner (especially on a parade); ***the recruits are learning how to march***

march past [ˌmɑːtʃ 'pɑːst] *verb* to march in ceremonial order past an officer or a saluting base; ***the battalion marched past to the tune of 'The British Grenadiers'***

marchpast ['mɑːtʃpɑːst] *noun* parade where troops march past a saluting base; ***the Queen took the salute at the marchpast***; *compare* FLYPAST

Marder ['mɑːdə] *noun* German-designed infantry fighting vehicle (IFV)

marine [mə'riːn] **1** *adjective* relating to the sea; ***he has a diploma in marine engineering*** **2** *noun* **(a)** infantry soldier serving with the navy, but trained to fight on land; **Royal Marines (RM)** = British units of marines; **United States Marine Corps (USMC)** = American units of marines **(b)** *GB* lowest non-commissioned rank in the marines (equivalent of a private in the army)

> COMMENT: in most armed forces, marines have the same or a similar rank structure to the army, but they follow the customs and traditions of the navy. In the British armed forces, a marine's rank has a higher status than the same rank in the army. Thus, a captain in the Royal Marines is considered to be the equivalent of a major in the army

maritime ['mærɪtaɪm] *adjective* relating to the sea and ships

mark [mɑːk] **1** *noun* **(a)** anything which is drawn, painted, written on, placed on or cut into the surface of an object or the ground, in order to convey a meaning; **aiming mark** = point at which one aims, in order to hit a target **(b)** any cut or indentation in the surface of an object or the ground as a result of damage; ***we could see the marks made by the shrapnel*** **(c)** model or type; ***Mark II*** *or* ***Mk. II fragmentation grenade*** (NOTE: can be shortened to **Mk** in this meaning) **2** *verb* **(a)** to make a mark on something; ***he marked the tree with an 'X'; the map was marked with all the enemy positions and minefields*** **(b)** *(of targets, landing zones, etc.)* to use a sign, light or coloured smoke, so that other people can see it; ***we will use yellow smoke to mark the LZ***

marker ['mɑːkə] *noun* anything which is used as a sign, in order to convey meaning, or to draw other people's attention to a location or object; ***he used a piece of mine tape as a marker***

marking ['mɑːkɪŋ] *noun* **(a)** act of making a mark; ***the general has forbidden the marking of maps because of security*** **(b)** numbers, letters, symbols or insignia, which are painted or printed on an object as a means of identification; ***high explosive shells are usually painted dark green, with yellow markings***

marksman ['mɑːksmən] *noun* soldier who is very good at shooting; ***they positioned marksmen on the roofs of surrounding buildings***

marsh [mɑːʃ] *noun* large area of permanently wet ground

marshal ['mɑːʃəl] **1** *noun* most senior army rank in certain armies; *GB* **field marshal (FM)** = most senior officer rank in the army; **Marshal of the Royal Air Force** = most senior officer rank in the RAF; *see also* AIR MARSHAL, AIR CHIEF MARSHAL, AIR VICE MARSHAL **2** *verb* to direct and organize vehicles at an assembly point or any other place where a lot of vehicles are gathered; ***we'll need some NCOs for marshalling the vehicles***; **marshalling area** = location where vehicles assemble before moving to another location or before deploying into formation; **marshalling yard** = railway yard where trains are assembled (NOTE: marshalling, marshalled; US marshaling, marshaled)

marshland ['mɑːʃlænd] *noun* terrain consisting mainly of marsh

MASH [mæʃ] *noun US* = MOBILE ARMY SURGICAL HOSPITAL field hospital

mask [mɑːsk] **1** *noun* face covering worn as protection or as a means of disguise; ***the terrorists were wearing masks*** **2** *verb* **(a)** to conceal; ***that hedge will mask our withdrawal*** **(b)** mask up =; **to put on respirators response to a chemical threat**

mass [mæs] **1** *noun* **(a)** large quantity of something; ***there is a huge mass of refugees at the frontier; there are masses of refugees at the frontier*** (NOTE: can be used in singular *or* plural form) **(b)** *(as a principal of war)* concentration of troops and firepower at a decisive point **2** *verb* to concentrate in large numbers; ***enemy troops are massing on the border***

massacre ['mæsəkə] **1** *noun* incident involving the killing of a large number of people; ***we are receiving reports of a massacre in Malmédy*** **2** *verb* to kill a large number of people (in one incident); ***the villagers were massacred by the retreating troops***

mast [mɑːst] *noun* tall metal structure, usually situated on high ground, for supporting communications equipment (eg. television, radio, mobile telephone technology)

master gunnery sergeant [ˌmɑːstə 'gʌnəri ˌsɑːdʒənt] *noun US* senior non-commissioned officer (SNCO) in the marines

master chief petty officer [mɑːstə ˌtʃiːf 'peti ˌɒfɪsə] *noun US* senior non-commissioned officer (SNCO) in the navy

master sergeant ['mɑːstə ˌsɑːdʒənt] *noun US* senior non-commissioned officer (SNCO) in the army, marines or air force

material [mə'tɪərɪəl] *noun* **(a)** any substance from which things can be made; ***this armour is made of steel and other materials*** **(b)** fabric or cloth; ***the new uniforms are made of flame-resistant material*** **(c)** information or data; ***this material is classified secret***

materiel [məˌtɪəri'el] *noun* equipment and supplies (as opposed to personnel);

logistics involves the purchase, transport and storage of materiel

> COMMENT: the department responsible for materiel in a headquarters is known as G4

Matterhorn ['mætəhɔːn] *noun* type of military boot made of soft leather with a Gortex lining

maul [mɔːl] *noun* to inflict a lot of casualties; ***the brigade was badly mauled*** (NOTE: this verb is normally used in the passive)

Maverick ['mævərɪk] *noun* American-designed air-to-ground missile (AGM)

MAW [ˌem eɪ 'dʌb(ə)ljuː] = MEDIUM ANTI-TANK WEAPON

MBA [ˌem biː 'eɪ] = MAIN BATTLE AREA

MBT [ˌem biː 'tiː] = MAIN BATTLE TANK

MCCP [ˌem siː siː 'piː] = MOVEMENT CONTROL AND CHECK-POINT

MCT(S) [ˌem siː tiː 'es] *noun GB* = MILAN COMPACT TURRET (SPARTAN) variant of Spartan armoured personnel carrier, adapted to fire Milan ATGW

means [miːnz] *noun* radio terminology for a type of communications system (eg: radio *or* telephone *or* fax *or* E-mail, etc); ***Hello 22, this is 2, change to secure means, over***

MEB [ˌem iː 'biː] *noun US* = MARINE EXPEDITIONARY BRIGADE divisional-scale combined-arms grouping of BLTs, helicopter squadrons and logistics units

mech [mek] = MECHANIZED INFANTRY

mechanic [mɪ'kænɪk] *noun* person who repairs and services machinery

mechanized ['mekənaɪzd] *adjective* equipped with machinery, especially transport; **mechanized battalion** = infantry battalion equipped with armoured personnel carriers (APCs) or infantry fighting vehicles (IFVs); **mechanized infantry** = infantry equipped with armoured personnel carriers (APCs) or infantry fighting vehicles (IFVs)

> COMMENT: infantry equipped with infantry fighting vehicles (IFVs) are often referred to as armoured infantry

medal ['medl] *noun* insignia (usually consisting of a metal cross, disk or star suspended from a piece of coloured fabric), which denotes the wearer's participation in a campaign or tour of operational duty, or that the wearer has received an award for bravery or for an outstanding achievement (usually during operational service); **medal ribbon** = piece of coloured fabric, which is worn on the breast of a uniform, to show that the wearer has received a medal (NOTE: also called **decoration**)

media ['miːdiə] *noun* general term for the television, radio and newspapers, and the reporters who work for them

medic ['medɪk] *noun* person who is not a doctor, but is trained to give medical treatment

medical ['medɪkl] *adjective* relating to the treatment of illness and injury; **medical officer (MO)** = doctor attached to a unit

Médicines Sans Frontières (MSF) [ˌmedɪsænz sɒnz ˌfrɒntɪ'eəz] *noun* non-governmental organization (NGO), dedicated to providing medical treatment to all casualties, regardless of nationality, during an armed conflict

medium ['miːdiəm] *adjective* **(a)** neither light nor heavy; **medium anti-tank weapon (MAW)** = man-portable launcher designed to fire a powerful anti-tank rocket or missile **(b)** neither small nor large

MEF [ˌem iː 'ef] *noun US* = MARINE EXPEDITIONARY FORCE corps-scale combined-arms grouping of BLTs, helicopter squadrons and logistics units

megaton ['megətʌn] *noun* unit of explosive power, corresponding to that produced by one million tons of TNT

MEL [ˌem iː 'el] *noun* = MOBILE-ERECTOR-LAUNCHER vehicle designed to carry and launch a surface-to-surface missile (SSM); *see also* TEL

Mentioned in Dispatches (MID) [ˌmenʃənd ɪn dɪs'pætʃɪz] *noun* British award in recognition of achievement on operational service

mercenary ['mɜːsənri] *noun* person who serves in the armed forces of another state for payment; ***the rebels are led by foreign mercenaries***

> COMMENT: the term **mercenary** is not usually applied to soldiers serving in officially recruited foreign units, such as the Gurkhas, or the French and Spanish Foreign Legions

merchant navy [ˌmɜːtʃənt 'neɪvi] *noun* a state's commercial shipping

Merkava [mɜː'kɑːvə] *noun* Israeli-designed 1980s-era main battle tank (MBT)

mess [mes] *noun* **(a)** place where servicemen of equal or similar rank eat and relax together; ***officers' mess***; ***sergeants' mess*** **(b)** meal; ***he's at mess***; **mess hall** = building where servicemen eat their meals; **mess tin** = metal container designed for cooking food in the field and also for use as a plate or dish; **mess-kit** = ceremonial uniform for evening wear.; **mess night** = formal dinner in a mess, where mess-kit is worn and certain customs and traditions are observed

message ['mesɪdʒ] *noun* verbal or written instruction, request, question or statement, which is passed from one person to another

met [met] = METEOROLOGICAL, METEOROLOGY

metalled *US* **metaled** ['metəld] *adjective; (of roads)* covered with a prepared surface of gravel *or* small stones (NOTE: also used, in a general sense, to describe a surface of **asphalt** *or* **tarmac**)

meteorological [ˌmiːtiərə'lɒdʒɪkl] *adjective* relating to meteorology; **meteorological office (met office)** = British government establishment for the study and prediction of weather conditions

meteorology [ˌmiːtiə'rɒlədʒi] *noun* study of the weather

meter ['miːtə] *noun US* metre

METT-T [met 'tiː] *noun US* = MISSION, ENEMY, TERRAIN, TROOPS AVALIABLE, TIME factors which a commander must consider when making his appreciation

MEU [ˌem iː 'juː] *noun US* = MARINE EXPEDITIONARY UNIT brigade-scale combined-arms grouping, based on a battalion landing team (BLT), a squadron of helicopters (attack and utility) and a logistics battalion, which can be reinforced by more BLTs as the tactical situation requires; **MEU (SOC)** = MARINE EXPEDITIONARY UNIT (SPECIAL OPERATIONS CAPABLE)

> COMMENT: At any time, the USA usually has at least three MEUs at sea in various parts of the world, ready to respond to any crisis in which American interests might be at risk

MFC [ˌem ef 'siː] = MORTAR FIRE CONTROLLER

MG [ˌem 'dʒiː] = MACHINE-GUN

MGRS [ˌem dʒiː ɑː 'es] *noun* = MILITARY GRID REFERENCE SYSTEM world-wide series of maps which are compatible with the GPS satellite navigation system

Mi [ˌem 'aɪ] Soviet-designed series of helicopters; **Mi-6** = transport helicopter known to NATO as the Hook; **Mi-8** = transport helicopter known to NATO as the Hip; **Mi-17** = updated version of the Hip Mi-8 transport helocopter; **Mi-24** = attack helicopter known to NATO as the Hind; **Mi-28** = attack helicopter known to NATO as the Havoc; **Mi-35** = updated version of the Hind Mi-24 attack helicopter

MIA [ˌem aɪ 'eɪ] = MISSING IN ACTION

MICLIC ['mɪklɪk] *noun US* = MINE-CLEARING LINE CHARGE apparatus for clearing a lane through a minefield, consisting of a long length of flexible tube filled with explosive, which is fired into the minefield by means of a rocket, and then detonated; *see also* GIANT VIPER

microphone ['maɪkrəfəʊn] *noun* **(a)** instrument (forming part of an intercom, radio or telephone mouthpiece) which converts sound into electromagnetic waves **(b)** mouthpiece of an intercom or radio

MID = MENTIONED IN DISPATCHES

mid-air ['mɪd eə] *noun, adjective & adverb* while flying, in the air; ***the helicopters collided in mid-air***; ***a mid-air collision***

middle distance [ˌmɪdl ˈdɪstəns] *noun* area half way between an observer's location and the horizon

midshipman [ˈmɪdʃɪpmən] *noun GB* lowest officer rank in the navy (an officer in training)

MiG [mɪg] *noun* Soviet-designed series of fighter aircraft; **MiG-21** = fighter aircraft with secondary ground attack role known to NATO as Fishbed; **MiG-23** = multirole fighter known to NATO as Flogger; **MiG-25** = high-altitude interceptor known to NATO as Foxbat; **MiG-27** = updated version of Flogger; **MiG-29** = multirole fighter aircraft known to NATO as Fulcrum; **Mig-31** = strategic interceptor known to NATO as Foxhound

Mike [maɪk] thirteenth letter of the phonetic alphabet (Mm)

mike [maɪk] *noun (informal)* **(a)** microphone **(b)** mouthpiece of an intercom or radio

Mil = MILITARY

mil [mɪl] *noun* unit of measurement for angles or bearings (NOTE: there are 6,400 mils in a circle. 6,400 mils are the equivalent of 360 degrees)

> COMMENT: many armies use **mils** instead of **degrees** in order to measure bearings, because they offer greater precision

Milan [mɪˈlæn] *noun* French/German-designed wire-guided anti-tank missile (ATGW)

Milbank bag [ˈmɪlbæŋk bæg] *noun* fabric bag, which is designed to filter dirty water

mile [maɪl] *noun* unit of linear measurement, corresponding to 1,760 yards or 1,609 metres; **nautical** *or* **sea mile** = unit of linear measurement at sea, corresponding to 2,025 yards or 1,852 metres

mileage [ˈmaɪlɪdʒ] *noun* number of miles travelled

MILES [maɪlz] *noun* = MULTIPLE INTEGRATED LASER ENGAGEMENT SYSTEM American-designed battle-simulation technology which uses harmless laser beams to simulate the firing of weapons, and laser detectors to record hits

militarize [ˈmɪlɪtəraɪz] *verb* to adapt for military use

military [ˈmɪlɪtri] **1** *adjective* relating to the armed forces (such as air force, army and navy); **military grid reference system (MGRS)** =world-wide series of maps which are compatible with the GPS satellite navigation system; **military hospital** = hospital for wounded military personnel; **military police** = organization responsible for police duties within the armed forces; **military policeman (MP)** = member of the military police; ***a military policeman directed us to the Brigade RV***; **military service** = service in one of the armed forces; ***all men of 18 years and over and required to do two years' military service*** **2** *noun* **the military** = the armed forces (in general); ***the new government proposals are being resisted by the military***

militia [mɪˈlɪʃə] *noun* **(a)** *(historical)* military force which is raised to supplement the regular army in the defence of a state's sovereign territory, and which does not normally serve overseas **(b)** military-style police force (mainly responsible for maintaining public order)

MILOB [ˈmaɪlɒb] = MILITARY OBSERVER

mine [maɪn] **1** *noun* **(a)** explosive device which is buried in or placed on the surface of the ground, and is designed to detonate when a person steps on it or a vehicle drives over it (NOTE: also called a **land mine**); **mine-detector** = device designed to locate mines; **mine plough (US mine plow)** = implement which is fitted to the front of an armoured vehicle and is designed to break up the ground, in order to dig up or detonate mines; **mine tape** = white or florescent tape, designed for marking lanes through a minefield or for marking a boundary; **anti-personnel mine** = mine designed to injure or kill a person; **anti-tank mine** = mine designed to damage or destroy an armoured vehicle; **off-route mine** = explosive device which is placed at the side of a road or track, and is designed to fire an anti-tank projectile into a passing vehicle automatically **(b)** explosive device which is placed into or under water, and is designed to detonate when a boat or ship hits it or passes over it **(c)** tunnel which is dug in order to detonate an explosive charge under an enemy fortification **(d)** tunnel or large hole, which is dug into the ground in order to extract minerals **2** *verb* **(a)** to lay mines in the ground or in water; ***the road has been mined*** **(b)** to tunnel under an enemy fortification; ***the engineers are mining under the forward enemy trench positions*** **(c)** to extract minerals from the ground

minefield ['maɪnfiːld] *noun* area of ground or sea in which mines have been laid

minelayer ['maɪnleɪə] *noun* **(a)** aircraft which scatters mines onto the ground **(b)** ship which places mines into the water **(c)** vehicle which places mines into or on top of the ground

minesweeper ['maɪnswiːpə] *noun* ship which is designed to destroy floating and underwater mines

miniflare ['mɪnɪfleə(r)] *noun* small illumination flare, produced in several different colours, which is fired from a simple hand-held launcher and is used mainly for signalling

Ministry of Defence (MOD) [ˌmɪnɪstri əv dɪ'fens] *noun* British government department dealing with the armed forces; ***the Ministry of defence has refused to comment on the situation***; ***he works at the MOD*** (NOTE: the American equivalent is the **Department of Defense** *or* **Pentagon**)

minute ['mɪnɪt] *noun* unit of time, corresponding to sixty seconds or a sixtieth part of one hour

Mirage [mɪ'rɑːʒ] *noun* French-designed fighter aircraft

MIRV [mɜːv] *noun* = MULTIPLE INDEPENDENTLY-TARGETED RE-ENTRY VEHICLE ballistic missile which carries several warheads, each of which is delivered to a different target

misdemeanour [ˌmɪsdɪ'miːnə(r)] *noun* doing something illegal *or* in contravention of regulations (but usually not very serious); ***he's always going in front of his company commander for minor misdemeanours***

misfire [mɪs'faɪə] **1** *noun (of ammunition or weapons)* act of not firing; ***the battalions have reported a large number misfires with the new ammunition*** **2** *verb (of ammunition or weapons)* to fail to fire; ***his rifle misfired***

miss [mɪs] **1** *noun* act of missing a target; ***the FOO has reported a miss on the last fire mission*** **2** *noun* **(a)** to fail to hit a target; ***he missed the tank*** **(b)** to fail to keep an appointment; ***he missed the briefing***

missile ['mɪsaɪl *US* 'mɪsəl] *noun* explosive projectile, containing its own propellant and usually equipped with a guidance system to control its flight onto the target; ***the plane was brought down by an enemy missile***; **missile launcher** = apparatus or vehicle from which a missile is fired; **National Missile Defence (NMD)** = projected American defence shield against hostile ballistic missile attacks, involving the use of anti-missile-missiles and missile-destroying lasers carried in aircraft *or* deployed in space satellites; *see also* GUIDED MISSILE, HEAT-SEEKING MISSILE

misinformation [ˌmɪsɪnfə'meɪʃən] *noun* deliberate use of incorrect information in order to deceive *or* mislead

missing ['mɪsɪŋ] *adjective* separated from your unit during operational duty, and possibly dead, wounded or captured by the enemy; ***he has been reported missing***; **missing in action (MIA)** = referring to a serviceman who has gone missing during a battle

mission ['mɪʃn] *noun* specific task assigned to a tactical grouping; ***our mission is to capture the bridge***; **force-orientated mission** = mission with the principal aim of destroying enemy forces; **mission analysis** = process where a subordinate examines his mission in order to identify the intentions of his commander and thereby determine what action he should take or be prepared to take so that those intentions will be achieved; **mission creep** = alteration to the original mission, in response to a changing situation (eg. troops being deployed to a disaster area to help in the rescue work, and then having to assist the local authorities in restoring order because rioting has broken out over a shortage of food); **terrain-orientated mission** = mission with the principal aim of either capturing or holding ground; *see also* MOPP

> COMMENT: mission analysis is an essential part of directive command

mist [mɪst] *noun* **(a)** cloud of water vapour suspended in the air, close to the ground **(b)** something which looks like mist

mist up [ˌmɪst 'ʌp] *verb (of glass)* to become obscured by water vapour, so that visibility is obscured; ***my weapon sight keeps misting up***

misty ['mɪsti] *adjective* obscured by mist

Mk = MARK

MLRS [ˌem el ɑːr 'es] *noun* = MULTIPLE-LAUNCH ROCKET SYSTEM American-designed multiple rocket launcher (MRL)

MNC = MAJOR NATO COMMAND, MAJOR NATO COMMANDER

MO [ˌem 'əʊ] = MEDICAL OFFICER

mobile ['məʊbaɪl] *adjective* designed for movement; *US* **mobile army surgical unit (MASH)** = field hospital; **mobile defence** = defensive doctrine employing the principles of manoeuvre warfare; **mobile training team (MTT)** = US small group of instructors sent to give military training to the armed forces of another country; *compare* POSITIONAL DEFENCE, STATIC DEFENCE

> COMMENT: in mobile defence, the holding of ground is of secondary importance, since the primary objective is to destroy the cohesion of an attacking force through the aggressive use of manoeuvre

mobility [mə'bɪləti] *noun* ability of a military force to move from one place to another

mobilization [ˌməʊbəlaɪ'zeɪʃn] *noun* a state's preparations for war by bringing soldiers together and organizing equipment, ammunition, etc. so that armed forces are trained and ready for action; ***the government ordered a general mobilization***

mobilize ['məʊbəlaɪz] *verb* to prepare armed forces for war by bringing them together; ***they are worried because the neighbouring states have mobilized***; *compare* IMMOBILIZE

MOD [ˌem əʊ 'diː] *GB* = MINISTRY OF DEFENCE

model ['mɒd(ə)l] *noun* representation (roughly to scale) of the ground over which an operation will take place, which is used as an aid to briefing the participants

> COMMENT: an efficient commander will normally carry a simple modelling kit, consisting of lengths of ribbon, coloured card and other suitable objects (eg. houses *or* hotels from a Monopoly™ set) to assist him in this task

MOD Police [ˌem əʊ ˌdiː pə'liːs] *noun* *GB* civilian police force responsible for dealing with offences committed by civilians on property owned by the Ministry of Defence (eg. airfields, barracks, depots, training areas, etc)

Molotov cocktail [ˌmɒlətɒf 'kɒkteɪl] *noun* improvised incendiary device consisting of a bottle filled with petrol and fitted with a wick of fabric, which is lit and then thrown at a target; *see also* PETROL BOMB

momentum [mə'mentəm] *noun* forward movement of an advance or attack; ***if we do not take that position immediately, the whole attack will lose momentum***

monarch ['mɒnək] *noun* king or queen

monitor ['mɒnɪtə] **1** *verb* **(a)** *(of radio)* to listen to other people's radio transmissions, in order to know what is happening; ***we are monitoring B Company's net*** **(b)** *(of incidents or situations)* to find out what is happening, without getting involved; ***UN observers are monitoring the ceasefire*** **2** *noun* part of a surveillance or detection system which the operator looks at or listens to (eg. television screen in a CCTV system); ***this monitor is not working***

monsoon [mɒn'suːn] *noun* season of heavy rain in southern Asia

monument ['mɒnjuːmənt] *noun* large ornamental structure, usually built to commemorate an important person *or* historical event

moor ['mʊə] **1** *noun* large area of flat, uncultivated high ground, usually covered by heather or coarse grass; ***we advanced across the moor*** **2** *verb* to secure a boat or ship to an anchor or an object on the shore, in order to stop it drifting; ***they moored the ship near the entrance to the harbour***

moorland ['mʊələnd] *noun* terrain consisting mainly of moors

MOPP [mɒp] *noun US* = MISSION-ORIENTED PROTECTIVE POSTURE state of readiness against chemical attack, which determines what protective measures should be taken

morale [mə'rɑːl] *noun* mental attitude, in relation to happiness and confidence; ***morale is very high at the moment***; ***the enemy is suffering from low morale***

morphine ['mɔːfiːn] *noun* drug used to relieve pain

Morse (code) [mɔːs] *noun* international code, in which letters of the alphabet are represented by combinations of short signals (dots) and long signals (dashes); ***he sent the message in Morse***

mortal ['mɔːtl] *adjective (of wounds)* causing death; ***the wound is probably mortal***

mortally ['mɔːtəli] *adverb (of wounding)* causing death; ***he was mortally wounded in the battle***

mortar ['mɔːtə] **1** *noun* simple indirect-fire weapon, which is designed to fire projectiles at very high trajectories; **mortar fire controller (MFC)** = non-commissioned officer (NCO) from the mortar platoon, who is attached to a rifle company in order to direct mortar fire; **mortar line** = fire position used by several mortars **2** *verb* to fire at a target with a mortar; ***'Hullo 2 this is 22, am being mortared, wait out'***

> COMMENT: mortars are normally used by the infantry, and provide a unit with its own indirect fire support

mosque [mɒsk] *noun* building used for religious worship by Muslims, the followers of Islam

motorized ['məʊtəraɪzd] *adjective; (of troops)* equipped with vehicles

motor-rifle ['məʊtə ˌraɪfl] *adjective; (of infantry)* Soviet term for mechanized; ***we have a motor-rifle regiment facing us***

Motor Transport Officer *see* MTO

motorway ['məʊtəweɪ] *noun* large road, with several lanes, which is designed for fast traffic

mount [maʊnt] **1** *noun* supporting structure for a weapon or other equipment **2** *verb* **(a)** *(of attacks, campaigns, operations, etc.)* to plan, prepare and carry out; ***this is the biggest operation which has ever been mounted by the allies*** **(b)** *(of weapons and other equipment)* to attach to supporting structure; ***this machine-gun can be mounted on a tripod*** **(c)** *(historical)* to get onto a horse; ***he mounted and rode away*** **(d)** to get into a high vehicle; ***he was shot as he was mounting his tank***

mountain ['maʊntən] *noun* very high land, rising much higher than the land surrounding it; ***the plane crashed in the mountains***

mountainous ['maʊntənəs] *adjective; (of terrain)* consisting mainly of mountains

mounting ['maʊntɪŋ] *noun* **(a)** supporting structure for a weapon or other equipment; ***the gun has come loose from its mounting*** **(b)** action of preparing an operation, including the assembly of forces, embarkation into transport, etc.; **mounting area** = place where forces are assembled ready to be moved to the scene of an operation

mouseholing ['mæushəʊlɪŋ] *noun* FIBUA tactic of blowing a hole into the wall of a building in order to provide an entry for infantrymen; ***we cleared each house by mouseholing through the attic and then checking every room from top to bottom***

MOUT [maʊt] = *US* MILITARY OPERATIONS IN URBAN TERRAIN

movement ['muːvmənt] *noun* action of moving personnel, equipment or supplies from one place to another; **movement control** = planning and scheduling of the movement of personnel, equipment or supplies from one place to another

MP [ˌem 'piː] = MILITARY POLICEMAN

MP-5 [ˌem piː 'faɪv] *noun* German-designed 9mm sub-machine-gun; **MP-5K** = shortened version of the MP-5

MPH *or* **mph** [ˌem piː 'eɪtʃ] = MILES PER HOUR

MPI [ˌem piː 'aɪ] *noun* = MEAN POINT OF IMPACT centre of a grouping of shots fired at a target (especially when zeroing); ***your MPI is 4cm to the left of the aiming mark***

MRE [ˌem ɑː 'iː] = MEAL READY-TO-EAT American-produced individual ration of food

MRL [ˌem ɑːr 'el] = MULTIPLE ROCKET LAUNCHER

MSC = MAJOR SUBORDINATE COMMAND

MSF [ˌem es 'ef] = MÉDICINES SANS FRONTIÈRES

MSR [ˌem es 'ɑːr] = MAIN SUPPLY ROUTE

MST [ˌem es 'tiː] = MOBILE SURGICAL TEAM

MT [ˌem 'tiː] *noun GB* = MOTOR TRANSPORT all vehicles held by a unit; ***the battalion's MT will move to the exercise area by rail***; **MT Platoon** = platoon consisting of the drivers of a unit's wheeled transport vehicles

MT-LB [ˌem tiː el 'biː] *noun* Soviet-designed multi-purpose armoured personnel carrier (APC)

MTO [ˌem tiː 'əʊ] *noun GB* = MOTOR TRANSPORT OFFICER officer with overall responsibility for a unit's vehicles

MTT ['em tiː tiː] *noun US* = MOBILE TRAINING TEAM small group of instructors sent to give military training to the armed forces of another country

mud [mʌd] *noun* wet soil

muddy ['mʌdi] *adjective* **(a)** covered with mud **(b)** *(of ground)* consisting mainly of mud

mudflat ['mʌdflæt] *noun* wide area of mud along the edge of a river *or* lake *or* sea

mufti ['mʌfti] *noun (informal)* civilian clothing; ***he was in mufti***

mujahideen [ˌmuːdʒəhə'diːn] *noun Arabic* muslim guerilla, who believes that he is fighting a "holy war" in defence of the Islamic faith; *see also* JIHAD

mule [mjuːl] *noun* four-legged animal produced by mating a horse with a donkey, which is suitable for carrying loads over rough terrain

> COMMENT: mules were used extensively as transport-animals during the Second World War

muleteer [ˌmjuːlɪ'tɪə(r)] *noun* person in charge of a mule; ***the muleteers will be attached to Headquarter Company***

mullah ['mʊlə(r)] *noun* muslim priest; ***mullahs have been inciting the crowd***

multinational [ˌmʌltɪ'næʃnl] *adjective; (of military forces)* consisting of contingents from several different nations; ***the multinational force is made up of British and German divisions***; **multinational logistics** = supporting operations which are made up of forces from different nations

multiple ['mʌltɪpl] **1** *adjective* consisting of several parts; **multiple integrated laser engagement system** = *see* MILES; **multiple-launch rocket system** = *see* MLRS; **multiple rocket launcher (MRL)** = rocket laucher which fires several rockets at the same time **2** *noun GB* patrol, consisting of several teams which move along different routes, but are always able to support each other

multi-purpose [ˌmʌlti'pɜːpəs] *adjective* having several different purposes

multirole [ˌmʌlti'rəʊl] *adjective* having several different roles or functions; ***the Americans are developing a multirole fighter***

munition [mjuː'nɪʃn] *noun* any type of explosive device which is used as a weapon (such as a bomb, grenade, mine, projectile, etc.); ***the bomb hit a munitions store***; **munitions factory** = factory where bombs, shells, etc., are made

murder ['mɜːdə] **1** *noun* unlawful killing of a person; ***he was accused of murder*** **2** *verb* to kill a person without legal justification; ***the mayor of the town has been murdered***

Murphy's Law ['mɜːfɪz lɔː] *noun* popular idea that if anything can go wrong, then it probably will go wrong; ***you forgot to take Murphy's Law into account***; *see also* SOD'S LAW

musket ['mʌskɪt] *noun (historical)* obsolete firearm, similar to a rifle, where the bullet and propellant are loaded through the muzzle

mustard gas ['mʌstəd ˌgæs] *noun* gas which causes blisters on exposed skin

muster ['mʌstə] **1** *noun* act of assembling troops; ***the muster will take place at 0800hrs***; **zulu muster** = location in the field where vehicles are kept when not in use; **muster parade** = assembly of all soldiers at the beginning of the day to receive instructions; **muster-roll** = official list of all people serving in a unit **2** *verb* **(a)** to assemble troops; ***they mustered all available forces*** **(b)** *(of troops)* to come together; ***B Company will muster at 0800hrs***

mutilate ['mjuːtɪleɪt] *verb* **(a)** to inflict a wound which causes severe physical damage (especially the loss of a limb or organ); ***his face was mutilated by shrapnel*** **(b)** to deliberately inflict several serious wounds on a person or dead body; ***the women's bodies had been mutilated***

mutineer [ˌmjuːtɪ'nɪə] *noun* serviceman who mutinies

mutiny ['mjuːtɪni] **1** *noun* rebellion by servicemen against the military authorities; ***the mutiny has spread to other units*** **2** *verb* to take part in a mutiny; ***units of the navy have mutinied***

mutual support [ˌmjuːtʃʊəl sə'pɔːt] *noun* ability of two *or* more defensive positions *or* groupings *or* vehicles to give fire support to each other; ***this is a bad position because there is no mutual support between the three platoons***

mutually supporting [ˌmjuːtʃʊəlɪ sə'pɔːtɪŋ] *adjective* able to provide mutual support; ***the enemy was well dug-in in mutually supporting positions***

muzzle ['mʌzl] *noun* open end of a gun's barrel; **muzzle velocity** = speed of a projectile, at the moment when it leaves the muzzle of a weapon

NOVEMBER - Nn

NAAFI ['næfi] *noun GB* = NAVY, ARMY AND AIR FORCE INSTITUTES **(a)** organization responsible for servicemen's welfare **(b)** bar, canteen or shop which is run by the NAAFI; **NAAFI break** = break to have a cup of tea or coffee (NOTE: the American equivalent is the **PX**)

NAIAD ['naɪæd] *noun* = NERVE AGENT IMMOBILISED ENZYME ALARM AND DETECTOR instrument which detects and warns of the presence of a nerve agent

napalm ['neɪpɑːm] *noun* jelly made from petroleum, which is used as an incendiary weapon

NAPS [næps] *plural noun* = NERVE AGENT PRE-TREATMENT SET tablets, which are taken at regular intervals when the chemical threat is high, in order to give the body some resistance against the effects of nerve agents; ***we've been ordered to start taking NAPS***

nation ['neɪʃn] *noun* community of people, united by a common language and history, who form a state

national ['næʃənl] *adjective* relating to a nation or state; **national logistic support** = logistic support given by a nation to its own forces, or as part of a multinational force; **national service** = compulsory military service

nationalist ['næʃənəlɪst] **1** *adjective* seeking national independence **2** *noun* person who is seeking national independence

nationality [ˌnæʃə'nælɪti] *noun* being a citizen of a particular state; ***his nationality is Czech***

NATO ['neɪtəʊ] *noun* = NORTH ATLANTIC TREATY ORGANIZATION military alliance consisting of the USA, Canada and most western European states, which was originally formed in order to counter the growing threat from the Soviet Union, following the end of the Second World War, and which now acts in a peacekeeping role in support of the United Nations; **NATO accounting unit** = fictitious currency used to calculate costs of construction, manpower, etc., which are funded from a common NATO source; **NATO commander** = commander of part of a NATO force; *see also* NON-NATO COMMANDER; **NATO military authority** = organization or HQ which is part of a NATO command

NATO - T [ˌneɪtəʊ 'tiː] *noun* method of illuminating a helicopter landing zone at night, by positioning five torches in the shape of a capital letter T, with the stem pointing downwind

nautical mile [ˌnɔːtɪkl 'maɪl] *noun* unit of linear measurement at sea, corresponding to 2,025 yards or 1,852 metres; *see also* SEA MILE

naval ['neɪvl] *adjective* **(a)** relating to a navy; ***the British naval capacity in the Atlantic*** **(b)** relating to war at sea; ***it was the biggest naval engagement of the war***; **naval gunfire support (NGS)** = indirect fire provided by warships in support of ground forces

navigable ['nævɪgəbl] *adjective; (of canals, estuaries, rivers, etc.)* allowing the passage of boats and ships

navigate ['nævɪgeɪt] *verb* to find your way from one location to another

navigation [ˌnævɪ'geɪʃn] *noun* act or process of navigating

navigator ['nævɪgeɪtə] *noun* aircrew member who is responsible for navigation

navy ['neɪvi] *noun* branch of a state's armed forces which operates at sea; ***ships from several NATO navies were patrolling the area***; **the Royal Navy (RN)** = the British navy; **the United States Navy (USN)** = the navy of the USA

NBC [ˌen biː 'siː] *adjective* = NUCLEAR, BIOLOGICAL AND CHEMICAL relating to nuclear, biological and chemical warfare; **NBC state** = degree of possibility or probability that the enemy will mount a nuclear, chemical or biological attack; **NBC suit** = special clothing, which provides protection from radioactive fallout and biological and chemical weapons (NOTE: also called a **Noddy suit**)

NCO [ˌen siː 'əʊ] = NON-COMMISSIONED OFFICER; ***he sent a squad of men with an NCO to investigate*** (NOTE: plural is **NCOs** [ˌen siː 'əʊz])

ND [ˌen 'diː] = NEGLIGENT DISCHARGE

need-to-know [ˌniːd tə 'nəʊ] *adjective; informal* relating to secret information which is only given to those people who actually need to know it; ***I'm afraid that's need-to-know at the moment, and you don't need to know it***

negative ['negətɪv] **1** *adjective* incorrect; ***that is negative*** **2** *adverb (radio terminology)* that is incorrect; ***'Hullo 23D this is 2, are you in position, over?' - '23D, negative, over'***; *compare* AFFIRMATIVE

negligence ['neglɪdʒəns] *noun* **(a)** lack of attention or proper care; ***the device malfunctioned because of negligence by the operator***; ***the accident was due to the negligence of the safety staff*** **(b)** failure to carry out correct procedure; ***the officer was accused of negligence***

negligent ['neglɪdʒənt] *adjective* behaving with negligence; **negligent discharge (ND)** = unintentional firing of a weapon

nerve agent ['nɜːv ˌeɪdʒənt] *noun* chemical agent designed to attack a person's central nervous system; ***the enemy are using a nerve agent***

net [net] *noun* **(a)** woven material made out of knotted cord, with large holes; **camouflage-net** = net decorated with pieces of fabric, which is used to conceal a vehicle, a piece of equipment or structure **(b)** net used as a container (especially for underslung loads) **(c)** *(of radio)* group of radio users (for example sub-units of a battalion, company or platoon), who are transmitting and receiving on the same frequency; ***we are monitoring the battalion net***

neutral ['njuːtrəl] *adjective (of states)* not allied to or supporting either of two opposing sides in a war or conflict

neutrality [njuː'trælɪti] *noun* state of being neutral; **armed neutrality** = use of armed force by a neutral state, in order to prevent interference by the military forces of other states which are involved in a war

neutralization [ˌnjuːtrəlaɪ'zeɪʃn] *noun* act of neutralizing

neutralize ['njuːtrəlaɪz] *verb* to fire at an enemy, in order to prevent him using his weapons; *see also* SUPPRESS

> COMMENT: when neutralizing an enemy, it is not necessary to kill him. The object is simply to make him keep his head down

neutron bomb ['njuːtrɒn bɒm] *noun* tactical nuclear weapon which produces high levels of radiation but little blast, thereby causing high loss of life but comparatively little damage to buildings and installations

next of kin [ˌnekst əv 'kɪn] *noun* person's closest living relative, who must be informed in the event of death *or* injury *or* other misfortune

> COMMENT: for most married people, the next of kin is their husband *or* wife. For most unmarried people, it is one of their parents.

NGO [ˌen dʒiː 'əʊ] *noun* = NON-GOVERNMENTAL ORGANIZATION humanitarian organization, which is not sponsored by any particular government, and can thus claim neutral status in a war zone; ***several NGOs are already established in the region***

NGS [ˌen dʒiː 'es] = NAVAL GUNFIRE SUPPORT

nickname ['nɪkneɪm] *noun* **(a)** name (often humorous), which is given to *or* used by a person instead of his real name; ***his***

nickname is Ferret **(b)** codename **(c)** codeword

nicknumber ['nɪknʌmbə] *noun* number used to denote a location on the ground

night [naɪt] *noun* period of darkness between sunset and sunrise; **night-observation device** *or* **night-viewing device** *or* **night-vision device** = optical equipment which utilizes night-viewing technology; **night-viewing technology** = optical equipment which makes it possible to see things at night (for example image intensification, infrared, thermal imaging, etc.); **night-sight** = weapon sight which utilizes night-viewing technology

Nighthawk ['naɪthɔːk] *see* F-117

Nimrod ['nɪmrɒd] *noun* British-designed multi-purpose aircraft, based on a passenger airliner; **Nimrod MR MK 2** = maritime patrol and anti-submarine aircraft; **Nimrod R MK1** = electronic intelligence (ELINT) aircraft

Nissen hut ['nɪsən ˌhʌt] *noun GB* semi-permanent structure with a curved roof of corrugated iron which extends down to ground level; ***your objective is the group of Nissen huts at the northern end of the airfield***

nitroglycerine [ˌnaɪtrəʊ'glɪsəriːn] *noun* chemical compound, used to make dynamite

NLT [ˌen el 'tiː] = NOT LATER THAN; ***A Company requests fuel replen NLT 1645hrs***

NMD [ˌen em 'diː] *noun US* = NATIONAL MISSILE DEFENCE projected American defence shield against hostile ballistic missile attacks, involving the use of anti-missile-missiles and missile-destroying lasers carried in aircraft or deployed in space satellites

> COMMENT: NMD is a new development of the unrealised **Strategic Defence Initiative (SDI)** *or* **Star Wars programme**

NOD [nɒd] = NIGHT OBSERVATION DEVICE

noddy suit ['nɒdi ˌsuːt] *noun (slang) GB* NBC suit

NODUF *or* **noduf** ['nəʊdʌf] *adverb* radio terminology indicating that the message refers to a real situation rather than an exercise scenario; ***hello 2 this is 22, NODUF, casualty with gunshot wound at grid 332598, request CASEVAC, over !; 22 has just sent a noduf message saying that someone has been shot***

NOE [ˌen əʊ 'iː] = *US* NONCOMBATANT EVACUATION OPERATION

no-fly zone [nəʊ 'flaɪ ˌzəʊn] *noun* airspace defined by a state or by international agreement, which the aircraft of another state are not allowed to enter

no-go area [nəʊ 'gəʊ ˌeərɪə(r)] *noun* area which is too dangerous for routine patrolling by security forces *or* peacekeepers; ***the eastern part of the town is now a complete no-go area***

nomad ['nəʊmæd] *noun* member of an ethnic group which has no permanent home, but travels around from one place to another (often with herds of animals)

nomadic [nəʊ'mædɪk] *adjective* relating to nomads

no-man's-land ['nəʊmænz lænd] *noun* area of ground between the forward positions of two opposing forces; ***a patrol was sent out into no-man's-land***

non-com [nɒn'kɒm] *noun (informal)* non-commissioned officer; *see also* NCO

noncombatant [nɒn'kɒmbətənt] *noun* person who has no military role in a war or conflict (ie a civilian); **noncombatant status** = status of servicemen who are not directly involved in the fighting or the support of those who fight (such as chaplains and medical personnel) (NOTE: the opposite is **combatant**)

non-commissioned officer (NCO) [ˌnɒnkəmɪʃənd 'ɒfɪsə] *noun* serviceman who holds a supervisory rank, but is not a commissioned officer (such as a corporal, sergeant, colour sergeant, etc.); ***he sent a squad of men with an NCO to investigate***

non-NATO commander [nɒnˌneɪtəʊ kə'mɑːndə] *noun* commander from a country which is not a member of NATO but who is part of the command structure of a NATO force

non-persistent [nɒnpə'sɪstənt] *adjective; (of chemical agents)* designed to disperse in the air after a few minutes

north [nɔːθ] **1** *noun* **(a)** one of the four main points of the compass, corresponding to a bearing of 0 degrees or 0 mils **(b)** area to the north of your location; ***the enemy are approaching from the north*** **(c) the North** = the northern part of a country **2** *adjective* relating to north; ***the landings took place on the north coast of France***; **north wind** = wind blowing from the north **3** *adverb* towards the north; ***the enemy is moving north***

North Atlantic Treaty [ˌnɔːθ ətˌlæntɪk 'triːti] *noun* treaty signed in Washington in 1949 by which NATO was set up

North Atlantic Treaty Organization [ˌnɔːθ ətˌlæntɪk 'triːti ɔːgənaɪˌzeɪʃn] *see* NATO

northbound ['nɔːθbaʊnd] *adjective* moving or leading towards the north; ***a northbound convoy***

northerly ['nɔːðəli] *adjective* **(a)** towards the north; ***they set off in an northerly direction*** **(b)** *(of wind)* from the north **(c)** situated towards the north; ***the most northerly point of a country***

northern ['nɔːðən] *adjective* relating to the north; ***the northern part of the country***

northing ['nɔːðɪŋ] *noun* **(a)** horizontal line of a map grid **(b)** one of the coordinates running from bottom to top across a map; *compare* EASTING

northward ['nɔːθwəd] **1** *adjective* towards the north; ***a northward direction*** **2** *adverb US* towards the north; ***the army is moving northward***

northwards ['nɔːθwədz] *adverb* towards the north; ***the army is moving northwards***

notice ['nəʊtɪs] **1** *noun* **(a)** written document which is displayed in order to pass on information or a warning; ***notices about the curfew have been displayed throughout the town*** **(b)** specified period of time before something happens; **notice to move (NTM)** = period within which a person or unit must be ready to move; ***we were at five minutes' notice to move***; **at short notice** = with little warning; ***the platoon must be ready to move at short notice*** **2** *verb* to become aware of something; ***he noticed that a tank was moving up the road***

notification [ˌnəʊtɪfɪ'keɪʃn] *noun* act of informing someone

notify ['nəʊtɪfaɪ] *verb* to inform someone; ***we were not notified of the change in plan***

November [nə'vembə] fourteenth letter of the phonetic alphabet (Nn)

NTM [ˌen tiː 'em] = NOTICE TO MOVE

nuclear ['njuːkliə] *adjective* relating to the use of nuclear energy; **nuclear energy** = energy produced by a nuclear reaction; **nuclear-powered** = driven or propelled by nuclear power; **nuclear submarine** = submarine driven by nuclear power; **nuclear warfare** = warfare involving the use of nuclear weapons; **nuclear weapon** = bomb or missile or other device which utilises the release of nuclear energy; *see also* NBC, TACTICAL

> COMMENT: although their meanings are not identical, the word **nuclear** has now superseded **atomic** for most general contexts

nuke [njuːk] *(slang)* **1** *noun* nuclear weapon **2** *verb* to attack with nuclear weapons; ***let's nuke them!***

NVG [ˌen viː 'dʒiː] = NIGHT-VIEWING GOGGLES

OSCAR - Oo

oasis [əʊ'eɪsɪs] *noun* place in a desert where water can be found; ***after three days they came to an oasis*** (NOTE: plural is **oases** [əʊ'eɪsiːz])

obey [əʊ'beɪ] *verb* to carry out a command or order; ***the soldiers were court-martialled for refusing to obey orders***

objective [əb'dʒektɪv] *noun* **(a)** something which must be accomplished or achieved; ***our first objective is to improve radio security throughout the brigade*** **(b)** location or position which must be destroyed or captured; ***the platoon will reorganize on the objective***

obscure [əb'skjʊə] *verb* to make something difficult to see; ***the objective is obscured by smoke***

observation [ˌɒbzə'veɪʃn] *noun* act of observing

observation post (OP) [ˌɒbzə'veɪʃn ˌpəʊst] *noun* **(a)** covert position from which an area of ground may be observed; ***an enemy OP has been located at grid 882014*** **(b)** troops occupying an observation post; ***all of the OP were captured***

observe [əb'zɜːv] *verb* to look at or watch something; ***we can observe the road from that hill***

observer [əb'zɜːvə] *noun* person who observes; **military observer** = person (usually a serviceman), who observes the activities of another state's armed forces (especially on operations)

obsolescent [ˌɒbsə'lesənt] *adjective* no longer in general use (because it is in the process of being replaced by something more modern); ***the enemy reserves are equipped with obsolescent tanks*** (NOTE: although it is not correct, many people use the word **obsolete** in this context)

obsolete ['ɒbsəliːt] *adjective* not used any more (because it has been replaced by something more modern); ***that tank is now obsolete*** (NOTE: when something is becoming obsolete, but is still capable of being used (and repaired), the correct term is actually **obsolescent.** However, many people use the word **obsolete** in this context as well.)

obstacle ['ɒbstəkl] *noun* natural or man-made feature, which hinders or obstructs the movement of a person or vehicle

> COMMENT: **natural obstacles** include features such as woods, rivers and high ground, while **man-made obstacles** include features such as built-up areas, canals and railway embankments. In addition, obstacles can be specially constructed: for example craters, barbed-wire entanglements and minefields

obstruct [əb'strʌkt] *verb* **(a)** to make it difficult or impossible for a person or vehicle to pass; ***the road was obstructed by a burning tank*** **(b)** to make it difficult or impossible for a person to carry out a task or duty; ***he was constantly obstructed by his platoon sergeant***

obstruction [əb'strʌkʃn] *noun* **(a)** act of obstructing; ***your platoon commander has accused you of obstruction*** **(b)** something which obstructs; ***we used explosives to clear the obstruction***; *see also* OBSTACLE

OC [ˌəʊ 'siː] *noun GB* = OFFICER COMMANDING term usually applied to an officer commanding a company or equivalent-sized grouping

occupation [ˌɒkju'peɪʃn] *noun* use of military forces to take possession of and then control territory belonging to another state; **forces of occupation** = military forces, which occupy territory belonging to another state; **foreign occupation** = occupation of territory by military forces of another state

occupy ['ɒkju:paɪ] *verb* **(a)** *(of tactical positions)* **(i)** to move into a position; **(ii)** to be in a position; ***we will occupy the position at 1700hrs***; ***the Coldstream Guards are occupying the forward positions*** **(b) to occupy a territory** = **(i)** to use military force to take possession of territory belonging to another state; **(ii)** to use military forces to control territory belonging to another state; ***Ruritania has occupied the neighbouring country***; ***the town has been occupied by NATO troops***

offence *US* **offense** [ə'fens] *noun* **(a)** aggressive military action (such as advance, attack, invasion, etc.); ***the country has been condemned for its use of offence to control the indigenous population***; *compare* DEFENCE **(b)** illegal act; ***he has committed several offences*** (NOTE: **offensive** is more common for describing aggressive military action)

offense [ə'fens] *see* OFFENCE

offensive [ə'fensɪv] **1** *adjective* relating to aggressive military action (such as advance, attack, invasion, etc.); ***their strategy concentrates mainly on offensive operations*** **2** *noun* aggressive military action (such as advance, attack, invasion, etc.); ***the rebels are planning a new offensive***; **to go or to move onto the offensive** = to change from defence to offence; *compare* DEFENSIVE; ***The transition from the defensive to the offensive is one of the most delicate operations in war - Napoleon***

office ['ɒfɪs] *noun* **(a)** room used for administrative and clerical work; ***the CO is in his office*** **(b)** administrative or supervisory position within an organization; ***he is unsuitable for this office*** (NOTE: a person who holds an office in the armed forces is known as an **officer**, while a person who holds an office in a civilian organization (especially one involved in government or local administration) is usually known as an **official**)

officer ['ɒfɪsə] *noun* **(commissioned) officer** = a serviceman with a supervisory rank, who derives his authority from a commission (such as a lieutenant, captain, major, etc.); ***two officers led the assault on the enemy position;*** *GB* **officer commanding (OC)** = term usually applied to an officer commanding a company or equivalent-sized grouping; *GB* **commanding officer (CO)** = term usually applied to an officer commanding a battalion or equivalent-sized grouping; **duty officer** *or* **officer of the day** *or* **orderly officer** = officer assigned by his unit to deal with incidents and carry out various routine tasks during a specified period; **field officer** = army officer of any rank above captain and below general; **non-commissioned officer (NCO)** = serviceman who holds a supervisory rank, but is not a commissioned officer (such as a corporal, sergeant, colour sergeant, etc.); *see also* PETTY OFFICER, WARRANT OFFICER (NOTE: despite their titles, non-commissioned officers, petty officers and warrant officers are not officers)

official [ə'fɪʃl] **1** *adjective* approved or authorized by someone who holds an office in an organization **2** *noun* person who holds an office in a civilian organization (especially one involved in government or local administration)

off limits [ˌɒf 'lɪmɪts] *adjective* prohibited (to the persons specified); ***this pub is off limits to officers and NCOs***

offr = OFFICER

off-route mine [ˌɒf ru:t 'maɪn] *noun* explosive device which is placed at the side of a road or track, and is designed to fire an anti-tank projectile into a passing vehicle automatically

O Group ['əʊ ˌgru:p] *noun* = ORDERS GROUP **(a)** meeting, where a commander issues operational orders to his subordinate commanders; ***the O Group is at 2200hrs*** **(b)** people who attend an O Group; ***the O Group was informed that the CO had been killed***

oil [ɔɪl] **1** *noun* **(a)** thick liquid refined from petroleum, which is used to lubricate machinery and protect metal from rust and corrosion **(b)** petroleum, a liquid mineral substance which is extracted from the ground and then refined to produce petrol, diesel, kerosene and lubricating oil; **oilfield** = area where petroleum is extracted from the ground; **oil rig** = structure which supports equipment for extracting petroleum from an oil well; **oil slick** = large patch of oil *or* petroleum floating on water (usually released from a damaged ship as a result of an accident *or* enemy action); **oil well** = hole in the ground from which petroleum is extracted **2** *verb* to apply oil to an object weapons should be cleaned and oiled before they are returned to the armoury

OK [ˌəʊ ˈkeɪ] *adverb* **(a)** I have understood your instruction **(b)** all right or satisfactorily; ***that went OK*** **(c)** *(as a question)* **(i)** do you understand? **(ii)** is everything all right?

OMG [ˌəʊ em ˈdʒiː] = OPERATIONAL MANOEUVRE GROUP

one-up [ˌwʌn ˈʌp] *adverb* tactical formation in which one sub-unit is leading as point, and the other two are following abreast of each other; ***we'll be advancing one-up***; *compare* TWO-UP

> COMMENT: this formation is suitable for an advance to contact

OOB = OUT OF BOUNDS

OOM = ORDER OF MARCH

OOTW = OPERATIONS OTHER THAN WAR

OP [ˌəʊ ˈpiː] **1** = OBSERVATION POST **2** = OUTPOST

op [ɒp] *noun (informal)* operation; **op order** = document, containing detailed instructions for a military operation (NOTE: short for **operation order**)

OP/Ack [ɒp ˈæk] *noun GB* assistant to a forward observation officer (FOO); *also known as a* SURVEYOR

open city [ˌəʊpən ˈsɪtɪ] *noun* city which is abandoned to the enemy, in order to avoid the serious destruction and loss of life, which would result from trying to defend it; ***Vienna has been declared an open city***

OPCON [ˈɒpkɒn] = OPERATIONAL CONTROL

open fire [ˌəʊpən ˈfaɪə] *verb* to start shooting; ***the guerillas opened fire on our platoon***

operate [ˈɒpəreɪt] *verb* **(a)** to carry out military activity; ***enemy special forces are operating in this area*** **(b)** *(medical)* to carry out surgery; ***we will have to operate in order to remove the bullet*** **(c)** to work the controls of an apparatus or device; ***he operates the ship's sonar equipment***; ***he has not been trained to operate this equipment*** **(d)** *(of an apparatus or device)* to work properly; ***the mechanism failed to operate***

operation [ˌɒpəˈreɪʃn] *noun* **(a)** act of operating; ***he has not been trained in the operation of this equipment*** **(b) operation (op)** = planned military task; ***this will be a covert operation***; **operations (ops)** = moving troops, equipment, etc., as part of a planned military task; **combined operations** = **(i)** operations involving various branches of the armed forces (army and marines, for example); **(ii)** *US* operations carried out in conjunction with the armed forces of other states (NOTE: the Americans refer to combined arms operations as joint operations); **operations (ops) officer** = officer responsible for the coordination and administration of a unit or grouping's operational tasks; **concept of operations** = general outline of how an operation is intended to proceed; **on operations** = on operational service **(c)** *(medical)* act of surgery; ***he will need an operation to remove the bullet***

> COMMENT: the department responsible for operations in a headquarters is known as **G3**

operational [ˌɒpəˈreɪʃnl] *adjective* relating to military operations; **operational command** = authority given to a commander to organize tasks, deploy personnel, etc., as he feels necessary to carry out an operation; **operational control** = authority given to a commander to direct the forces under his command so as to carry out the mission that has been assigned to him; **operational manoeuvre group (OMG)** = Soviet armoured grouping designed to exploit a breakthrough; **operational mobility** =

ability of forces to move rapidly from place to place; **operational service** = service involving the possibility of real combat (as opposed to peacetime soldiering) (NOTE: in the British armed forces, the term **operational service** is used to describe counter-insurgency and peacekeeping operations. When a state of war exists, the term **active service** is used instead)

operator ['ɒpəreɪtə] *noun* person who operates an apparatus or device; ***he is the platoon commander's radio-operator***

OPFOR [ˌɒp 'fɔː] *noun* = OPPOSING FORCES enemy on a training exercise

oppo ['ɒpəʊ] *noun (informal)* = OPPOSITE NUMBER colleague who holds the same position as you in another unit *or* sub-unit; ***my oppo in 6 Platoon was killed***

oppose [ə'pəʊz] *verb* **(a)** to be hostile to someone **(b)** to offer resistance; ***we were opposed by a group of well-armed snipers*** **(c)** to disagree with someone; ***the general is opposed to our plan***

opposition [ˌɒpə'zɪʃn] *noun* **(a)** act of opposing; ***there was some opposition to the CO's plan*** **(b)** resistance; ***the brigade has met with little opposition so far***

ops [ɒps] = OPERATIONS

ops officer ['ɒps ˌɒfɪsə] = OPERATIONS OFFICER

Ops Room ['ɒps rʊm] *noun* = OPERATIONS ROOM command post in a permanent base location

OPSCHED ['ɒpʃed *or* 'ɒpsked] = OPERATION SCHEDULE

OPSEC ['ɒpsek] = OPERATIONAL SECURITY

Opso ['ɒpsəʊ] *noun GB* air-force operations officer

optic ['ɒptɪk] **1** *adjective* relating to the eyes or vision; **optic sight** = weapon sight which gives the firer a magnified image of an aiming mark or target **2** *noun* **optics** = optical equipment (such as imaging equipment, optic sights, periscopes, etc.); ***the tank's optics were damaged by artillery fire***

optical ['ɒptɪkl] *adjective* relating to the eyes and vision; **optical device** = device which a person looks through (usually as a means of improving visibility)

option ['ɒpʃn] *noun* one of two or more alternative courses of action; ***we have no option but to withdraw***

OPV [ˌəʊ piː 'viː] *noun* = OBSERVATION POST VEHICLE armoured personnel carrier (APC) used by a forward observation officer (FOO)

OR [ˌəʊ 'ɑː] = OTHER RANK

ORBAT ['ɔːbæt] = ORDER OF BATTLE

order ['ɔːdə] **1** *noun* **(a)** instruction or command; ***I gave you an order!***; **op order** = document containing detailed instructions for a military operation; **standing orders** = set of rules and regulations relating to duties and discipline; **warning order** = message which warns a unit or sub-unit of a future operation or task, and provides sufficient information for the unit to start making its preparations **(b)** position or sequence in which things or events are arranged; **order of battle (ORBAT)** = arrangement of people, vehicles or sub-units as a tactical grouping; *see also* ORGANIZATION; **order of march (OOM)** = sequence in which the sub-units of a grouping move (either on foot or by vehicle) from one location to another **(c)** good behaviour; ***there has been a breakdown of law and order in the town*** **(d)** state in which things are correct; ***is everything in order?*** **(e)** style of dress or equipment; **fighting order** = equipped with webbing only; **marching order** = equipped with webbing and bergen **(f)** tidiness; ***you are in bad order*** **2** *verb* to tell someone to do something; ***he ordered the platoon to load their weapons***

orderly ['ɔːdəli] **1** *adjective* **(a)** disciplined or under control; ***the brigade carried out an orderly withdrawal*** **(b)** relating to the execution of orders; **orderly officer** = duty officer; **orderly room** = administrative office of a unit **2** *noun* **(a)** serviceman who delivers messages and carries out various simple tasks in a headquarters **(b)** serviceman who cleans an officer's kit; *see also* BATMAN

orders ['ɔːdəz] *noun* **(a)** detailed instructions for an operation, given by a commander to his subordinates; **orders group (O Group)** = meeting, where a commander issues operational orders to his subordinate commanders **(b)** unit or sub-unit parade, where disciplinary matters are dealt with; ***Company Commander's orders are at 1000hrs*** **(c)** daily document produced by a unit or sub-unit, containing a programme of the day's events and any other information which may be important; ***it was written on battalion orders***

ordnance ['ɔːdnəns] *noun* **(a)** military equipment and supplies in general **(b)** weapons and munitions; **explosive ordnance** = general term for any projectile or device which contains an explosive substance or which uses an explosive substance as its propellant

Org = ORGANIZATION

organization [ˌɔːgənaɪ'zeɪʃn] *noun* **(a)** act of organizing; ***he is responsible for the organization of our sports competition*** **(b)** arrangement of people, vehicles or sub-units as a grouping; ***the organization of an infantry battalion consists of three rifle companies, a support weapons company, a headquarter company and a battalion headquarters***; *see also* ORDER OF BATTLE (ORBAT) **(c)** group of people who are arranged into smaller groups or departments and are given different responsibilities and tasks, so that they can work together for a common purpose; ***how long have you been working for this organization?***

Organization for Security and Cooperation in Europe (OSCE) international organization of 55 member states, including all European countries and the USA and Canada, with the aim of reducing tension and solving international problems within Europe

organize ['ɔːgənaɪz] *verb* **(a)** to arrange a group of people into smaller groups or departments with individual responsibilities and tasks, so that they can work together for a common purpose; ***an infantry battalion is usually organized into three rifle companies, a support weapons company, a headquarter company and a battalion headquarters*** **(b)** *(of events)* to make all the necessary arrangements, so that an event can take place; ***he organized the battalion sports day***

orient ['ɔːrɪənt] *verb* *US* **(a) to orient yourself** = to establish your exact location **(b)** *(maps)* to hold a map, so that the top of the sheet is pointing towards north (NOTE: **orient - oriented - orientation**; British English is **orientate - orientated - orientation**)

orientate ['ɔːriənteɪt] *verb* *GB* **(a) to orientate yourself** = to establish your exact location **(b)** *(maps)* to hold a map, so that the top of the sheet is pointing towards north (NOTE: **orientate - orientated - orientation**; American English is **orient - oriented - orientation**)

orientation [ˌɔːriən'teɪʃn] *noun* **(a)** being aware of your exact location **(b)** action of orienting or orientating yourself **(c)** map-reading race, where competitors navigate their way from one location to another

Orion [ə'raɪən] *noun* American-designed multi-purpose aircraft, which is based on a passenger airliner; **Orion AEW & C** = airborne early warning and control aircraft, with a large disk-like antenna (radome) mounted on the fuselage; **Orion EP-3** = electronic intelligence (ELINT) aircraft; **Orion P-3** = anti-submarine and maritime patrol aircraft

Orthodox Church [ˌɔːθədɒks 'tʃɜːtʃ] *noun* eastern European form of Christianity, found in Greece, Russia, parts of the Balkans and the Near East; *compare* PROTESTANT CHURCH, ROMAN CATHOLIC CHURCH

Oscar ['ɒskə] fifteenth letter of the phonetic alphabet (Oo)

OSCE [ˌəʊ es siː 'iː] *noun* = ORGANIZATION FOR SECURITY AND COOPERATION IN EUROPE

OSO ['əʊsəʊ] *noun* *US* = OFFENSIVE SYSTEMS OFFICER aircrew member on a bomber who operates the aircraft's weapons systems; *compare* DSO

other rank (OR) [ˌʌðə 'ræŋk] *noun* *GB* serviceman who is not an officer (NOTE: American English is **enlisted man**)

Otomat ['ɒtəmæt] *noun* French/Italian-designed long-range anti-ship missile (ASM)

out [aʊt] *adverb (radio terminology)* this is the end of the conversation; ***2, roger, out.***; *compare* OVER; *see also* WAIT OUT

COMMENT: although 'over and out' is often heard in films, it is not correct radio procedure

outbrief ['aʊtbriːf] *noun* final briefing before aircrew get into their aircraft, including an update of weather conditions, last minute changes to situation, and equipment checks

outfit ['aʊtfɪt] *noun US* unit

outflank [aʊt'flæŋk] *verb* to manoeuvre around an enemy's flank; ***the enemy are outflanking us on the left***

outmaneuver [ˌaʊtmə'nuːvə] *see* OUTMANOEUVRE

outmanoeuvre *US* **outmaneuver** [ˌaʊtmə'nuːvə] *verb* to manoeuvre successfully against an enemy force which is trying to manoeuvre against you

out of area operation [ˌaʊt əv ˌeəriə ɒpə'reɪʃn] *noun* military operation conducted outside the area in which a state's armed forces usually operate (eg British troops operating in South America)

out of bounds (OOB) [ˌaʊt əv 'baʊndz] *adverb* where one is not allowed to go; ***that pub is out of bounds to troops***

outpost ['aʊtpəʊst] *noun* small detachment of troops placed at a distance from the main force, in order to provide warning of an approaching enemy; ***all the outposts have withdrawn to the main position***

outrange [aʊt'reɪndʒ] *verb* to be able to shoot further than another weapon; ***during the Gulf War, British and American tanks outranged the Soviet-designed tanks used by the Iraqis***

outrank [aʊt'ræŋk] *verb* to hold a higher rank than someone else; ***a naval lieutenant outranks a lieutenant in the army***

over ['əʊvə] *adverb (radio terminology)* it is your turn to speak; ***'Hullo 2, this is 22, what is your location, over?'***; *compare* OUT

overalls ['əʊvərɔːlz] *plural noun* **(a)** garment combining jacket and trousers, which is worn over other clothes in order to protect them from dirt, mud, oil, etc **(b)** skin-tight trousers worn by cavalry soldiers and members of certain supporting arms, as part of a ceremonial uniform *or* mess kit

overboard ['əʊvəbɔːd] *adverb* from a ship or boat, into the water; ***he fell overboard***

overhead [ˌəʊvə'hed] **1** *adverb* directly above you; ***helicopters were flying overhead*** **2** *adjective* positioned above you; **overhead protection** = roof constructed over a trench, in order to provide protection from shrapnel and chemical weapons

overlay ['əʊvəleɪ] *noun* piece of transparent paper or plastic, marked with boundaries, positions, routes, and other information relating to an operation, which is designed to be placed over a map as a means of briefing the participants (NOTE: also called a **trace**)

overrun [ˌəʊvə'rʌn] *verb* to fight your way onto an enemy position; ***we are being overrun***; ***the enemy easily overran our defences*** (NOTE: **overrunning - overran - have overrun**)

overseas [ˌəʊvə'siːz] **1** *adverb* in or to a foreign country; ***he was posted overseas*** **2** *adjective* located in a foreign country; ***he wants an overseas posting***

COMMENT: for American and British servicemen, most foreign countries are indeed located **overseas.** The word would be less appropriate for French soldiers serving in Germany, for example, since the two countries are not separated by a sea

over-watch ['əʊvə ˌwɒtʃ] *noun* role in which troops *or* tanks observe and give covering fire if necessary two squadrons deployed into over-watch positions

PAPA - Pp

P-15 [ˌpiː fɪf'tiːn] *noun* Soviet-designed long-range anti-ship missile (ASM) (NOTE: known to NATO as **Styx**)

PAA [ˌpiː eɪ 'eɪ] *noun* = PRIMARY AIRCRAFT AUTHORIZED number of aircraft allocated to a unit for the performance of its operational role (as opposed to training aircraft, spares, aircraft under maintenance)

pace [peɪs] **1** *noun* **(a)** single movement of a foot when walking; ***the squad took two paces forward*** **(b)** distance which a person's foot moves when walking one pace; ***an officer should march four paces in front of the parade*** **(c)** speed (especially when walking); **to keep pace with someone** = to move at the same speed as another person or vehicle **2** *verb* to measure distance by counting your paces; ***he paced the distance to the river***

pace-stick ['peɪs stɪk] *noun* giant set of mathematical dividers, traditionally carried by drill instructors in order to determine the length of pace for marching; ***he was charged for hitting a recruit with his pace-stick***

pacifism ['pæsɪfɪzm] *noun* belief that war is the wrong way to settle disputes

pacifist ['pæsɪfɪst] **1** *noun* person who believes that war is the wrong way to settle disputes; ***he is a committed pacifist*** **2** *adjective* referring to pacifism; ***he became a conscientious objector because of his pacifist beliefs***

pack [pæk] *noun* **(a)** large fabric container, designed to be carried on a person's back; *see also* BERGEN, RUCKSACK **(b)** paper *or* cardboard *or* plastic container; ***a pack of cigarettes; a 24 hour ration pack***

package ['pækɪdʒ] *noun* any object which is wrapped in a protective covering of paper *or* plastic *or* fabric; ***the package might be a bomb***

package formation ['pækɪdʒ fɔːˌmeɪʃn] *noun* large aerial attack force made up of different types of aircraft (eg. attack, escort, reconnaissance, SEAD, EW)

pack-animal ['pæk ˌænɪməl] *noun* mule *or* horse used to carry ammunition *or* equipment *or* supplies

PACOM ['pækɒm] *noun US* = PACIFIC COMMAND department of the US forces responsible for defending American national interests in the Pacific

paddle ['pædl] **1** *noun* instrument like a short oar, designed to propel an assault boat or canoe by hand; ***he dropped his paddle into the water*** **2** *verb* to propel a boat with a paddle; ***they paddled quietly up to the castle walls***

paddy-field ['pædi fiːld] *noun* field which is submerged in water, in order to cultivate rice

padre ['pɑːdreɪ] *noun* Christian army chaplain

pagoda [pə'gəʊdə] *noun* tall building used for religious worship in China, Korea, Japan, etc.

pain [peɪn] *noun* unpleasant physical sensation, caused by illness or injury

painkiller ['peɪnkɪlə] *noun* drug (for example morphine) designed to stop pain

paint [peɪnt] **1** *noun* liquid substance which is applied to an object in order to colour it (as for camouflage), to provide protection from water and damp or to provide resistance to certain types of surveillance equipment (such as infrared) **2** *verb* **(a)** to apply paint to an object; ***they painted the vehicles white*** **(b)** *(informal)* to illuminate a target with a laser target designator

pair [peə] *noun* two people or things acting or being used together; ***the section assaulted the position in pairs***; ***I saw a pair of tanks by the wood***

pallet ['pælɪt] *noun* wooden platform, designed to provide a firm base for a heavy load

pallisade [ˌpælɪ'seɪd] *noun* barrier or fortification constructed from wooden stakes, which are positioned vertically in the ground

Paludrin ['pæljʊdrɪn] *noun* drug which provides resistance to malaria

panga ['pæŋgə] *noun (East Africa)* long broad-bladed knife designed for clearing vegetation and often used as a weapon; *see also* MACHETE, PARANG

panic ['pænɪk] **1** *noun* loss of self-control as a result of fear or anxiety; ***the civilian population fled in panic as the soldiers arrived*** **2** *verb* to be affected by panic; ***he panicked when the enemy opened fire*** (NOTE: **panicking - panicked**)

panoramic sketch [pænəˌræmɪk 'sketʃ] *noun* simple drawing of an area of ground, made by an FOO *or* MFC, with likely targets and other useful information marked on it

panzer ['pænzə] *German noun* armour; **panzer grenadier** = German armoured infantryman

Papa ['pɑːpə] sixteenth letter of the phonetic alphabet (Pp)

para ['pærə] *noun (informal)* paratrooper; ***British paras have taken the village***

parachute ['pærəʃuːt] **1** *noun* apparatus consisting of a fabric canopy and a suspension harness which allows a person, vehicle or load to descend safely from an aircraft in flight; ***he was killed when his parachute failed to open*** **2** *verb* **(a)** to descend by parachute; ***the group parachuted behind enemy lines*** **(b)** to drop something by parachute; ***they parachuted supplies into the village***

parachutist ['pærəʃuːtɪst] *noun* someone who descends by parachute

parade [pə'reɪd] **1** *noun* **(a)** action of assembling at a specified time and place in order to be inspected before the commencement of an operation or period of duty; **muster parade** = assembly of all soldiers at the beginning of the day to receive instructions; **parade ground** = large area near a barracks, where troops can parade **(b)** ceremonial occasion (usually involving marching and military music); ***the passing-out ceremony was followed by a parade*** **2** *verb* to assemble for inspection at a specified time and place; ***the platoon will parade at the armoury at 0745 hours***

parados ['pærədɒs] *noun* raised mound of earth protecting the rear of a trench; *compare* PARAPET

paraffin oil ['pærəfɪn ɔɪl] *see* KEROSENE

paramedic [ˌpærə'medɪk] *noun* serviceman or civilian, with a high level of medical training, who is qualified to perform emergency treatment on serious casualties

paramilitary [ˌpærə'mɪlɪtri] *adjective* organized like an army (and possibly armed); ***there are several paramilitary groups operating in the region***

parang ['pæræŋ] *noun (Malaya)* long broad-bladed knife designed for clearing vegetation and often used as a weapon; *see also* MACHETE, PANGA

parapet ['pærəpet] *noun* raised mound of earth protecting the front of a trench; *compare* PARADOS

parasite ['pærəsaɪt] *noun* organism or insect which lives on or inside another animal and feeds off that animal (for example a louse, or worm); ***the disease is transmitted by a parasite***

parasitic [ˌpærə'sɪtɪk] relating to a parasite; ***the disease is transmitted by a parasitic worm***

paratrooper ['pærətruːpə] *noun* infantryman or member of a supporting arm who deploys into a war zone by parachute

> COMMENT: paratroopers of most armies undergo an exceptionally hard training programme and are consequently considered to be elite troops

paratroops ['pærətru:ps] *plural noun* paratroopers; ***enemy paratroops landed near the village***

parole [pə'rəʊl] **1** *noun* release of a prisoner on the condition of a promise made by that prisoner (for example that he will no longer fight against the army which captured him); ***he was released on parole*** **2** *verb* to release a prisoner on parole; ***he has been paroled***

partisan [ˌpɑ:tɪ'zæn] *noun* irregular soldier fighting against regular troops; ***the convoy was ambushed by a group of partisans***; *see also* GUERILLA

pass [pɑ:s] **1** *verb* **(a)** to go past something; ***we passed the fuel dump an hour ago*** **(b)** to move on; ***we passed through Hildesheim without stopping*** **(c)** to hand something to another person; ***pass me that map, please*** **(d)** to approve or allow something; ***I have passed your application to join the Mortar Platoon*** **(e)** to complete an examination or test successfully; ***you have passed the sergeants' exam*** **2** *noun* **(a)** document authorizing the holder to do something; ***show me your leave pass, please*** **(b)** narrow route through mountainous country; ***the enemy had blocked all the passes through the mountains*** **(c)** approach flight towards a target made by an attacking aircraft; ***we hit the tank on our second pass***

passage of lines [ˌpæsɪdʒ əv 'laɪnz] *noun* process whereby a unit or grouping moves through the positions of another unit or grouping

passive ['pæsɪv] *adjective* relating to night-viewing devices which do not require an external source of infrared (IR) light in order to operate; *compare* ACTIVE

pass out [ˌpɑ:s 'aʊt] *verb* to successfully complete an officer training course; ***he passed out of Sandhurst in December***; ***parents of cadets were invited to the passing-out parade***

password ['pɑ:swɜ:d] *noun* words, letters or numbers used as a verbal recognition signal, usually in the form of a challenge and a reply; ***the sentry opened fire because the patrol did not give the correct password***; *see also* COUNTERSIGN

path [pɑ:θ] *noun* **(a)** small track, which has been made artificially, or simply by people walking along it over a long period; ***the guide led the group along steep mountain paths*** **(b)** line along which something travels; ***the path of a missile***

pathfinder ['pɑ:θˌfaɪndə(r)] *noun* **(a)** aircraft which travels ahead of the main attack force in order to test enemy air defences and to mark targets **(b)** soldier (especially paratrooper *or* marine) who deploys ahead of the main force in order to reconnoitre, secure and mark a DZ *or* LZ *or* beach-landing site; *see also* ITG

Patriot ['peɪtriət] *noun* American-designed surface-to-air missile (SAM)

> COMMENT: during the Gulf War in 1991, the Patriot proved itself to be highly effective at shooting down Iraqi Scud missiles

patrol [pə'trəʊl] **1** *noun* **(a)** detachment of soldiers or vehicles sent out by a larger unit to carry out a specific task; ***the patrol was ambushed as it entered the village*** **(b)** covert or overt task carried out by a small detachment of soldiers or vehicles; ***he was killed on a patrol***; ***we will have to mount a patrol to check the route***; **patrol base** = covert base established in no-man's-land or enemy territory from which patrols can be mounted; **ambush patrol** = large well-armed patrol sent out to lay an ambush; **clearing patrol** = patrol sent out from a patrol base or defensive position in order to check the surrounding area for enemy; **fighting patrol** = large well-armed patrol sent out on an offensive operation (eg. snatching a prisoner for interrogation); **recce patrol** = small covert patrol sent out to gather information or reconnoitre ground; **standing patrol** = patrol sent out to occupy a covert position in no-man's-land in order to provide warning of enemy activity **(c)** act of walking or driving around an area on a regular basis in order to deter or prevent illegal or hostile activity; ***there are regular enemy patrols along this route***; **border patrol** = patrol sent out to prevent or provide warning of border incursions **2** *verb* to carry out a patrol; ***soldiers patrolled the streets during the curfew*** (NOTE: **patrolling - patrolled**)

pattern ['pætən] *noun* **(a)** decorative design, which is printed or painted on fabric or other materials; **disruptive pattern** = camouflage pattern, which is designed to break up the outline of an object **(b)** design of clothing, footwear or webbing; ***he was wearing the latest pattern of combat boot***

Pave Tack ['peɪv tæk] *noun* American-designed airborne laser target designator

Paveway ['peɪvweɪ] *noun* American-designed laser-guided bomb (LGB)

pay [peɪ] **1** *noun* money which a person receives for doing his job; ***he has difficulty bringing up his family on a corporal's pay*** **2** *verb* to give someone his pay; ***the troops mutinied because they had not been paid for months*** (NOTE: **paying - paid**)

payload ['peɪləʊd] *noun* **(a)** ordnance and equipment carried by an aircraft; ***this fighter is capable of carrying an enormous payload; the payload of this aircraft includes laser-guided bombs and heat-seeking missiles*** **(b)** type of ordnance delivered by a missile (such as bomblets, chemical agent, high explosive, etc.); ***this missile is designed to carry a variety of payloads***

paymaster ['peɪmɑːstə] *noun* unit officer who is responsible for pay

PB pill [piː 'biː pɪl] *noun* pill taken to counter the effects of nerve agents

P Company ['piː ˌkʌmpənɪ] *noun GB* mandatory parachute training course for airborne troops; ***he failed P Company***

> COMMENT: P Company is an exceptionally difficult course to pass and requires an extremely high level of physical fitness. Only those soldiers who have successfully completed this course are entitled to wear the famous red beret of the Parachute Regiment.

PD [ˌpiː 'diː] *noun* = PHENYLDICHLOROARSINE type of blister and vomiting agent

PE [ˌpiː 'iː] = PLASTIC EXPLOSIVE

peace [piːs] *noun* **(a)** state of not being involved in a war or armed conflict; ***after the end of the war, Europe enjoyed two decades of peace*** **(b)** ending of a war or armed conflict; **peace talks** = negotiations aimed at ending a war or armed conflict; **peace treaty** = signed agreement between opposing sides at the end of a war, stating the conditions under which they agree to exist in peace with each other; ***the peace treaty was signed at Versailles***

peacekeeper ['piːskiːpə(r)] *noun* serviceman, who is a member of a peacekeeping force; ***several UN peacekeepers have been injured***

peacekeeping ['piːskiːpɪŋ] **1** *noun* deployment, usually by the United Nations, of a neutral military force into an area where two sides are, or have recently been, engaged in armed conflict, in order to prevent or deter further military action by either side **2** *adjective* referring to peacekeeping; ***the UN is deploying a peacekeeping force in the region; Ireland is very active in its UN peacekeeping role***

peacetime ['piːstaɪm] *noun* period during which a state is not involved in a war or armed conflict; ***the army was disbanded and the troops returned to their peacetime occupations***

peak [piːk] *noun* **(a)** sharp summit of a mountain **(b)** moment when something is at its worst, best, etc.; ***when the bombardment was at its peak, hundreds of shells were falling on the city every minute***

peat [piːt] *noun* type of soil, composed of decayed vegetation

> COMMENT: during the Falklands Conflict in 1982, many artillery rounds failed to explode because they landed in soft peat

penetrate ['penətreɪt] *verb* **(a)** to force a way through the surface of something; ***this round is capable of penetrating most modern types of armour***; *see also* PIERCE **(b)** to force your way through a fortification or line of defence; ***the enemy have penetrated our perimeter in several places*** **(c)** to find a way through a barrier or obstacle; ***enemy engineers have penetrated the minefield***

penetration [ˌpenə'treɪʃn] *noun* act of penetrating something; ***we have managed to contain the enemy penetration in the south***

Penguin ['peŋgwɪn] *noun* Norwegian-designed anti-ship missile (ASM)

peninsula [pə'nınsjʊlə] *noun* long narrow strip of land projecting into a sea or lake; ***the enemy advanced along the peninsula***

pennant ['penənt] *noun* small triangular flag

Pentagon ['pentəgən] *noun* national headquarters of the US Department of Defense

> COMMENT: the Pentagon is named after the five-sided building in which the Defense Department is housed

pepper-pot ['pepə pɒt] *verb* GB *(informal)* to skirmish (ie. use fire and manoeuvre); ***the patrol pepper-potted across the open ground***

percussion [pə'kʌʃən] *noun* act of one object striking another object; **percussion cap** = small explosive charge, designed to ignite the propellant of a projectile, when struck by the firing mechanism of a weapon

perimeter [pə'rımıtə] *noun* outer boundary of a fortified or defended area; ***the guerillas broke through the northern perimeter of the camp***; ***they strengthened the perimeter fence with barbed wire***; **perimeter lights** = lights round the edge of a helicopter landing area

peripheral [pə'rıfərəl] *adjective* on the edge of an area or in the surrounding area; ***the base has suffered some peripheral damage***

periphery [pə'rıfəri] *noun* edge of an area or its surrounding area

periscope ['perıskəʊp] *noun* optical instrument, which enables an observer on a lower level (for example in a submerged submarine or at the bottom of a trench) to see things on a higher level (such as on the surface of the sea or ground)

permission [pə'mıʃn] *noun* consent or authorization given by one person, which allows another person to do something; ***he left the barracks without permission***; ***permission to carry on, Sir?***

permit 1 ['pɜːmıt] *noun* document which authorizes someone to do something; ***you will need a permit to get into the camp*** **2** [pə'mıt] *verb* to allow or authorize someone to do something; ***this pass permits two people to visit the HQ***

persistent [pə'sıstənt] *adjective; (of chemical agents)* designed to remain effective for several hours or days; **non-persistent** = designed to disperse in the air after a few minutes

> COMMENT: **persistent agents** are normally used against targets in rear areas, whereas **non-persistent agents** are used against a forward position, shortly before an assault is mounted

personnel [ˌpɜːsə'nel] *noun* people who are employed by an organization; **personnel carrier** = vehicle (usually armoured) designed to carry troops; *see also* ANTI-PERSONNEL

> COMMENT: the department responsible for personnel in a headquarters is known as **G1**

petrol ['petrəl] *noun* liquid fuel made from petroleum, used by motor vehicles; **petrol bomb** = improvised incendiary device consisting of a bottle filled with petrol and fitted with a wick of fabric, which is lit and then thrown at a target; *see also* MOLOTOV COCKTAIL (NOTE: American English is **gasoline** in many other languages it is **benzin**)

petroleum [pə'trəʊliəm] *noun* liquid mineral substance which is extracted from the ground and then refined to produce petrol, diesel, kerosene and lubricating oil

petty officer (PO) ['peti ˌɒfısə] *noun* GB & US non-commissioned officer (NCO) in the navy; GB & US **chief petty officer (CPO)** = senior non-commissioned officer (SNCO) in the navy

Pfc [ˌpiː ef 'siː] *US* = PRIVATE FIRST CLASS

PGM [ˌpiː dʒiː 'em] = PRECISION GUIDED MUNITION

Phalanx ['fælænks] *noun* American-designed radar-controlled 20mm naval anti-aircraft cannon (CIWS), which automatically detects, tracks and engages targets (NOTE: Phalanx is based on the M-61A1 Vulcan)

Phantom ['fæntəm] *see* F-4

phase [feɪz] *noun* specific stage in a planned or predicted sequence of events; ***this will be a five-phase operation***; **phase line** = imaginary line (often defined by a topographical feature, such as a road) used as a reference point or objective during movement; ***the squadron crossed Phase Line Charlie at 1345 hours***

phonetic alphabet [fəˌnetɪk 'ælfəbet] *noun* alphabet consisting of words (such as Alpha, Bravo, Charlie), which is designed to avoid confusion between similar letters, when speaking on the radio

Phosgene ['fɒzdʒiːn] *see* CG

Phosgene Oxime [ˌfɒzdʒiːn 'ɒksiːm] *see* CX

phosphorus ['fɒsfərəs] *see* WHITE PHOSPHORUS (WP)

photo ['fəʊtəʊ] *noun (informal)* photograph

photograph ['fəʊtəgrɑːf] **1** *noun* picture produced by a camera; **air photograph** = picture of an area of ground, taken from an aircraft; **satellite photograph** = picture of an area of ground, taken from a satellite **2** *verb* to take a photograph with a camera; ***he was arrested while he was photographing the base***

photo-interpreter [ˌfəʊtəʊ ɪn'tɜːprɪtə(r)] *noun* person who studies air *or* satellite photographs

physical training (PT) [ˌfɪzɪkl 'treɪnɪŋ] *noun* activities and exercises designed to improve or maintain physical fitness; **physical training instructor (PTI)** = serviceman who is trained to supervise physical training

pick [pɪk] **1** *noun* simple tool consisting of a curved metal bar with a point at one end and a blade at the other, attached at right angles to a long wooden handle; designed to break up hard ground; **pick helve** = pick handle; sometimes used as a baton; *see also* PICKAXE **2** *verb* to select; ***the sergeant picked two soldiers to carry the ammunition***; **pick off** = to shoot systematically at selected targets; ***the snipers were ordered to pick off the enemy commanders***

pickaxe *US* **pickax** ['pɪkæks] *noun* simple tool consisting of a curved metal bar with a point at one end and a blade at the other, attached at right angles to a long wooden handle; designed to break up hard ground; *see also* PICK

picket ['pɪkɪt] **1** *noun* **(a)** small group of soldiers sent out to watch for the enemy or to cover the activities of other troops; ***it was necessary to place pickets along the route***; *see also* OUTPOST, PICQUET, STANDING PATROL **(b)** metal stake used in the revetting of trenches and other fortifications **2** *verb* to deploy a picket; ***it will be necessary to picket the high ground***

pick up [ˌpɪk 'ʌp] *verb* **(a)** to lift an object off the ground (usually by hand); ***you should not pick up unexploded bombs*** **(b)** to collect people or things with an aircraft, boat or vehicle, in order to transport them to another location; ***the patrol was picked up by helicopter***

pick-up point (PUP) ['pɪk ʌp ˌpɔɪnt] *noun* location where people or things are picked up

picquet ['pɪkɪt] *see* PICKET; **picquet officer** = duty officer

pierce [pɪəs] *verb* to force a way through the surface of something; ***the shell pierced the tank's armour***; *see also* PENETRATE

pillbox ['pɪlbɒks] *noun* small bunker constructed from reinforced concrete

pilot ['paɪlət] **1** *noun* **(a)** person who operates the flying controls of an aircraft; *GB* **pilot officer (PO)** = lowest officer rank in the air force **(b)** person employed to take control of ships which are entering or leaving a harbour, or passing through a waterway **2** *verb* to act as a pilot on an aircraft or ship

pincer movement ['pɪnsə ˌmuːvmənt] *noun* tactical manoeuvre, in which two groupings attack an enemy force at the same time from different directions

pin down [ˌpɪn 'daʊn] *verb* to direct so much fire at an enemy that he is unable to move (in any direction); ***we were pinned down for over an hour***

P-INFO [ˌpiː 'ɪnfəʊ] *noun* = PUBLIC INFORMATION department responsible for dealing with the media

Pink [pɪŋk] *noun GB* **the Pink** = document containing the planned or probable sequence of events for a military exercise

pioneer [ˌpaɪəˈnɪə] *noun* infantry soldier who is trained to carry out field engineering tasks for his unit (preparing fortifications, clearing obstacles, etc.)

pip [pɪp] *noun GB* insignia in the form of a star, which is used in certain badges of rank by the British Army; *see also* STAR

> COMMENT: a second lieutenant has one pip; a lieutenant has two; a captain has three; a lieutenant colonel has a pip and a crown

pipe [paɪp] *noun* **(a)** tube made of concrete, metal or plastic, which is used to convey gas or liquid; **pipe-bomb** = homemade grenade, consisting of a piece of metal pipe filled with explosive; **pipe range** = indoor shooting range constructed from a length of large-diameter concrete pipe, which is used for the zeroing of weapons **(b) pipes** = bagpipes; **pipes and drums** = band of pipers and drummers, belonging to a battalion or regiment

pipeline [ˈpaɪplaɪn] *noun* huge pipe built to convey water, oil *or* gas over long distances; ***the pipeline will be a serious obstacle for our tanks***

piper [ˈpaɪpə] *noun* musician who plays the bagpipes

pistol [ˈpɪstəl] *noun* small gun designed to be held in one hand; **pistol-grip** = handle shaped like a pistol, which is mounted behind to the trigger of a machine-gun *or* **assault weapon;** *see also* HANDGUN

pit [pɪt] *noun* wide deep man-made hole in the ground; **weapon pit** = pit dug as a fire-position for a large weapon, which offers concealment and protection from enemy fire

pitch [pɪtʃ] **1** *noun* **(a)** *(of aircraft and ships)* movement up and down by the front and rear alternately; ***the elevator is used to control pitch*** **(b)** area of ground where a sport is played; ***the helicopter landed on the football pitch*** **2** *verb* to erect a tent; ***we pitched our tents in a small field***

plague [pleɪg] *noun* = BUBONIC PLAGUE highly infectious and potentially fatal disease, which is often carried by rat fleas and can be transmitted to man, sometimes resulting in widespread epidemics

> COMMENT: caused by a bacterium *Pasteurella pestis,* its symptoms are fever, shivering and swellings on the lymph nodes. Certain nations are known to have developed plague for use as a biological weapon

plain [pleɪn] *noun* large comparatively level area of ground (usually with few trees); ***the division advanced rapidly across the north German plains***

plan [plæn] **1** *noun* procedure, decided after consideration by a person or group, by which a mission or task will be carried out; ***the Company Commander is making his plan***; **according to plan =** in the same way as it was planned; ***the operation is running according to plan*** **2** *verb* to make a plan; ***they planned the operation very carefully***; ***we plan to attack under cover of darkness*** (NOTE: **planning - planned**); ***"No plan survives contact with the enemy." Moltke***

plane [pleɪn] *noun* fixed-wing aircraft; *see also* AEROPLANE, AIRPLANE

planner [ˈplænə] *noun* person involved in the making of a plan; ***military planners have prepared the invasion in great detail***

planning [ˈplænɪŋ] *noun* process of making a plan; ***he was involved in the planning of the operation***

plastic [ˈplæstɪk] *noun* man-made material; **plastic bullet** = large projectile made of plastic or rubber which is fired from a special gun and is designed to knock a person over but not to cause a serious injury (NOTE: also called **baton round**); **plastic explosive (PE)** = soft explosive substance which can be moulded into a required shape by hand

plasticuff [ˈplæstɪkʌf] *noun* flexible plastic strip with a self-locking catch, which is designed to bind a prisoner's hands together; ***the prisoners were secured with plasticuffs***

> COMMENT: the advantage of plasticuffs is that they are light and disposable; thus each man can carry several and he doesn't have to worry about retrieving them

plateau ['plætəʊ] *noun* wide level area of high ground

platoon [plə'tu:n] *noun* **(a)** tactical and administrative infantry grouping of three or more sections or squads (ie. about 30 men) **(b)** tactical and administrative armoured grouping of three or more tanks or armoured reconnaissance vehicles

> COMMENT: platoons are usually commanded by lieutenants or second lieutenants. In the British Army, platoon-sized groupings of tanks, artillery and certain supporting arms (such as engineers) are known as **troops**

platform ['plætfɔ:m] *noun* **(a)** raised structure alongside the railway line at a railway station, which enables the passengers to get into the train **(b)** firm surface *or* structure to which a weapon *or* other device is fitted **(c)** aircraft *or* ship *or* vehicle upon which a weapon *or* weapons system is fitted and used

plot [plɒt] **1** *noun* secret plan to carry out an illegal act **2** *verb* **(a)** to make a secret plan to carry out an illegal act; ***they were plotting a coup*** **(b)** to plan a course or route on a chart or map; ***he plotted a course between the islands***

plotter ['plɒtə(r)] *noun* circular board, printed with a grid of squares and fitted with a revolving cursor, which is used to calculate the bearing required for an artillery piece *or* mortar to hit a target

> COMMENT: in most armies, plotters have been replaced by computerized fire-control data systems

plume [plu:m] *noun* tall decoration of coloured feathers *or* animal hair, which is attached to a ceremonial headdress; ***the Coldstream Guards wear red plumes in their bearskins***

PMC [ˌpi: em 'si:] = PRIVATE MILITARY COMPANY

PNG [ˌpi: en 'dʒi:] *noun* = PASSIVE NIGHT GOGGLES night-viewing device, similar to a pair of binoculars, which is normally fitted to a person's face so that his hands are free

PO [ˌpi: 'əʊ] = PETTY OFFICER, PILOT OFFICER

pod [pɒd] *noun* external container suspended under an aircraft (used to carry electronic equipment *or* fuel *or* weapon systems *or* munitions); ***the aircraft is fitted with a radar-jamming pod under its starboard wing and a chaff-dispensing pod under the port wing.***

point [pɔɪnt] **1** *noun* **(a)** sharp or tapered end; ***the point of my pencil has broken*** **(b)** location; ***they found a suitable point to dump the ammunition***; **point of main effort** = part of the battlefield or operational theatre which a commander identifies as the place to maximize his efforts in order to fulfil his mission; **pick-up point (PUP)** = location where people or things are picked up; **water point** = location where water may be replenished; **point defence** = naval anti-air warfare (AAW) term for a warship's use of its short-range surface-to-air missiles (SAM) and other weapons (eg. CIWS) for self-defence **(c)** precise moment; ***it was at this point that he decided to withdraw*** **(d)** reason or purpose; ***there was no point in continuing the patrol*** **(e)** important factor; ***the point is that we are short of ammunition*** **(f)** leading soldier, vehicle or unit in a formation; ***the point was killed in the first exchange of fire*** **(g)** role of leading a formation; ***we were on point for the first phase of the advance*** **2** *adjective* leading; ***the point tank was destroyed by a mine*** **3** *verb* **(a)** to indicate with finger, hand or other object; ***he pointed at the wood***; **to point out** = to draw someone's attention to an object or fact; ***I pointed out the fact that the bridge was too narrow for tanks*** **(b)** to direct or aim a weapon; ***he pointed his gun at the officer***

point-blank range [ˌpɔɪnt blæŋk 'reɪnʒ] *noun* very close range; ***he was shot at point-blank range***

POL [ˌpi: əʊ 'el] = PETROL, OIL, LUBRICANTS; **POL point** = location where vehicles are refuelled

police [pə'li:s] *noun* civil organization responsible for the maintenance of law and order within a state; **military police** = organization responsible for police duties within the armed forces

policeman [pə'li:smən] *noun* member of a police force; **military policeman**

(MP) = member of the military police; ***a military policeman directed us to the Brigade RV***

policy ['pɒlɪsi] *noun* decisions taken by a government or military command on the general way something should be done

poncho ['pɒntʃəʊ] *noun* waterproof cape; **poncho liner** = lightweight quilt used for bivouacking

pond [pɒnd] *noun* very small lake

pongo ['pɒŋgəʊ] *noun GB (air-force slang)* member of the army

pontoon bridge ['pɒntu:n 'brɪdʒ] *noun* temporary bridge supported by boats (pontoons)

POR [ˌpi: əʊ 'ɑ:(r)] = POST-OPERATION REPORT

port [pɔ:t] *noun* **(a)** harbour, containing docks and other facilities for the loading and unloading of ships **(b)** left-hand side of an aircraft, boat or ship; ***enemy fighters approaching port!***; *compare* STARBOARD

portable ['pɔ:təbl] *adjective* able to be carried easily; *see also* AIR-PORTABLE, MAN-PORTABLE

Portakabin ['pɔ:təkæbɪn] *noun* small building, which is easy to assemble and can be transported by vehicle (designed for use as a temporary office or shelter on building sites, etc.)

port arms [ˌpɔ:t 'ɑ:mz] *verb* to hold a rifle diagonally across the chest; **for inspection port arms** = to hold a weapon with the working parts pulled to the rear, so that it can be inspected to make sure that it is clear of ammunition

position [pə'zɪʃn] **1** *noun* **(a)** place occupied by troops or equipment for tactical purposes; ***the enemy positions were clearly visible in the satellite photograph*** **(b)** situation; ***our position is now critical*** **2** *verb* to place in a specific location; ***he positioned the mortar line in a shallow gully***

positional defence [pəˌzɪʃənl dɪ'fens] *noun* defensive doctrine which relies on static defensive positions and the use of attrition to halt an enemy advance; *also known as* STATIC DEFENCE; *compare* MOBILE DEFENCE

positional warfare [pəˌzɪʃənl 'wɔ:feə] *noun* military doctrine which places emphasis on the possession of ground and its denial to the enemy; *compare* MANOEUVRE WARFARE

posn = POSITION

post [pəʊst] **1** *noun* **(a)** place where a serviceman is stationed **(b)** military base or installation **(c)** tactical position; **firing-post** = missile launcher; **observation post (OP)** = covert position from which an area of ground may be observed; *see also* LISTENING POST **(d)** job or position in an organization **2** *verb* **(a)** to assign a serviceman to a new grouping or location; ***he has been posted to HQ 7 Brigade***; ***I've been posted to Belize*** **(b)** to position soldiers for a task; ***the sentries have been posted***; **Last Post** = bugle-call blown in barracks and bases at bedtime (usually around 2200hrs) and also at military funerals

posting ['pəʊstɪŋ] *noun* assignment of a serviceman to a new grouping or location; ***my next posting is in Germany***

postpone [pəʊst'pəʊn] *verb* to make an event happen at a later time than originally planned; ***H-Hour has been postponed until 1530hrs***

post-traumatic stress disorder [ˌpəʊst trɔ:ˌmætɪk 'stres dɪsˌɔ:də] *noun* mental collapse, as a result of a frightening or horrific experience; *see also* BATTLE FATIGUE, SHELLSHOCK

potable ['pɒtəbl] *adjective US* fit for drinking; ***it was impossible to find any potable water in the contaminated area***

pouch [paʊtʃ] *noun* webbing or leather container, which is attached to a soldier's belt and is designed to hold ammunition or equipment

pound [paʊnd] *noun* unit of weight, corresponding to 0.4536 kilograms

POW [ˌpi: əʊ 'dʌb(ə)lju:] = PRISONER OF WAR

power ['paʊə] **1** *noun* **(a)** mechanical or electrical energy; ***the village has been deprived of power for ten days***; ***the engine keeps losing power***; **power lines** = wires, which convey large quantities of electricity from one location to another; **power station** = installation which produces electricity; **nuclear power** = energy produced by a nuclear reaction **(b)** military strength; ***does the gun have the power to knock out that battery?***; **firepower** = destructive capability of weapons; **sea power** = warships and weaponry used at sea **(c)** state with international influence which is based upon military strength; ***there is a danger of involvement by foreign powers***; **nuclear power** = state which possesses nuclear weapons; *see also* SUPERPOWER **(d)** authorization to do something; **power of arrest** = legal authority to arrest a person **2** *verb* to supply an apparatus or machine with mechanical or electrical energy; ***this submarine is powered by nuclear energy***

powerpack ['paʊəpæk] *noun* engine of an armoured fighting vehicle; *also known as* POWERPLANT

powerplant ['paʊəplɑːnt] *noun* engine of an armoured fighting vehicle; *also known as* POWERPACK

PR [ˌpiː 'ɑː] = PUBLIC RELATIONS

precaution [prɪ'kɔːʃn] *noun* action or procedure which is carried out in advance, in order to counter or prevent danger or failure; ***the accident happened because he failed to take the correct precautions***; ***wear your helmet as a precaution***

precautionary [prɪ'kɔːʃnəri] *adjective; (of actions or procedure)* as a precaution; ***this is just a precautionary measure***

precision [prɪ'sɪʒn] *noun* accuracy; **precision guided munition (PGM)** = bomb or missile, containing an automatic guidance system (such as an anti-radar missile, cruise missile, laser-guided bomb, etc.); *see also* SMART BOMB

Predator ['predətə] *noun* American-designed hand-held anti-tank missile

pre-emptive strike [priˌemptɪv 'straɪk] *noun* act of attacking a potential enemy before he attacks your own forces or territory

preparation [ˌprepə'reɪʃn] *noun* act of preparing for something; ***the operation failed because of poor preparation***; ***we were still making our preparations when the enemy attacked***; **artillery preparation** = bombardment of an objective, prior to an assault

preparatory [prɪ'pærətri] *adjective* as a preparation; **preparatory bombardment** = bombardment of an objective, prior to an assault

prepare [prɪ'peə] *verb* to make yourself or a thing ready for something; ***he prepared the weapon for firing***; ***Prepare to move!***

present ['prezənt] **1** *adverb* at this location; ***the company is present and ready for your inspection, Sir*** **2** *noun* this moment in time; ***at present, we are unaware of the enemy's intentions*** **3** *verb* [prɪ'zent] **a** to offer *or* give; ***he was presented with a painting of a Warrior***; **present arms** = to salute someone by holding a rifle in front of the body in a vertical position

pressel-switch ['pres(ə)l ˌswɪtʃ] *noun* switch, connected to *or* forming part of a microphone, which is pressed in order to transmit a radio message; ***no-one will hear you if you don't keep the pressel-switch pressed down***; ***three clicks on the pressel-switch means that the enemy is approaching the ambush***

prevent [prɪ'vent] *verb* **(a)** to stop something happening; ***it was impossible to prevent the accident*** **(b)** to stop someone doing something; ***we managed to prevent the civilians attacking the prisoner***

primary aircraft authorized *see* PAA

prime [praɪm] *verb* to prepare an explosive device for detonation; ***he is priming grenades***; ***the missile primes itself when fired***

primer ['praɪmə] *noun* small explosive charge used to detonate the explosive of a bomb or other explosive device; ***he removed the primer from the bomb***

principle ['prɪnsɪpl] *noun* general rule which is the basis for political or military action; **principles of war** = mass, manoeuvre, surprise, security, simplicity, objective, offensive, economy of force, unity of command

prison ['prɪzn] *noun* secure location, where people (especially convicted criminals) are confined; **prison camp** = camp, where prisoners of war are confined

prisoner ['prɪznə] *noun* **(a)** person who is confined in a prison **(b)** person who has been arrested or captured; **prisoner of war (POW** *or* **PW)** = serviceman captured by the enemy during a war; **to take someone prisoner** = to capture someone; ***A prisoner of war is someone who tries to kill you and fails, and then asks you not to kill him - Sir Winston Churchill***

private ['praɪvət] *noun (short for* PRIVATE SOLDIER) *GB & US* lowest rank in the army

> COMMENT: in certain regiments and corps of the British Army, privates are known by different titles: for example craftsman, guardsman, gunner, trooper, etc.

private first class (Pfc) [ˌpraɪvət fɜːst 'klɑːs] *noun US* **(a)** experienced private soldier in the army, with certain supervisory responsibilities (equivalent to a lance corporal in the marines) **(b)** lowest rank in the marines

private military company (PMC) [ˌpraɪvət ˌmɪlɪtəri 'kʌmpəni] *noun* private business organization, which markets military training and expertise, and in some cases, personnel and equipment, to foreign clients (eg. governments of developing nations), usually in a counter-insurgency *or* internal security situation

> COMMENT: this is an emotive subject, and many critics, including journalists and government officials, have dismissed PMCs as *mercenaries*, using the term in a derogatory context. PMCs, which operate within the constraints of international law and the Geneva Convention, would argue that they provide a legitimate service, especially to the governments of poorer nations, whose own military assets are inadequate and who, for political *or* strategic reasons, have been unable to obtain military assistance from the international community (eg. UN peacekeeping forces)

PRO [ˌpiː ɑː 'əʊ] = PUBLIC RELATIONS OFFICER

proactive [prəʊ'æktɪv] *adjective; (of actions or policy)* taking the initiative in order to prevent problems occurring; ***we need a more proactive response to the increase in terrorism***; *compare* REACTIVE

probe [prəʊb] **1** *verb* to attack in order to test the enemy defences and locate any weak points; ***the enemy has been probing our line throughout the night*** **2** *noun* act of making a probing attack; ***the attack was just a probe***; **refuelling probe** = long rigid tube, fitted to the front of an aircraft, which is inserted into a receptacle called a "basket", in order to carry out air-to-air refuelling

procedure [prə'siːdʒə] *noun* series of actions, which are carried out in specific sequence or manner; **standard operating procedure (SOP)** = set of instructions, produced by an arm, grouping or unit, which explain exactly how various duties and tasks should be carried out

production logistics [prə'dʌkʃn ləˌdʒɪstɪks] *noun* the design and manufacture of materiel for use by the armed forces

projectile [prə'dʒektaɪl] *noun* **(a)** anything which is fired at a target (for example a bullet, missile, shell, etc.); **projectile velocity** = the speed at which a bullet, shell, etc., travels **(b)** anything which is thrown or propelled by other means at a target; ***the defenders ran out of ammunition and threw stones and other projectiles at the attacking forces***

promote [prə'məʊt] *verb* to raise a serviceman to a higher rank; ***he was promoted to sergeant***; *compare* DEMOTE

promotion [prə'məʊʃn] *noun* act of promoting someone; ***on his promotion to sergeant he decided to get married***

prone [prəʊn] *adjective* lying on your stomach; ***he was shooting from the prone position***

prong [prɒŋ] *noun* one of two or more pointed parts of a fork

pronged [prɒŋd] *adjective* like prongs; **two-pronged attack** = attack mounted on two different parts of the enemy's line at the same time

propaganda [ˌprɒpə'gændə] *noun* selective release of information (whether true or untrue), in order to influence public opinion

propellant [prə'pelənt] *noun* explosive charge, which is used to fire a projectile

protect [prə'tekt] *verb* **(a)** *(of people)* to prevent injury or capture **(b)** *(of things)* to prevent damage or capture

protection [prə'tekʃn] *noun* **(a)** act of protecting; ***I am responsible for the protection of these civilians*** **(b)** something which protects; ***eye-protection should be worn when using this device***

protective [prə'tektɪv] *adjective* designed to protect; ***you should wear protective clothing***

protractor [prə'træktə] *noun* mathematical instrument used to calculate grid bearings

Protestant Church [ˌprɒtɪstənt 'tʃɜːtʃ] *noun* western European form of Christianity, promoting some beliefs which differ widely from those held by the Roman Catholic Church; *compare* ORTHODOX CHURCH, ROMAN CATHOLIC CHURCH

provocation [ˌprɒvə'keɪʃn] *noun* act which provokes

provoke [prə'vəʊk] *verb* to say or do something, which causes another person to act offensively; ***the soldiers have been ordered not to react when they are provoked***

provost [prə'vəʊ] *adjective* relating to the military police; **provost company** = company of military police; **Provost - Marshal** = senior administrative appointment in the military police; **Provost Sergeant** = NCO in charge of the Regimental Police (RP)

PROWLER ['praʊlə] *see* EA-6

proxy bomb ['prɒksi bɒm] *noun* terrorist bombing tactic, where an innocent civilian is forced by the terrorists to carry an explosive device *or* drive a car containing an explosive device up to a target (eg. security force base); the device is then initiated by a timer *or* by remote control.

> COMMENT: the most effective way of forcing a person to deliver a proxy bomb, is by taking members of his family hostage and threatening to harm them if he does not comply.

PSG pill [ˌpiː es 'dʒiː pɪl] *noun* pill taken to counter the effects of nerve agents

psychological [ˌsaɪkə'lɒdʒɪkl] *adjective* relating to a person's mental state; **psychological operations (PSYOPS)** = activities designed to lower the enemy's morale (such as the use of leaflets, radio transmissions, etc.); **psychological warfare (PSYWAR)** = use of psychological operations in wartime

PSYOPS ['saɪɒps] = PSYCHOLOGICAL OPERATIONS

PSYWAR ['saɪwɔː] = PSYCHOLOGICAL WARFARE

PT [ˌpiː 'tiː] = PHYSICAL TRAINING

PT-76 [ˌpiː tiː ˌsev(ə)nti 'sɪks] *noun* Soviet-designed light tank

PTI [ˌpiː tiː 'aɪ] = PHYSICAL TRAINING INSTRUCTOR

PTP [ˌpiː tiː 'piː] *noun US* = PETROLEUM TRANSFER POINT location where fuel is transferred from large tanker vehicles into smaller tanker vehicles which are capable of advancing with the fighting units

public ['pʌblɪk] **1** *adjective* **(a)** relating to people in general; **public order** = state of law and order within a community; **public relations** = management of an organization's image and reputation; **public relations officer (PRO)** = person whose job it is to communicate with the public, in order to improve the image of an organization **(b)** open to the general public; ***this is not a public road*** **2** *noun* ***the general public*** = people in general; ***the army base is not open to the general public***

public duties [ˌpʌblɪk 'djuːtiz] *plural noun GB* ceremonial duties in London *or* at one of the other royal residences (eg. guarding Buckingham Palace, Trooping the Colour, etc)

pulka ['pʊlkə(r)] *noun* lightweight sledge designed to be towed by a man on skis, and used to carry extra equipment *or* stores

pull back [ˌpʊl 'bæk] *verb* to withdraw; ***we were forced to pull back***

pull out [ˌpʊl 'aʊt] *verb* to abandon a position; ***the enemy is pulling out***

pull-through ['pʊlθruː] *noun* length of cord with a weight at one end and a strip of flannelette at the other, which is pulled through the barrel of a weapon in order to clean it; ***I've broken my pull-through***

Puma ['pju:mə] *noun* French-designed transport helicopter

punish ['pʌnɪʃ] *verb* to make someone suffer, because they have done something wrong; ***he was punished for being absent without leave***

punishment ['pʌnɪʃmənt] *noun* act of punishing a person

punitive ['pju:nətɪv] *adjective* intended as a punishment; ***the government is mounting a punitive expedition against the rebels***

punji ['pʌndʒi] *noun* *(Vietnam)* sharpened stick concealed in a shallow pit in order to injure the foot of anyone who steps on it

PUP [ˌpi: ju: 'pi:] = PICK-UP POINT

purification tablet [ˌpjʊərɪfɪ'keɪʃən ˌtæblɪt] *noun* tablet designed to make water fit to drink by killing bacteria; ***you must use purification tablets with this water***

Puritabs™ ['pjʊərɪˌtæbs] *noun* type of water purification tablet

pursue [pə'sju:] *verb* to follow a retreating or withdrawing enemy force, with the intention of destroying or capturing as much of it as possible; ***we were too exhausted to pursue the enemy into the mountains***; **to pursue by fire** = to shoot at a retreating enemy until he is no longer visible or beyond the effective range of your weapons

pursuit [pə'sju:t] *noun* act of pursuing (especially after a successful attack); **hot pursuit** = pursuit of terrorists, guerillas, etc., who have just attacked someone; ***we are only allowed to cross the border in the event of a hot pursuit***

push [pʊʃ] **1** *noun* **(a)** act of pushing; ***the car's stuck in the mud, can you give us a push?*** **(b)** large-scale offensive operation; ***they made a push to capture the capital before Christmas*** **2** *verb* to use physical force on an object, in order to move it away from one; ***we had to push the truck off the road***; ***we pushed the enemy back to their original positions***

push on [ˌpʊʃ 'ɒn] *verb* to move forwards as fast as possible; ***our aim is to push on as fast as possible to capture the capital***

puttee ['pʌti:] *noun* strip of cloth (similar to a bandage) which is wrapped around the ankle and lower leg in order to provide support and to prevent small stones and other loose objects going into your boots; ***when I joined the army, we still wore puttees***

PW [ˌpi: 'dʌb(ə)lju:] = PRISONER OF WAR

PWO ['pi:wəʊ] *noun GB* = PRINCIPAL WARFARE OFFICER officer on a warship who coordinates the sea battle; *compare* AWO

PX [ˌpi: 'eks] *noun US* = POST EXCHANGE shop or shopping centre on an American military base or camp (NOTE: the British equivalent is the **NAAFI**)

pylon ['paɪlən] *noun* metal structure, designed to support power lines above the ground

pyrotechnic [ˌpaɪrə'teknɪk] **1** *adjective* **(a)** designed to burn, in order to produce light for illumination or signalling purposes **(b)** designed to explode, in order to produce noise for battle simulation **2** *noun* pyrotechnic device (for example a trip-flare, thunderflash, Very light, etc.)

pyrrhic victory [ˌpɪrɪk 'vɪktərɪ] *noun* victory in which the losses suffered by the winning side are so high, that they outweigh the advantages gained by winning the battle

PZ [ˌpi: 'zed US ˌpi: 'zi:] = PICK-UP ZONE

QUEBEC - Qq

Q-fever [ˌkjuː 'fiːvə(r)] *noun* disease of cattle and sheep, which is transmissible to humans

> COMMENT: caused by *Coxiella burnetti*. The symptoms are fever, cough and headaches. Certain nations are known to have developed Q-fever for use as a biological weapon

Q-5 [ˌkjuː 'faɪv] *noun* Chinese-designed close-support fighter aircraft (NOTE: known to NATO as the **Fantan**)

QM [ˌkjuː 'em] = QUARTERMASTER

Q-matters [kjuː'mætəz] *noun* logistics at battalion level (ie those dealt with by the quartermaster or company quartermaster sergeants)

QMG [ˌkjuː em 'dʒiː] = QUARTERMASTER GENERAL

QRA [ˌkjuː ɑːr 'eɪ] *noun* = QUICK REACTION ALERT air-force state of readiness, where aircraft are prepared to take off and go into action at very short notice.; ***the squadron is on QRA***

QRF [ˌkjuː ɑː 'ef] = QUICK REACTION FORCE

quarry ['kwɒri] *noun* place where stone is extracted from the ground; ***the RAP was located in a quarry***

quarter ['kwɔːtə] **1** *noun* **(a)** house or flat provided for a married serviceman and his family; ***we've got a lovely quarter*** **(b)** **quarters** = living accommodation on a military base; ***he was confined to his quarters***; ***he's in his quarters***; **married quarters** = place where married servicemen and their families live **(c)** sparing an enemy's life on condition that he surrenders; ***quarter was neither expected nor given***; **no quarter !** = take no prisoners ! **2** *verb* ***to quarter someone on someone*** = to provide accommodation for someone; ***he was quartered on the local priest***

quartermaster (QM) ['kwɔːtəmɑːstə] *noun* officer (usually a captain) responsible for the logistics of a battalion or equivalent-sized grouping; *GB* **quartermaster commission** = commission held by an officer who has been promoted from the ranks, instead of undergoing normal officer selection and training which results in a Queen's Commission; *GB* **company quartermaster sergeant (CQMS)** = non-commissioned officer (usually a colour sergeant or staff sergeant) responsible for the logistics of a company *GB*; **regimental quartermaster sergeant (RQMS)** = warrant officer who assists the quartermaster of a battalion or equivalent-sized grouping; **quartermaster general (QMG)** = senior army officer, responsible for quartering; *GB* **technical quartermaster (TQM)** = officer (with a quartermaster commission) responsible for all technical equipment and machinery held by a battalion or equivalent-sized grouping; **technical quartermaster sergeant (TQMS)** = warrant officer who assists the technical quartermaster of a battalion or equivalent-sized grouping

> COMMENT: in the British Army, a quartermaster commission has a lower status than a Queen's Commission. Thus, a captain holding a quartermaster commission is considered to be junior to a second lieutenant holding a Queen's Commission

quay [kiː] *noun* structure built alongside or into water, which is used for the loading and unloading of ships

Quebec [kwɪ'bek] seventeenth letter of the phonetic alphabet (Qq)

Queen's Commission [ˌkwiːnz kə'mɪʃn] *noun GB* commission held by an officer who has undergone normal officer selection and training (NOTE: when the British monarch is a king, the commission becomes the **King's Commission**)

Queen's Regulations [ˌkwiːnz regjʊ'leɪʃnz] *plural noun GB* set of rules and regulations governing the conduct of the British Army (NOTE: when the British monarch is a king, the regulations become **King's Regulations**)

quick reaction force (QRF) [ˌkwɪk ri'ækʃn ˌfɔːs] *noun* detachment of soldiers on stand-by to deal with any incident which might occur within their unit's tactical area of responsibility (TAOR); *compare* RAPID REACTION FORCE

ROMEO - Rr

RA [ˌɑː 'eɪ] = ROYAL ARTILLERY

rabble ['ræb(ə)l] *noun* disorganized crowd of people; ***the enemy force disintegrated into a rabble***

rabid ['ræbɪd] *adjective* infected with rabies; ***a rabid fox came into the camp***

rabies ['reɪbiːz] *noun* fatal virus affecting mammals (especially dogs, foxes and wolves but also man), which is transmitted by the animal's saliva

> COMMENT: wild animals infected with rabies often lose their fear of man.

RAC [ˌɑː ˌeɪ 'siː] = ROYAL ARMOURED CORPS

radar ['reɪdɑː] *noun* system for detecting aircraft, vehicles, ships or other objects, through the transmission of high-frequency electromagnetic waves which are reflected back by the object; ***the aircraft flew low in order to avoid enemy radar***; **radar-absorbent material (RAM)** = substance which does not reflect radar waves, used as a covering on earlier types of stealth aircraft, in order to make them invisible to enemy radar equipment; **radar-absorbent structural material (RAS)** = substance which does not reflect radar waves, used in the construction of stealth aircraft, in order to make them invisible to enemy radar; **radar cross-section (RCS)** = shape of an object in relation to its ability to reflect radar waves; **radar guided** = equipped with a guidance system which uses radar to acquire its targets; **radar scan** = the movement of a radar beam as it turns and searches for objects; **anti-radar missile (ARM)** = missile designed to home in on an enemy radar transmission

radiation [ˌreɪdi'eɪʃn] *noun* emission of energy as rays of heat, light or electromagnetic waves; **radiation sickness** = illness caused by exposure to radiation from a radioactive substance; **initial nuclear radiation** = harmful rays of energy given off by a nuclear explosion; **thermal radiation** = rays of heat and light given off by a nuclear explosion; **residual nuclear radiation** = harmful rays of energy given off by radioactive substances left after a nuclear explosion

radio ['reɪdɪəʊ] *noun* **(a)** transmission and reception of audible signals, using electromagnetic waves; **radio-controlled** = controlled by radio signals; **radio silence** = state when no one is allowed to transmit on the radio; **to impose radio silence** = to start radio silence; **to lift radio silence** = to end radio silence; **radio watch** = period of duty, which is spent listening to a radio **(b)** apparatus designed to transmit and receive radio signals; **radio check** = radio transmission to ensure that the radios are working and on the correct frequency; ***'hullo 2 this is 22, radio check, over' - '2, OK, out'***

radioactive [ˌreɪdiəʊ'æktɪv] *adjective* giving off radiation in the form of harmful rays

radioactivity [ˌreɪdiəʊæk'tɪvɪti] *noun* radiation given off by a radioactive substance

radome ['reɪdəʊm] *noun* large disk-like antenna, which is mounted on the fuselage of airborne early warning and control (AEW & C) aircraft such as A-40, E-3, Orion AEW&C

RAF [ˌɑː eɪ 'ef] = ROYAL AIR FORCE

raid [reɪd] **1** *noun* military operation in which a small force enters enemy territory in order to cause casualties, destroy equipment or take prisoners, and then withdraws back to its own lines again; ***we must keep a look out for possible raids by guerillas***; **air raid** = attack by aircraft against a target on the ground (usually with bombs); **artillery raid** = tactic using artillery, where the guns move into enemy territory to attack a specific target and then withdraw before the enemy can retaliate **2** *verb* to carry out a raid; ***the enemy raided our supply depot***

rail [reɪl] *noun* **(a)** one of many long metal bars, which are fitted together to form a railway line **(b)** transport by train; ***the battalion is moving by rail***

railhead ['reɪlhed] *noun* point on a railway, where troops leave their trains and continue their journey using other forms of transport

railroad ['reɪlrəʊd] *noun US* railway

railway *or* **railway line** ['reɪlweɪ] *noun* two parallel lines of rails, along which a train travels

rain [reɪn] **1** *noun* drops of water, which fall from the sky; ***heavy rain delayed the start of the offensive***; ***the tropical rain turned the paths to mud*** **2** *verb (of rain)* to fall from the sky; ***it is raining heavily and this makes any immediate attack unlikely***

rainy season ['reɪni ˌsiːzən] *noun* time of year when there is a lot of rain; ***with the end of the rainy season, military operations will restart***; *compare* DRY SEASON, MONSOON

raise [reɪz] *verb* **(a)** to lift *or* put an object into a higher position; ***the tank raised its gun barrel*** **(b)** to promote to a higher rank; ***he was raised to corporal*** **(c)** *(radio terminology)* to establish radio contact with another call-sign; ***I can't raise 22B***

rake [reɪk] *verb* to fire over a wide area with an automatic weapon; ***they raked the street with machine-gun fire***

rally ['ræli] *verb* to reassemble after a period of confusion; ***the platoon rallied at the edge of the village***; ***the sergeant rallied his men*** (NOTE: **rallying - rallied**)

RAM [ˌɑː ˌeɪ 'em] = RADAR-ABSORBENT MATERIAL

ramp [ræmp] *noun* inclined surface designed to enable people *or* vehicles to move onto *or* off a vehicle *or* aircraft *or* ship

R & D; = RESEARCH AND DEVELOPMENT

R & R [ˌɑː ən 'ɑː] *noun* = REST AND RECUPERATION period of holiday taken during operational duty; ***American troops come to the resort for a period of R & R***

range [reɪndʒ] *noun* **(a)** maximum distance that a weapon can fire; ***this missile has a range of one hundred kilometres***; **in** *or* **within range** = within the effective range of a weapon; ***that tank is in range***; **out of range** = beyond the effective range of a weapon; ***that tank is out of range*** **(b)** distance between a weapon and its target; ***he destroyed the tank at a range of two hundred metres***; **range card** = card showing topographical features or targets and the distance to them from a specific location; *see also* POINT-BLANK RANGE **(c)** area of ground used for shooting practice; **field firing range** = area of open ground, where soldiers can practice shooting in battle conditions; **gallery range** = formal shooting range, consisting of a firing point, where several people can shoot side-by-side, and the butts, where targets are positioned; **pipe range** = indoor shooting range constructed from a length of large-diameter concrete pipe, which is used for the zeroing of weapons

rangefinder ['reɪndʒfaɪndə] *noun* device designed to calculate the exact distance to an object; **laser rangefinder** = rangefinder which utilizes a laser beam

Rangers ['reɪndʒəz] *noun* US Army special forces organization

range-tables ['reɪndʒ ˌteɪblz] *plural noun* set of figures displayed in columns, showing the elevations required by artillery *or* mortars to hit targets at specific distances

> COMMENT: in most armies, range-tables have been replaced by computerized fire-control data systems

rank [ræŋk] *noun* **(a)** official title, indicating a serviceman's position in the hierarchy (such as corporal, sergeant, lieutenant, etc.); **field rank** = any army rank above captain and below general; **other rank (OR)** = serviceman who is not an officer (NOTE: American English is **enlisted man**) **(b) the ranks** ordinary soldiers as a group; ***he rose from the ranks to become a general*** **(c)** parade formation, in which men stand side-by-side in a line; ***form three ranks!***; **pull rank** = to invoke the authority of your rank when disagreeing with a subordinate; ***I'll have to pull rank on you over this matter***

ranker ['ræŋkə] *noun* soldier, who is not an officer

RAP [ˌɑː eɪ 'piː] = REGIMENTAL AID POST

rape [reɪp] **1** *noun* act of raping a woman; ***two soldiers were accused of rape*** **2** *verb* to force a woman to have sex; ***the marines were accused of raping a girl they had met in a bar***

rapid ['ræpɪd] *adjective* quick; **rapid fire** = to fire several shots in quick succession; **rapid reaction force (RRF) = (i)** combined-arms force, which is ready to deploy to an area of operations at very short notice **(ii)** multinational combined-arms force under command of the European Union (EU), with a primary role of peacekeeping; **Allied Rapid Reaction Corps (ARRC)** = British-led NATO force designed to react at short notice to any crisis involving NATO countries; *compare* QUICK REACTION FORCE

Rapier ['reɪpɪə] *noun* British-designed surface-to-air missile (SAM)

rappel [ræ'pel] *verb* to descend, using a rope; ***the patrol rappelled into the jungle from a helicopter***

Raptor ['ræptə(r)] *see* F-22

RAS [ˌɑː eɪ 'es] = RADAR-ABSORBENT STRUCTURAL MATERIAL

rate of fire [ˌreɪt əv 'faɪə] *noun* number of rounds fired by a weapon in one minute

rating ['reɪtɪŋ] *noun GB* junior non-commissioned rank in the navy; **able rating** = lowest non-commissioned rank in the navy; **leading rating** = junior non-commissioned officer in the navy (equivalent to a corporal in the army, marines or air force) (NOTE: the ranks of **ordinary rating** and **junior rating** were abolished in April 1999)

ration ['ræʃn] **1** *noun* **(a)** regulation amount of food, fuel or other resource, which is issued or available to a person in times of shortage; ***the water ration is one litre per man*** **(b) rations** = food issued to a soldier on operations or exercise; ***the company will draw rations at 1400hrs*** **2** *verb* to limit the amount of food, fuel or other resource that a person is allowed to have; ***petrol is being rationed***

rationing ['ræʃnɪŋ] *noun* policy of restricting supplies, especially during wartime; ***the government is going to introduce petrol rationing***

rats = RATIONS

Raven ['reɪvən] *see* EF-111A

ravine [rə'viːn] *noun* deep narrow valley, with steep sides

raw [rɔː] *adjective* **(a)** *(of meat)* uncooked; ***the refugees were eating raw horse meat*** **(b)** *(of soldiers)* very inexperienced; ***the reserves consisted of raw recruits***

RCS [ˌɑː siː 'es] = RADAR CROSS-SECTION

Rct = RECRUIT

RE [ˌɑː 'iː] = ROYAL ENGINEERS

react [ri'ækt] *verb* to do something in response to an incident or situation

reaction [ri'ækʃn] *noun* **(a)** act of reacting to something; ***his immediate reaction was one of shock*** **(b)** action a person takes when reacting to something; ***the general's reaction was to order further bombing raids***; **reaction force** = small mobile multinational force whose role is to provide immediate aid in a crisis situation; **quick reaction force (QRF)** = detachment of soldiers on stand-by to deal with any incident which might occur within their unit's tactical area of responsibility (TAOR); **rapid reaction force (RRF) = (i)** combined-arms force, which is ready to deploy to an area of operations at very short notice; **(ii)** multinational combined-arms force under command of the European Union (EU), with a primary role of peacekeeping; *see also* ALLIED RAPID REACTION CORPS

reactive [ri'æktɪv] *adjective (of actions or policy)* dealing with a problem once it has started to occur; ***most of our countermeasures seem to be purely reactive***; *compare* PROACTIVE

readiness ['redɪnəs] *noun* being ready to do something; **readiness state** *or* **state of readiness** = period of time in which a person *or* grouping must be ready to do something; ***the squadron is currently on a readiness state of fifteen mimutes***

ready ['redi] *adverb* fully prepared and in a position to do something; **to make ready** = to cock a loaded weapon, so that it is ready to fire (NOTE: American English is **lock and load**)

real estate ['riːəl ɪsˌteɪt] *noun* *US* ground (especially in a logistics *or* administration context)

reallocate [riː'æləkeɪt] *verb* to allocate in a different way; ***the national forces which have been provided to NATO may be reallocated to different sectors***

reallocation [ˌriːˌæləʊ'keɪʃn] *noun* the action of reallocating; **reallocation authority** = authority given to a NATO commander to reallocate multinational forces under his command

rear [rɪə] **1** *adjective* **(a)** moving or located at the back of a formation or position; ***we engaged the rear platoon*** **(b)** located behind the forward positions; **rear headquarters** = primary logistical headquarters for a large tactical grouping (normally located well to the rear of the front line) **2** *noun* **(a)** back of a formation or position; ***he was moving at the rear*** **(b)** area behind the front line; ***we are moving to the rear*** **(c) Rear** = rear headquarters; ***Rear is located at grid 453654***

rear-admiral [ˌrɪə 'ædmərəl] *noun* *GB* senior officer in the navy

rear admiral [ˌrɪə 'ædmərəl] *noun* *US* senior officer in the navy

rearguard ['rɪəgɑːd] *noun* small military force, positioned at the rear of a withdrawing force, in order to fight off the enemy pursuit

rearm [riː'ɑːm] *verb* to equip with new weapons

rearmament [riː'ɑːməmənt] *noun* act of rearming

rearmost ['rɪəməʊst] *adjective* furthest in the rear

rearward ['rɪəwəd] **1** *adjective* towards the rear; ***all rearward movement has been delayed by the chemical attack*** **2** *adverb* towards the rear; ***the enemy is moving rearward***

rebel 1 ['rebəl] *noun* person who uses armed force to oppose the established government; ***the rebels have captured the barracks***; *see also* FREEDOM FIGHTER, INSURGENT, REVOLUTIONARY **2** [rɪ'bel] *verb* to oppose the established government with armed force; ***some mountain tribes have rebelled against the provincial government***

rebellion [rɪ'beljən] *noun* armed resistance to the established government or to the army command; ***the rebellion spread rapidly to neighbouring provinces***; *see also* INSURGENCY, INSURRECTION

rebro ['riːbrəʊ] = REBROADCAST

rebroadcast [riː'brɔːdkɑːst] **1** *noun* act of rebroadcasting a radio signal (NOTE: called **rebro** for short) **2** *verb* to receive a radio signal and then transmit it on to another receiver, which is too far away to receive the original signal; *see also* RELAY

recall [rɪ'kɑːl] *verb* to order a person or unit to return to a location or grouping; ***the patrol has been recalled***; ***the government is recalling reservists***

recapture [riː'kæptʃə] *verb* **(a)** to capture a position or location which has been captured by the enemy; ***we recaptured the positions we had lost the previous day*** **(b)** to capture a prisoner who has escaped; ***after three days' search, all the prisoners were recaptured***

recce ['reki] *GB* **1** *adjective* relating to reconnaissance; **recce group (R Group)** = small group (usually consisting of a commander and his subordinate commanders), which carries out a reconnaissance before planning an operation or task; **recce patrol** = small covert patrol sent out to gather information or reconnoitre ground **2** *noun* **(a)** reconnaissance; ***we need to carry out a recce of the bridge***; **shovel recce** = (slang) going to the toilet in the field; ***I'm just going on a shovel recce*** **(b)** reconnaissance units; ***enemy recce has been sighted to the east of Mikulov*** **3** *verb* to reconnoitre; ***we need to recce the bridge***

reception [rɪ'sepʃn] *noun* **(a)** act of receiving a radio signal; ***he acknowledged***

reception of the signal **(b)** quality of a radio signal; ***there is very poor reception in this area***

recharge [riː'tʃɑːdʒ] *verb* **(a)** to put electrical power into a dead battery **(b)** to put ammunition into an empty magazine or ammunition compartment

recognition [ˌrekəg'nɪʃn] *noun* act of identifying a person or thing; **recognition marking** = distinctive symbol painted on all vehicles, so that friendly forces will not mistake each other for the enemy; ***during the Gulf War, the coalition recognition marking was an inverted V***; **recognition signal** = verbal or other signal which identifies a person or unit as being friendly

recognize ['rekəgnaɪz] *verb* **(a)** to identify a person, place or thing because one remembers it; ***he recognized the crossroads*** **(b)** to acknowledge that something is correct or legal; ***I do not recognize your authority in this matter***; ***the EU countries have recognized the new government***

recoil **1** ['riːkɔɪl] *noun* backward movement of a gun when it is fired; ***this weapon has hardly any recoil*** **2** [rɪ'kɔɪl] *verb (of guns and firing mechanisms)* to move backwards when fired; ***he was injured when the gun recoiled***

recon ['riːkɒn] *US* **1** *adjective* relating to reconnaissance; **recon platoon** = platoon which specializes in reconnaissance **2** *noun* **(a)** reconnaissance; ***we need to do a recon of the bridge*** **(b)** reconnaissance units; ***enemy recon has been sighted to the east of Mikulov*** (NOTE: British English is **recce**)

reconnaissance [rɪ'kɒnɪsəns] **1** *adjective* relating to reconnaissance; **reconnaissance unit** = unit which moves ahead of the main body of an advancing force, in order to identify suitable routes, give warning of natural and man-made obstacles, and to locate the enemy **2** *noun* **(a)** examination or survey of ground or a specific location, in order to plan an operation or task; ***he carried out a detailed reconnaissance of the enemy position*** **(b)** act of examining terrain in order to identify suitable routes and give warning of natural and man-made obstacles **(c)** act of looking for the enemy; **reconnaissance by fire** = act of firing at likely enemy positions, in order to cause the enemy to reveal his location by moving or by returning fire; **reconnaissance in force** = reconnaissance carried out by a large well-armed grouping, which is strong enough to engage the enemy if necessary **(d)** reconnaissance units; ***enemy reconnaissance has been sighted to the east of Mikulov***; *see also* RECCE, RECON; ***Time spent in reconnaissance is seldom wasted - British Army Field Service Regulations 1912***

reconnoiter [ˌrekə'nɔɪtə] *see* RECONNOITRE

reconnoitre *US* **reconnoiter** [ˌrekə'nɔɪtə] *verb* to carry out reconnaissance; ***a patrol was sent out to reconnoitre the enemy position***; *see also* RECCE, SCOUT

record **1** ['rekɔːd] *noun* data or information, which is written down for future reference; ***units will submit records of all ammunition fired during the year***; ***we captured a large quantity of enemy records*** **2** [rɪ'kɔːd] *verb* to write down data or information, so that it may be used in the future; ***he recorded the information in his notebook***

recover [rɪ'kʌvə] *verb* **(a)** to become healthy after illness or injury; ***he is recovering from his wounds*** **(b)** to find and bring back; ***the crew of the helicopter which was brought down have all been recovered*** **(c)** *(of damaged or defective vehicles)* to collect and bring back for repair; ***the tank has been recovered***

recovery [rɪ'kʌvri] *noun* act of recovering a vehicle; ***'hullo 2, this is 22, request recovery at grid 559321, over'***; ***the recovery will have to be delayed until first light***

recruit [rɪ'kruːt] **1** *noun* newly-enlisted serviceman, who is undergoing basic training; ***a batch of raw recruits arrived at the barracks*** **2** *verb* to encourage or persuade people to join the armed forces, and then arrange for them to do so; ***they are running a TV advertising campaign to recruit for the marines***

recruitment [rɪ'kruːtmənt] *noun* the process of recruiting men and women for the armed forces; ***the marines are planning a recruitment drive***

red [red] *noun & adjective* the colour of blood; ***he wore red flashes on his collar***

> COMMENT: the positions of enemy forces are usually marked on a map in red, while those of friendly forces are marked in blue

red berets [ˌred ˈbereɪz] *noun (informal)* British paratroopers; *compare* BLUE BERETS, GREEN BERETS

redcap [ˈredkæp] *noun GB informal* military policeman; ***run for it boys ! The redcaps are here !***

Red Crescent [ˌred ˈkresənt] *noun* **(a)** international medical and relief organization in Islamic countries, which is closely connected to the Red Cross **(b)** internationally-recognized insignia, consisting of a red crescent-moon on a white background, used by the Red Crescent, displayed on military ambulances, hospital ships, and medical facilities, and worn by medical personnel in Islamic countries; *see also* RED CROSS

Red Cross [ˌred ˈkrɒs] *noun* **(a)** international organization, dedicated to providing assistance (especially medical) during times of war or natural disaster **(b)** internationally-recognized insignia, consisting of a red cross on a white background, used by the Red Cross, displayed on military ambulances, hospital ships, and medical facilities, and worn by medical personnel; *see also* RED CRESCENT

redeploy [ˌriːdɪˈplɔɪ] *verb* **(a)** to deploy to a new location; ***we redeployed to a position south of the hill*** **(b)** to deploy into a different tactical formation; ***they redeployed into columns***

redistribution [riːˌdɪstrɪˈbjuːʃn] *noun* action of distributing in a different way; **redistribution authority** = authority given to a NATO commander to redistribute forces which have been assigned by different nations in such a way that they are best used in the operations he is controlling

red-light district [redˈlaɪt ˌdɪstrɪkt] *noun* part of a town or city, containing a large number of brothels, sex clubs, etc.; ***the MPs went round the red-light district collecting drunken soldiers***

reduce [rɪˈdjuːs] *verb* to make smaller or less; ***the government is planning a reduction in defence expenditure***

reduction [rɪˈdʌkʃn] *noun* act of reducing

redundancy [rɪˈdʌndənsi] *noun* practice of maintaining additional military resources to replace those which are destroyed or neutralized

re-entrant [riːˈentrənt] *noun (topographical)* small valley cutting into the side of a hill or mountain (often between two spurs); ***the vehicles were concealed in a small re-entrant on the other side of the hill***

re-entry vehicle (RV) [riːˈentrɪ ˌviːɪkl] *noun* warhead of a surface-to-surface missile which is designed to travel through space on its way to its target; ***this missile is fitted with three nuclear re-entry vehicles***

> COMMENT: one missile might be fitted with several re-entry vehicles, each of which might be directed at a different target; *see also* MIRV

ref [ref] = REFERENCE

reference [ˈrefrəns] **1** *noun* **(a)** direction for obtaining information; ***Reference: Section 69 of the Army Act, 1955***; **reference point** = any object or feature on the ground, which is used to assist in the giving of directions or to draw another person's attention to a target or other object of interest; **grid reference** *or* **map reference** = six- or eight-figure reference, obtained from the coordinates of a map grid, used to denote an exact location on the map; ***the grid reference for the church is 656364*** **(b)** grid or map reference; ***what's your reference?*** **2** *preposition* in relation to; ***with reference to my letter dated 26 November***; ***reference the church; look left. Enemy tank moving towards the wood.***; ***'hullo 22, this is 2, reference my last order, cancel, over'***

reflect [rɪˈflekt] *verb (of heat, light, sound, radar waves, etc.)* to send something back towards its source; ***sonar detects underwater objects by transmitting sound waves which are reflected back by the object***

refuel [riː'fjuːəl] *verb* to put fuel into a vehicle's fuel tank; ***we made a refuelling stop before crossing the desert*** (NOTE: **refuelled - refuelling**; US **refueled - refueling**)

refuge ['refjuːdʒ] *noun* place of safety; ***the villagers sought refuge in the crypt of the local church***

refugee [ˌrefjuː'dʒiː] *noun* person who leaves his or her home in order to escape from danger (especially war), and looks for refuge elsewhere; ***the road is blocked with refugees***

refuse [rɪ'fjuːz] *verb* **(a)** to not agree to do something; ***he refused to obey my orders***; ***I refuse to believe that the enemy are beaten*** **(b)** (of a line of defence) to position one end of the line back at an angle to the main frontage, in order to meet the threat of a flanking attack *or* envelopment

regiment (Regt) ['redʒɪmənt] *noun* **(a)** tactical and administrative army grouping of two or more battalions; ***two enemy regiments have crossed the river*** **(b)** *GB* administrative grouping of one or more infantry battalions plus a separate regimental headquarters; ***he served in one of the Highland Regiments*** **(c)** *GB* tank battalion, consisting of three or more squadrons; ***the brigade consists of two infantry battalions and one armoured regiment*** **(d)** *GB* battalion-sized artillery grouping, consisting of three or more batteries; ***we have a regiment of artillery in support*** **(e)** *GB* battalion-sized grouping for certain supporting arms (such as engineers); ***the government is sending a regiment of engineers to assist in the rescue operation*** **(f)** *US* armored cavalry grouping of two or more squadrons; ***an American armoured cavalry regiment led the advance***

COMMENT: in the British Army, the use of the word **regiment** is rather confusing, since it no longer refers to a tactical grouping of two or more battalions (the British use a **brigade** of three or more battalions instead). Regiments continue to exist in name, however, because their histories and traditions are considered to be extremely valuable in promoting unit identity and esprit de corps. Infantry regiments are essentially administrative groupings, and most currently consist of only one battalion plus a regimental headquarters, which is responsible for recruiting, career planning and welfare. If a regiment does have more than one battalion, these do not normally serve together in the same brigade. Battalion-sized groupings of artillery, tanks and certain supporting arms (such as engineers) are known as **regiments** for different historical reasons. A British armoured brigade might consists of two armoured regiments and one armoured or mechanized infantry battalion or, alternatively, two infantry battalions and one armoured regiment, plus artillery and supporting arms. On operations, these units are broken down and combined into **battle groups.** As an example, an armoured battle group might consist of two squadrons of tanks and one infantry company, which are organized into two **squadron and company groups** and a **company and squadron group** under the command of the armoured regimental HQ. The exact composition will vary according to the tactical requirement at the time. In the US Army, a battle group is known as a task force, while company and squadron groups and squadron and company groups are known as company teams

regimental [ˌredʒɪ'mentəl] *adjective* **(a)** relating to a regiment; ***regimental headquarters (RHQ)***; *GB* **regimental lieutenant colonel** = officer commanding a regimental headquarters (usually a colonel) **(b)** *GB (in certain contexts only)* relating to a specific infantry battalion; **regimental aid post (RAP)** = battalion casualty clearing-station, where casualties are assessed and given emergency medical treatment, before being evacuated to a dressing station; **regimental medical officer (RMO)** = doctor attached to a battalion; **regimental police (RP)** = small group of NCOs assigned to carry out police duties within a battalion *or* equivalent-sized grouping; **regimental quartermaster sergeant (RQMS)** = warrant officer who assists the quartermaster of a battalion or equivalent-sized grouping; **regimental sergeant major (RSM)** = most senior warrant officer in a battalion, who assists the adjutant in disciplinary matters and the day-to-day administration of the battalion; **regimental signals officer (RSO)** = battalion officer responsible for communications **(c)** relating to service with a unit (as opposed to service on a staff or at a training establishment); ***he is an excellent regimental officer***

region ['ri:dʒən] *noun* large area of land with well-defined boundaries or distinctive features

register ['redʒɪstə] **1** *noun* **(a)** official record of information or data **(b)** book or document in which information or data is recorded **2** *verb* **(a)** to record in writing **(b)** *(of artillery targets)* to allocate a target with a target number, and then calculate and record the firing data, which must be applied to the guns in order to to hit it; **to register with fire** = to register an artillery target and then fire at it in order to confirm that the firing data is correct, making adjustments if necessary

registration [ˌredʒɪ'streɪʃn] *noun* act of registering something; **registration number** *or* **vehicle registration** = combination of numbers and letters, which is displayed on a vehicle as identification

regroup [ri:'gru:p] *verb* to stop an activity or operation temporarily, in order to reorganize; ***the battalion was ordered to regroup south of the town***

Regt = REGIMENT

regular ['regjʊlə] **1** *adjective* **(a)** *(of actions)* always happening at the same time or on the same day; ***this is a regular weekly inspection*** **(b)** *(of armed forces and servicemen)* relating to a full-time professional force, with a proper organization and rank structure and regular training; **regular soldier** = person who makes a career of soldiering **2** *noun* member of the regular armed forces (as opposed to a reservist, territorial or guerrilla); ***we were attacked by a battalion of regulars***

regulate ['regjuleɪt] *verb* to control the actions of a person or thing

regulation [ˌregju'leɪʃn] *noun* official directive, restriction or rule; **Queen's** (*or* **King's) Regulations** = set of rules and regulations governing the conduct of the British Army

rehearsal [rɪ'hɜ:səl] *noun* act of rehearsing; ***there will be a rehearsal at 1430hrs***; **dress rehearsal** = final rehearsal just before the operation starts

rehearse [rɪ'hɜ:s] *verb* to practise carrying out an operation or task, before doing it for real; ***the brigade rehearsed the passage of lines in some fields outside the town***

reinforce [ˌri:ɪn'fɔ:s] *verb* **(a)** to make something stronger **(b)** *(of military forces)* to send additional personnel and equipment to give assistance or to replace casualties; ***our right flank needs reinforcing***

reinforcement [ˌri:ɪn'fɔ:smənt] *noun* **(a)** act of reinforcing; ***before reinforcement, the brigade was down to approximately 1,500 men*** **(b)** **reinforcements** = men or units sent to reinforce another grouping; ***the division was unable to send any reinforcements***

rejoin [rɪ'dʒɔɪn] *verb* to return to a person or grouping, after being away from them; ***we were unable to rejoin our unit after the bridge was blown up***

relay **1** ['ri:leɪ] *noun* act of relaying a radio signal; ***we are setting up a relay*** **2** [rɪ'leɪ] *verb* **(a)** to receive a radio signal and then transmit it on to another receiver, who is too far away to receive the original signal; ***the signal was relayed to 7 Brigade HQ***; *see also* REBROADCAST **(b)** to pass on a message to another person; ***he received the message from HQ and immediately relayed it to the troops***

release [rɪ'li:s] **1** *noun* act of releasing; ***the aim of the operation is to secure the release of the hostages*** **2** *verb* **(a)** to let a prisoner go free; ***we captured the camp and released the prisoners*** **(b)** to remove a control or restriction; ***he was released from duty to attend the funeral***; ***he released the safety catch on his rifle***; **release point** = point on a route, where sub-units leave their parent unit and continue independently by different routes **(c)** to allow a substance to come out; ***the missile released a chemical agent over our position***

relief [rɪ'li:f] *noun* **(a)** act of providing assistance or support; **relief agency** = civilian organization which provides assistance to victims of war or disaster **(b)** act of destroying or driving off an enemy force which has surrounded another friendly force or is besieging or investing a friendly town or city; ***he took part in the relief of Mafeking***; **relief force** = military force, which is sent to relieve another friendly unit or place which is surrounded by the enemy **(c)** act of taking over a duty or task from another person or unit; ***my relief is at 2000hrs***

relief-in-place [rɪˌli:f ɪn 'pleɪs] *noun* act of taking over positions from another unit or grouping

relieve [rɪ'li:v] *verb* **(a)** to provide assistance or support; ***we requested reinforcements to relieve the pressure on the small garrison*** **(b)** to destroy or drive off an enemy force, which has surrounded another friendly force or is besieging or investing a friendly town or city; ***a force of marines was sent to relieve the town*** **(c)** to take over a duty or task from another person or unit; ***I'll send someone to relieve you at 2100hrs***

religion [rɪ'lɪdʒən] *noun* belief in a god, and the rituals which express that belief

religious [rɪ'lɪdʒəs] *adjective* relating to religion

reload [ri:'ləʊd] *verb* to load a weapon again after firing; ***he was shot as he was reloading***

relocate [ˌri:ləʊ'keɪt] *noun* to move to a new location; ***we'll have to relocate the RAP***; ***the sniper always relocates after every shot***

REME ['ri:mi] = ROYAL ELECTRICAL AND MECHANICAL ENGINEERS

remit ['ri:mɪt] *noun* instructions *or* guidelines *or* rules (usually in relation to a peacekeeping mission); ***that sort of task is not part of my remit***

remf [remf] *noun (slang)* soldier who is not serving in the front line (and is therefore not in any danger)

remote [rɪ'məʊt] *adjective* **(a)** distant; **remote control** = means by which a device or machine can be operated from a distance (eg. command wire, radio signals); ***the bomb was detonated by remote control*** **(b)** isolated; ***the village is very remote***

remotely piloted vehicle (RPV) [rɪˌməʊtli ˌpaɪlətɪd 'vi:ɪkl] *noun* small unmanned radio-controlled aircraft designed to carry surveillance equipment (NOTE: also known as an **unmanned aerial vehicle (UAV)** *or* **drone**)

rendezvous (RV) ['rɒndeɪvu:] **1** *noun* place where people have agreed to meet up with each other; ***the rendezvous is at grid 453213*** **2** *verb* to meet up with other people at a rendezvous; ***we will rendezvous at grid 654776***

reoccupy [ri:'ɒkjʊpaɪ] *verb* to occupy again, after being away; ***the enemy have reoccupied the village***

reorg [rɪ'ɔ:g] = REORGANIZE; REORGANIZATION

reorganize [rɪ'ɔ:gənaɪz] *verb* **(a)** to organize in a different way; ***the brigade has been reorganized*** **(b)** to carry out the reorganization phase of an attack; ***the platoon will reorganize on the objective***

reorganization [rɪˌɔ:gənaɪ'zeɪʃn] *noun* **(a)** act of reorganizing; ***the reorganization of the brigade has been completed*** **(b)** phase following a successful assault, during which the assaulting troops go firm, in order to redistribute ammunition, deal with any casualties or prisoners, and reassess their situation before continuing their task; ***there will be a resupply of ammunition during reorganization***

reorientate [ri:'ɔ:riənteɪt] *verb* to stop and confirm your exact location and the direction in which one should be heading, before continuing a journey

repair [rɪ'peə] **1** *noun* act of repairing; ***the vehicle must be sent back for repair*** **2** *verb* to mend something which is damaged or defective; ***the vehicle is being repaired***

repatriate [ri:'pætrɪeɪt] *noun* to return a prisoner to his own country; ***he was repatriated at the end of the war***

repeat [rɪ'pi:t] *verb* **(a)** to say something again; ***he repeated the question*** **(b)** to do something again; ***he was warned not to repeat his mistake*** (NOTE: on the radio, **repeat** is only used when you want the artillery or mortars to repeat a fire-mission. It is never used when you want someone to repeat a message. In that situation, **say again** is used instead)

repel [rɪ'pel] to fight and push back; ***they repelled all attempts to board their ship***

replace [rɪ'pleɪs] *verb* **(a)** to put something back in its original position; ***he replaced the handset*** **(b)** to obtain a new object to take the place of one that has been damaged or lost; ***we need to replace the firing pin*** **(c)** to take the place or job of another person; ***I am replacing Major Knight*** **(d)** to arrange for someone to take the place of another person; ***we will have to replace Sergeant Jones*** **(e)** *(in passive)* **to be replaced** = to be dismissed from a position or job; ***the general is being replaced***

replacement [rɪ'pleɪsmənt] *noun* person or thing which takes the place of another person or thing; **battle casualty replacement (BCR)** = soldier who remains on stand-by in order to take the place of a soldier who is killed or wounded

replen ['ri:plen] = REPLENISHMENT

replenish [rɪ'plenɪʃ] *verb* to provide someone with fresh supplies (such as ammunition, food and water, fuel, etc.)

replenishment [rɪ'plenɪʃmənt] *noun* act of providing fresh supplies (such as ammunition, food and water, fuel, etc.)

report [rɪ'pɔ:t] **1** *noun* **(a)** verbal or written information, which is given or sent to another person; **contact report** = information relating to a sighting of the enemy; **situation report (SITREP)** = information relating to the current situation; **report line** = real or imaginary line on the ground, with a code word which units use when they cross, so that a commander can monitor their progress **(b)** disciplinary action; ***I am putting you in the report*** **2** *verb* **(a)** to give or send information to another person; ***B Company report enemy recce to the east of Karlsbad*** **(b)** to take disciplinary action against someone; ***I am reporting you for this offence*** **(c)** to show that you are officially present; ***he reported for duty three days late***; **to report to a place** = to arrive officially at a place; ***new recruits are asked to report to the training camp***

reprisal [rɪ'praɪzl] *noun* severe punitive action (usually intended as a deterrent); ***ten villagers were shot as a reprisal for the attack on the convoy***

repulse [rɪ'pʌls] *verb* to beat off an attack; ***all enemy attacks were repulsed with considerable losses on both sides***

request [rɪ'kwest] **1** *noun* act of asking for someone or something; ***have you received my ammunition request?*** **2** *verb* to ask for something; ***'hullo 2, this is 22, request immediate assistance, over'***

re-route [ri:'ru:t] *verb* to make someone travel by a different route; ***the convoy was re-routed because of the chemical attack***

rescue ['reskju:] **1** *noun* act of saving someone; **air-sea rescue** = using aircraft and helicopters to save someone from the sea **2** *verb* **(a)** to save someone from a dangerous situation; ***we used a helicopter to rescue the sailors*** **(b)** to free someone from captivity; ***all the prisoners of war have been rescued***; ***the hostages were rescued by the SAS***

reserve [rɪ'zɜ:v] *noun* **(a)** units or sub-units which are held back from an engagement, so that they can be used to reinforce or support any unit which gets into difficulties; ***the general was unwilling to use his reserves***; **in reserve** = acting as a reserve; ***6 Platoon will be in reserve for Phase 1*** **(b) the Reserve** = manpower (usually consisting of ex-servicemen) which can be used to supplement the regular forces in times of war or national emergency; ***on leaving the armed forces, a serviceman is liable for a further seven years in the Reserve***

reservist [rɪ'zɜ:vɪst] *noun* person (usually an ex-serviceman) who is liable for service with the Reserve, in the event of war or national emergency; ***all reservists must report to their nearest police station***

reservoir ['rezəvwɑ:] *noun* natural or man-made lake, used as a supply of water

residual force [rɪˌzɪdjʊəl 'fɔ:s] *noun* small security force which remains in a war zone after hostilities have ended and the main force has withdrawn

resist [rɪ'zɪst] *noun* **(a)** to fight against something; ***we were surprised that the garrison resisted so strongly***; ***the snipers resisted all our attempts to dislodge them***

(b) to oppose something; ***the Army is resisting the new proposals***

resistance [rɪ'zɪstəns] *noun* **(a)** act of resisting; ***the enemy is putting up little resistance*** **(b) resistance (movement)** = secret organization which opposes (often with armed force) the established government or an occupying power; ***the Resistance attacked the railway lines***

resource [rɪ'zɔːs] **1** *noun* anything which is available for use (such as ammunition, equipment, manpower, etc.); ***the guerillas have enough resources to cause considerable damage*** **2** *verb* to allocate resources; ***he is responsible for resourcing within the Corps***

respirator ['respɪreɪtə] *noun* protective face-covering containing an apparatus to filter air, which is used to protect a person from chemical agents and radioactive contamination; *see also* GAS MASK

respond [rɪ'spɒnd] *verb* **(a)** to answer or reply; ***he did not respond to the accusation*** **(b)** to take action as a result of an incident or situation; ***they responded by shooting the hostages***

response [rɪ'spɒns] *noun* **(a)** act of responding; ***there has been no response to our demands*** **(b)** means or method of responding; **nuclear response** = use of nuclear weapons in order to defeat or punish an act of aggression by another state

responsibility [rɪˌspɒnsɪ'bɪlɪti] *noun* **(a)** obligation or duty; ***one of my responsibilities is the security of the ammunition compound***; *see also* TACTICAL AREA OF RESPONSIBILITY **(b)** person or thing for which one is responsible; ***these men are my responsibility*** **(c)** sensible behaviour; ***you should show more responsibility***

responsible [rɪ'spɒnsɪbl] *adjective* **(a)** having an obligation or duty to do something; ***you are responsible for maintaining these vehicles*** **(b)** behaving in a sensible manner; ***you should be more responsible***

restrict [rɪ'strɪkt] *verb* to put a limit on something; ***access to the training ground is restricted to certain personnel only***

restricted [rɪ'strɪktɪd] *adjective* lowest security classification for information or documents; relating to material which is restricted to members of the armed forces and may not be passed on to the media or general public

> COMMENT: information is classified according to its importance, eg: restricted, confidential, secret, top secret, etc.

restriction [rɪ'strɪkʃn] *noun* **(a)** act of restricting; ***gangrene can be caused by a restriction of the blood supply***; ***the new restrictions mean that there will be less ammunition for training***; ***all restrictions on movement have been lifted***; **restriction of privileges** = military punishment where a serviceman is required to parade at certain times of the day in a specified order of dress **(b)** something which affects or limits a person's choice of action; ***restrictions in time and space left us with no alternative but to mount a frontal assault***

restrictive control [rɪˌstrɪktɪv kən'trəʊl] *noun* doctrine of command and control which relies upon detailed planning and strict obedience to precise orders; *compare* DIRECTIVE COMMAND

resupply [ˌriːsʌ'plaɪ] **1** *noun* act of providing fresh supplies (such as ammunition, food and water, fuel, etc.); ***there will be a resupply of ammunition at 1800hrs*** **2** *verb* to supply something again; ***they resupplied the garrison with water***

retake [riː'teɪk] *verb* to capture a position or location which has been captured by the enemy; ***B Company have retaken the hill***

retaliate [rɪ'tælɪeɪt] *verb* to take action in response to an attack, insult or provocation; ***the soldiers have been ordered not to retaliate if they are insulted by the local population***

retaliation [rɪˌtælɪ'eɪʃn] *noun* act of retaliating; ***there has been no retaliation to the air strikes***; ***they killed three villagers in retaliation for the murder of the general***

retaliatory [rɪ'tæljətərɪ] *adjective* made in retaliation to an act *or* incident; ***NATO has authorized retaliatory air strikes***

retire [rɪ'taɪə] *verb* **(a)** to move away from the enemy; ***we retired to our own***

lines **(b)** to move back towards your own forces or territory; ***the enemy have retired***; *see also* RETREAT, WITHDRAW

retreat [rɪ'triːt] **1** *noun* **(a)** act of retreating; ***the enemy forces are in retreat***; ***the retreat from Moscow was accompanied by huge losses*** **(b)** (Retreat) ceremony to signify the close of the working day in barracks (usually around 1800hrs), when the flags are lowered; **beating the retreat** = ceremonial parade of music and marching, which is held in the evening**;** *GB* bugle call, blown in barracks to signify the end of the working day (usually around 1800hrs; should not be confused with Last Post) **2** *verb* **(a)** to move away from the enemy; ***we had to retreat when B Company's position was overrun*** **(b)** to move back towards your own forces or territory; ***the enemy are retreating***; *see also* RETIRE, WITHDRAW

> COMMENT: the word **retreat** is normally used when one is forced to move back (for example because one has been defeated or your position has become untenable), whereas **retire** *or* **withdraw** imply rearward movement as part of a planned manoeuvre or in order to occupy a better position. Consequently, **retire** *or* **withdraw** are sometimes used instead of **retreat** because they sound more positive COMMENT: if you are visiting a British barracks and you hear "Retreat" being blown on the bugle, it is good manners to stop what you are doing and stand still until the call has finished.

return [rɪ'tɜːn] *verb* **(a)** to come back; ***the patrol has returned***; ***all planes returned safely to base*** **(b)** to give something back; ***he returned the equipment***; **to return fire** = to shoot back, when shot at

reveille [rɪ'væli] *noun* **(a)** time at which troops are woken up; ***reveille at 0600hrs*** **(b)** bugle-call blown at reveille

reverse [rɪ'vɜːs] **1** *verb* to drive backwards; ***the tank reversed into a barn*** **2** *noun* failure *or* misfortune; ***after several reverses, the enemy withdrew*** **3** *adjective* opposite; **reverse slope** = far side of a hill, which is in dead ground to the enemy; ***the Duke of Wellington always preferred a reverse slope position***

revet [rɪ'vet] *verb* to strengthen the sides of a trench or other fortification, using corrugated iron, wood or other material; ***pickets are used in the revetting of trenches and other fortifications*** (NOTE: **revetting - revetted**)

review [rɪ'vjuː] **1** *noun* large-scale parade and inspection of troops; ***he is attending a review*** **2** *verb* to inspect a large number of troops, etc., on parade; ***the Queen came to review the fleet***

revolt [rɪ'vəʊlt] **1** *noun* armed resistance to the established government or to the army command; ***the revolt has spread to the neighbouring provinces*** **2** *verb* to oppose the established government with armed force; ***we expect members of the former president's bodyguard to revolt against military rule***

revolution [ˌrevə'luːʃn] *noun* overthrow of a government or social order by the use of force

revolutionary [ˌrevə'luːʃənri] **1** *adjective* relating to a revolution; **Revolutionary Guard** = elite troops in some communist armies **2** *noun* person who takes an active role in a revolution; ***most of the leading revolutionaries have been arrested***

revolver [rɪ'vɒlvə] *noun* hand-held gun, with a chamber which turns when a shot is fired, so that a fresh cartridge is ready for firing; ***he used his revolver to kill the civilian***

RF-4C [ɑː ˌef fɔː 'siː] *noun* reconnaissance version of the F-4 Phantom

R Gp = R GROUP

R Group ['ɑː ˌgruːp] *noun* = RECCE GROUP small group (usually consisting of a commander and his subordinate commanders), which carries out a reconnaissance before planning an operation or task

RHA [ˌɑː eɪtʃ 'eɪ] = ROLLED HOMOGENEOUS ARMOUR

RHQ [ˌɑː eɪtʃ 'kjuː] = REGIMENTAL HEADQUARTERS

ricochet ['rɪkəʃeɪ] **1** *noun* projectile (especially a bullet) which bounces off a surface; ***he was hit by a ricochet*** **2** *verb (of projectiles, especially bullets)* to hit the ground or some other hard object and bounce off (often in a different direction to the original line of flight); ***the bullet ricochetted around the room*** (NOTE: the past form is pronounced: ['rɪkəʃeɪd])

ride [raɪd] **1** *noun* wide path cut through a wood or forest; ***we came under fire as we were crossing a ride*** **2** *verb* **(a)** to travel on a horse or motorcycle; ***the general rode into the defeated city on a white horse*** **(b)** to travel, sitting on the top of a vehicle; ***the infantry were riding on tanks*** (NOTE: **riding - rode - has ridden**)

ridge [rɪdʒ] *noun* long narrow line of high ground, formed where two slopes meet each other; ***after several hours of hand-to-hand fighting we took the ridge***

rifle ['raɪfl] **1** *noun* hand-held firearm with a long rifled barrel and a butt, which is placed against the shoulder for firing; **rifle company** = normal company in an infantry battalion (as opposed to support company) **2** *verb* to cut spiral grooves in the barrel of a gun or artillery piece, in order to make the projectile spin during flight; ***the barrel is carefully rifled***

rifleman ['raɪflmən] *noun* infantry soldier armed with a rifle

> COMMENT: most modern riflemen carry assault weapons

rifling ['raɪflɪŋ] *noun* spiral grooves, cut into the barrel of a gun or artillery piece in order to make the projectile spin during flight

ring of steel [ˌrɪŋ əv 'stiːl] *noun* tactical manoeuvre in which tanks, having fought their way through an objective, then form a protective screen around the flanks and far side so that dismounted infantry can clear all the trenches and bunkers on the objective

riot ['raɪət] **1** *noun* violent public disturbance; ***there have been riots in all the major cities*** **2** *verb* to take part in a riot; ***the inhabitants rioted when the police arrested the editor of the local newspaper***

rioter ['raɪətə] *noun* person who takes part in a riot; ***three of the rioters were shot dead***

rip-cord ['rɪp kɔːd] *noun* device which is pulled by hand in order to open a parachute

rise [raɪz] *noun* high ground; ***the enemy position is just over the next rise***

river ['rɪvə] *noun* body of fresh water, which flows along a natural channel towards a sea or lake

RM [ˌɑː 'em] = ROYAL MARINES

RMO [ˌɑː em 'əʊ] = REGIMENTAL MEDICAL OFFICER

RN [ˌɑː 'en] = ROYAL NAVY

road [rəʊd] *noun* way with a prepared surface, designed for use by vehicles; ***only one road was suitable for tanks***; ***there are very few roads in the region***

roadblock ['rəʊdblɒk] *noun* **(a)** obstruction set up by troops or police in order to control the movement of vehicles; ***we set up a roadblock on the edge of the town*** **(b)** the troops or police manning a roadblock; ***the roadblock was attacked during the night***

rock [rɒk] *noun* **(a)** hard mineral substance, forming part of the earth's surface; ***the sappers were digging into the rock under the castle walls*** **(b)** piece of rock; ***rocks had rolled down the hillside and blocked the road***

rocket ['rɒkɪt] *noun* projectile which contains its own propellant; **rocket launcher** = apparatus or vehicle from which a rocket is fired

> COMMENT: the term **rocket** usually refers to a direct- or indirect-fire weapon, whereas a **missile** is normally equipped with its own guidance system, which controls its flight onto the target

rocket-booster ['rɒkɪt ˌbuːstə(r)] *noun* additional charge of propellant fitted to an artillery shell, which detonates in mid-air and increases its range

rocky ['rɒki] *adjective; (of terrain)* consisting mainly of rock

ROE [ˌɑː əʊ 'iː] = RULES OF ENGAGEMENT

roger ['rɒdʒə] *adverb (radio terminology)* **(a)** that is correct; ***'hullo 22, this is 2, confirm that you are at the RV, over' - '2, roger, out'*** **(b)** I have understood your instructions; ***'hullo 22, this is 2, move***

now, over' - '2, roger, out.'; *see also* AFFIRMATIVE, COPY

Roland ['rəʊlənd] *noun* French/German-designed short-range surface-to-air missile (SAM)

role [rəʊl] *noun* function or purpose for which a person or thing is used; ***the role of an anti-tank platoon is the destruction of enemy armour***; **role specialization** = situation where one nation is responsible for supplying one type of equipment or personnel for a multinational force; *see also* MULTIROLE

roll up [ˌrəʊl 'ʌp] *verb* to assault through an enemy position sideways, destroying or capturing it trench by trench; ***once we had gained a foothold on the position, we were able to roll it up from the right***

rolled homogeneous armour (RHA) [ˌrəʊld hɒməˌdʒiːniəs 'ɑːmə] *noun* armour composed of a single substance (such as steel alloy), which has been rolled to a uniform thickness (NOTE: 'homogeneous' is often written incorrectly as 'homogenous', pronounced [hə'mɒdʒənəs], which is actually a biological term describing organisms which are similar because they share a common ancestry. The error is so common that many dictionaries now treat the two words as interchangeable)

ROM [rɒm] = REFUELLING ON THE MOVE

Roman Catholic Church [ˌrəʊmən ˌkæθlɪk 'ʃɜːtʃ] *noun* original western European form of Christianity; *compare* ORTHODOX CHURCH, PROTESTANT CHURCH

Romeo ['rəʊmiəʊ] eighteenth letter of the phonetic alphabet (Rr)

romer ['rəʊmə(r)] *noun* simple mathematical instrument for calculating accurate grid references from a map, consisting of a piece of clear plastic printed with a grid which subdivides a grid square into ten northings and ten eastings; ***I need a romer for a 1:20,000 map***

> COMMENT: military protractors are usually printed with romers for the most common scales of map (ie. 1:50,000 and 1:25,000

rookie ['rʊki] *noun (slang)* recruit

rope [rəʊp] *noun* thick line of twisted fibres, normally used for pulling or suspending heavy objects

rotary-wing aircraft [ˌrəʊtəri wɪŋ 'eəkrɑːft] *noun* helicopter

rotate [rəʊ'teɪt] *verb* **(a)** to revolve around an axis; ***the rotor was not rotating properly*** **(b)** to complete one tour of duty and move on to the next one; ***I am rotating back to the States next month***

rotor ['rəʊtə] *noun* set of horizontally rotating blades, which gives a helicopter its upward lift

round [raʊnd] *noun* one projectile plus the propellant required to fire it; ***we require 2,000 rounds of 5.56mm ball***; ***ten rounds HE, fire for effect!***; **blank round** = piece of training ammunition, consisting of the propellant but no projectile, which is designed to simulate the firing of a weapon; ***we require 5,000 rounds of 7.62mm blank***; **live round** = piece of real ammunition (as opposed to a blank round); ***the militia fired live rounds into the crowd***

roundel ['raʊnd(ə)l] *noun* circular identification mark painted on aircraft (usually denoting nationality); ***British aircraft usually have roundels of red and blue***

rout [raʊt] **1** *noun* retreat following a defeat, where command and control has completely broken down; ***what was intended to be an orderly retreat turned into a rout*** **2** *verb* to force the enemy into a rout; ***the invaders were routed***

route [ruːt *US* raʊt] *noun* way from one location to another; ***the companies moved to the FUP by different routes***; **route card** = card showing the different stages of a journey, with locations, distances, bearings and other information; **route march** = long-distance march, designed to improve or maintain physical fitness

routine [ruː'tiːn] **1** *adjective* normal, which happens all the time; ***this is a routine inspection*** **2** *noun* regular programme of tasks or duties; ***that is not part of our routine***

royal ['rɔɪəl] *adjective* in the service of a king or queen; **Royal Air Force (RAF)** = British air force; **Royal Armoured Corps (RAC)** = tank regiments of the British Army; **Royal Artillery (RA)** = artillery of the British Army; **Royal Engineers (RE)** = engineers of the British Army; **Royal Electrical and Mechanical Engineers (REME)** = vehicle mechanics of the British Army; **Royal Marines (RM)** = British marines; **Royal Navy (RN)** = British navy

RP [ˌɑː ˈpiː] = REGIMENTAL POLICE

RPG-7 [ˌɑː piː dʒiː ˈsevən] *noun* Soviet-designed hand-held anti-tank rocket

RPK [ˌɑː piː ˈkeɪ] *noun* Soviet-designed light machine-gun (LMG), based on the AK-47 assault weapon

RPV [ˌɑː piː ˈviː] = REMOTELY PILOTED VEHICLE

RQMS [ˌɑː kjuː em ˈes] = REGIMENTAL QUARTERMASTER SERGEANT

RRF [ˌɑː ɑː ˈef] = RAPID REACTION FORCE

RSM [ˌɑː es ˈem] = REGIMENTAL SERGEANT MAJOR

RSO [ˌɑː es ˈəʊ] = REGIMENTAL SIGNALS OFFICER

RTU [ˌɑː tiː ˈjuː] *verb* = RETURNED TO UNIT to remove a person from a training course prematurely, usually as a result of injury *or* failure *or* misdemeanour; ***he's been RTU'd***

rubber bullet [ˌrʌbə ˈbʊlɪt] *noun* large projectile made of plastic or rubber which is fired from a special gun and is designed to knock a person over but not to cause a serious injury (NOTE: also known as **plastic bullet** *or* **baton round**)

rubble [ˈrʌbl] *noun* fragments from damaged or destroyed buildings; ***the streets are blocked with rubble***

rucksack [ˈrʌksæk] *noun* large fabric container, designed to be carried on a person's back; *see also* BERGEN, PACK

rudder [ˈrʌdə(r)] *noun* **(a)** vertical blade at the stern of a boat *or* ship which is used for steering. **(b)** vertical blade hinged to the tail of an aircraft, which is used for steering

run [rʌn] *verb* **(a)** to move quickly on foot; ***he ran to the latrine*** **(b)** to manage something; ***he is running the mortar course*** (NOTE: **running - ran - have run**)

runner [ˈrʌnə] *noun* soldier used to deliver verbal messages; ***a runner came up with a message from the general***

runway [ˈrʌnweɪ] *noun* prepared surface used by aircraft for take-off and landing; ***the runway has been captured by the enemy***; ***this runway if too short for transport aircraft***; **runway lights** = lights arranged along the sides of a runway or across it, to indicate where it is

Rupert [ˈruːpət] *noun GB (soldiers' slang)* officer; ***He got busted for hitting a Rupert***

rural [ˈrʊərəl] *adjective* relating to the countryside (as opposed to towns and cities); *compare* URBAN

ruse [ruːz] *noun* act of deception; ***the enemy withdrawal was just a ruse***

rush [rʌʃ] **1** *noun* sudden assault; ***the last enemy rush overran 6 Platoon's trenches*** **2** *verb* **(a)** to move suddenly and quickly towards something; ***they rushed towards the vehicles*** **(b)** to make a sudden assault; ***the enemy tried to rush our positions***

rust [rʌst] **1** *noun* harmful brown discolouration to iron or steel, caused by exposure to damp air or water; ***there is rust on your weapon*** **2** *verb* to be affected by rust; ***your weapon will rust if you don't oil it***; ***rusting equipment littered the courtyard of the former command HQ***

rusty [ˈrʌsti] *adjective* affected by rust; ***look at that gun - it's rusty***; ***the road through the desert was lined with rusty tanks***

rut [rʌt] *noun* deep mark made by wheels passing over damp ground; ***the lorry got stuck in a deep rut***

rutted [ˈrʌtɪd] *adjective; (of ground, especially roads and tracks)* affected by ruts; ***the track is badly rutted***

RV [ˌɑː ˈviː] **1** *noun* **(a)** rendezvous; ***the RV is at grid 453213*** **(b)** re-entry vehicle; ***the missile is fitted with three nuclear RVs*** **2** *verb* to meet at a rendezvous; ***we will RV at grid 453213***

RWR [ˌɑː dʌb(ə)ljuː ˈɑː(r) *or* rɔː] *noun* = RADAR WARNING RECEIVER device which warns a pilot that his aircraft is being hit by a radar beam

SIERRA - Ss

S-3 [ˌes 'θriː] *noun* American-designed sea-strike aircraft (NOTE: also known as **Viking**)

SA- [ˌes 'eɪ] *abbreviation* NATO prefix given to Soviet-designed surface-to-air missiles (SAM); **SA-1** = Guild surface-to-air missile; **SA-2** = Guideline surface-to-air missile; **SA-3** = Goa surface-to-air missile; **SA-4** = Ganef surface-to-air missile; **SA-5** = Gammon surface-to-air missile; **SA-6** = Gainful surface-to-air missile; **SA-7** = Grail surface-to-air missile; **SA-8** = Gecko surface-to-air missile; **SA-9** = Gaskin surface-to-air missile; **SA-10** = Grumble surface-to-air missile; **SA-13** = Gopher surface-to-air missile; **SA-14** = Gremlin surface-to-air missile; **SA-16** = Gimlet surface-to-air missile

SA-80 [ˌes eɪ 'eɪti] *noun* British-designed 5.56mm assault rifle (NOTE: plural is **SA-80s** [ˌes eɪ 'eɪtiz])

SAA 1 SMALL ARMS AMMUNITION **2** SKILL AT ARMS

Saab-35 [ˌsɑːb θɜːti 'faɪv] *noun* Swedish-designed multirole fighter aircraft (NOTE: also known as **Draken**)

Saab-37 [ˌsɑːb θɜːti 'sevən] *noun* Swedish-designed fighter aircraft, which is also suitable for ground attack (NOTE: also known as **Viggen**)

Saab-39 [ˌsɑːb θɜːti 'naɪn] *noun* Swedish-designed multirole fighter aircraft (NOTE: also known as **Gripen**)

saber ['seɪbə] *see* SABRE

sabkha *or* **sabqua** ['sæbkə(r)] *noun* *Arabic* firm crust on the surface of a dried-up lake bed, underneath which is soft mud; ***we got stuck in sabkha***

> COMMENT: sabkha is a serious hazard in the desert, because the outer margins are usually strong enough to support the weight of a vehicle, thus allowing it to get bogged down out in the middle of the lake bed from where it is extremely difficult to recover. It therefore provides a useful natural obstacle for the tactician. Good desert maps will show known areas of sabkha

sabot ['sæbəʊ *US* 'seɪbəʊ] *noun* **(a)** metal collar or sleeve, which is fitted to a long-rod penetrator in order to give it stability and extra kinetic energy as it travels up the barrel of the gun, and which is designed to fall away, once the projectile has left the muzzle of the weapon **(b)** any type of tank ammunition which consists of a long-rod penetrator and a discarding-sabot; ***load with sabot!***; *see also* ARMOUR-PIERCING DISCARDING-SABOT (APDS), ARMOUR-PIERCING FIN-STABILIZED DISCARDING-SABOT (APFSDS)

sabotage ['sæbətɑːʒ] **1** *noun* act of damaging or destroying an enemy installation or piece of equipment, so that it cannot be used **2** *verb* to carry out an act of sabotage; ***protesters tried to sabotage the missile installation***

saboteur [ˌsæbə'tɜː] *noun* person who carries out an act of sabotage

sabqua ['sæbkə] *see* SABKHA

sabre *US* **saber** ['seɪbə] *noun* heavy curved sword, traditionally used by cavalrymen; *GB* **sabre squadron** = squadron of tanks

sabre-rattling ['seɪbə ˌrætlɪŋ] *noun (informal)* increase in military activity during a period of international tension (deployment of troops on a border, recall of reservists, etc.); ***after a period of sabre-rattling by both sides, the situation was resolved by the United Nations***

SACEUR ['sækɜː] *noun* = SUPREME ALLIED COMMANDER EUROPE most senior NATO commander in Europe, one of the two MNCs

SACLANT ['sæklænt] *noun* = SUPREME ALLIED COMMANDER ATLANTIC most senior NATO commander in the Atlantic, one of the two MNCs

SACLOS ['sæklɒs] *noun* = SEMI-AUTOMATIC COMMAND LINE OF SIGHT missile guidance system, which relies upon the operator continuing to track the target after launching until the missile actually hits it

sacrifice ['sækrɪfaɪs] *verb* to accept the loss of one *or* more of your groupings in order to avoid greater loss elsewhere; ***5 Brigade was deliberately sacrificed so that the rest of the corps could withdraw***

saddle ['sædl] *noun* ridge joining the tops of two hills; ***the enemy opened fire as we were moving across the saddle***

safe [seɪf] **1** *adjective* **(a)** free from danger; ***you are safe now*** **(b)** *(of weapons)* loaded, but not cocked and with no round in the breech; *GB* **to make safe** = to fully unload a cocked weapon and then replace the loaded magazine back onto the weapon; **made safe** = with a loaded magazine fitted, but the weapon is not cocked and there is no round in the breech; **on safe** = with the safety catch applied **(c)** *(of explosive devices)* not armed; ***the bomb is now safe*** **2** *noun* strong container fitted with a lock, which is used to store secret documents, money, valuable property, etc.

safety ['seɪfti] *noun* state of being safe; **safety catch** = mechanism which prevents a weapon from being fired

Sagger ['sægə] *noun* Soviet-designed wire-guided anti-tank missile (ATGW)

sail [seɪl] *verb* to travel by ship; ***the German fleet sailed along the English Channel***

sailor ['seɪlə] *noun* **(a)** member of a ship's crew; ***sailors from the aircraft carrier came ashore*** **(b)** person serving in the navy (especially one who is not an officer); ***fights broke out when a group of soldiers went into a sailors' bar***

salient ['seɪliənt] *noun* part of an army's front line, which sticks out at an angle towards the enemy's front line

> COMMENT: a salient usually occurs in one of two ways: either when an attack has largely failed but a small part of the attacking force has managed to capture ground, or when an attack has largely succeeded but a small part of the defending force has managed to hold its positions. Either way, the troops occupying the salient are particularly vulnerable, because they can be attacked on two sides by the enemy

Salmonella [ˌsælmə'nelə] *noun* group of bacteria which cause food poisoning and typhoid fever

> COMMENT: certain nations are known to have developed Salmonella bacteria for use as a biological weapon

salute [sə'luːt] **1** *noun* **(a)** military greeting made between officers and other ranks, which is carried out by raising the right hand to the peak of the cap; **butt salute** = salute made by slapping the butt or handguard of the rifle; **to take the salute** = to salute and be saluted by marching troops on parade **(b) gun salute** = greeting made by firing guns; ***the Queen's birthday is celebrated by a 21-gun salute*** **2** *verb* to carry out a salute; ***ordinary ranks must salute officers***

> COMMENT: the British naval salute has the hand more or less flat and horizontal, and is similar to the American style of saluting. The British army and air force salute has the hand flat, but with the palm facing outwards

saluting base [sə'luːtɪŋ ˌbeɪs] *noun* small wooden stand for the officer or important person taking the salute at a military parade

salvo ['sælvəʊ] *noun* **(a)** firing of several large-calibre guns at the same time (especially at sea) **(b)** firing of several rockets at the same time

SAM [sæm] = SURFACE-TO-AIR MISSILE

Sam Browne [ˌsæm 'braʊn] *noun* leather belt with a diagonal shoulder strap, traditionally worn by army officers

sand [sænd] *noun* substance consisting of tiny grains of rock, which covers the ground on beaches and in deserts

sandbag ['sændbæg] *noun* small hessian sack, designed to be filled with sand or soil, and used in the construction of fortifications

sandbank ['sændbæŋk] *noun* deposit of sand, found in shallow water in rivers, estuaries and the sea

sangar ['sæŋgə] *noun* **(a)** field fortification, constructed by building a circular wall of rocks or sandbags, when the ground is too hard or too wet to dig trenches **(b)** reinforced position for a sentry, constructed on the perimeter of an army base or installation

Sandhurst ['sændhɜːst] *noun* = ROYAL MILITARY ACADEMY SANDHURST British army officer training establishment; ***he passed out of Sandhurst in 1980***

sanitation [ˌsænɪ'teɪʃn] *noun* **(a)** practice of keeping yourself and your surroundings clean, in order to prevent disease; ***sanitation is extremely important in refugee camps*** **(b)** washing and toilet facilities; *see also* HYGIENE

sanitize ['sænɪtaɪz] *verb* ***a*** to disinfect something **b** to destroy *or* neutralize enemy positions; ***once the tanks have sanitized the objective, they form a ring of steel on the far side***

sapper ['sæpə] *noun (traditional)* **(a)** engineer **(b)** title of the lowest rank in the Royal Engineers; ***Sapper Williams***

SAR [ˌes eɪ 'ɑː(r)] = SEARCH AND RESCUE

Saracen ['særəsən] *noun* British-made wheeled armoured personnel carrier (APC)

Sarin ['sɑːrɪn] *see* GB

SAS [ˌes eɪ 'es] *noun* = SPECIAL AIR SERVICE elite British Army special forces organization

satellite ['sætəlaɪt] *noun* unmanned spacecraft, which is positioned in the earth's orbit and is designed to carry communications, surveillance or other electronic equipment

SATNAV ['sætnæv] = SATELLITE NAVIGATION

savannah [sə'vænə] *noun* wide level area of grassland with few trees (in tropical regions)

save [seɪv] *verb* **(a)** to rescue a person from danger; ***the helicopter managed to save the crew of the ship*** **(b)** to rescue a person from captivity; ***the hostages were saved by the SAS*** **(c)** to avoid waste; ***save your ammunition*** **(d)** to keep something for future use; ***we are saving most of our mortar rounds for the main assault***

SAW [sɔː] = SQUAD AUTOMATIC WEAPON; *see* M-249

Saxon ['sæksən] *noun* British-designed wheeled armoured personnel carrier (APC)

say again [ˌseɪ ə'gen] *verb (radio terminology)* to repeat a message; ***'Hello 2, this is 22. Say again last message. Over.'*** (NOTE: on the radio, **repeat** is only used when you want the artillery or mortars to repeat a fire-mission. It is never used when you want someone to repeat a message)

SBS [ˌes biː 'es] *noun* = SPECIAL BOAT SERVICE elite British special forces organization, recruited from the Royal Navy and Royal Marines

scale [skeɪl] *noun* **(a)** ratio of size between a map and the area of ground which it represents; ***most military maps have a scale of 1:50,000*** **(b)** diagram representing distance, which is usually found on the key of a map; ***look at the scale to work out how far the village is from here*** **(c)** quantity of ammunition, equipment or weapons, which are allocated to a unit or grouping; ***special forces have larger scales of ammunition than normal units***; ***the new ammunition scale is 200 rounds per gun***

Scarab ['skæræb] *noun* Soviet-designed tactical surface-to-surface missile

scarper ['skɑːpə] *verb GB (slang)* to run away; ***the enemy have scarpered***

scatter ['skætə] *verb* **(a)** to throw or drop objects over a wide area; ***these shells are designed to scatter leaflets over the enemy lines*** **(b)** *(of a group or crowd)* to run away or take cover in different directions; ***the protesters scattered when we opened fire***

Schwerpunkt ['ʃveərpʊnkt] *noun German* point of main effort; ***the Schwerpunkt for this attack will be at Lingen*** (NOTE: German nouns are spelt with a capital letter)

Scimitar ['sɪmɪtə] *noun* British-made light tank (CVRT), equipped with a 30mm Rarden cannon

scorched earth [ˌskɔːtʃt 'ɜːθ] *noun* deliberate destruction of your own infrastructure and resources so that the enemy cannot use them; ***throughout history, the Russians have used scorched earth tactics to defeat invaders***

Scorpion ['skɔːpiən] *noun* **(a)** British-made light tank (CVRT), equipped with a 76mm gun **(b)** insect with two front claws and a long jointed tail which it uses to inflict a venomous sting

scout [skaʊt] **1** *noun* **(a)** person sent out on a reconnaissance; ***the scouts haven't returned yet*** **(b)** person or vehicle which moves ahead of a grouping, in order to find a suitable route or locate the enemy; ***the scout vehicle drove into a minefield*** **(c)** **Scout** = small British-made utility helicopter **2** *verb* **(a)** to act as a scout **(b)** to reconnoitre; ***6 Platoon is scouting the enemy position***

scramble ['skræmbl] *verb* **(a)** to move over rocky terrain, using your hands when necessary; **scramble net** = net used by soldiers to climb from a ship into a landing-craft **(b)** *(of fighter aircraft)* to take off quickly in order to go into action; ***two fighter squadrons scrambled*** **(c)** *(of communications)* to adapt a transmission electronically, so that it can only be understood by someone with the correct receiving equipment; ***all messages to HQ must be scrambled***

scrape [skreɪp] *noun* shallow pit dug to provide a hull-down position for an armoured fighting vehicle; ***the tank was hit as it was reversing out of the scrape***; *see also* SHELL-SCRAPE

scree [skriː] *noun* loose surface of a mountain slope, consisting of a thick layer of small stones; ***we heard the enemy patrol moving across the scree***

scrim [skrɪm] *noun* small pieces of fabric, used as camouflage on helmets and camouflage nets (NOTE: no plural)

scrub [skrʌb] *noun* **(a)** vegetation consisting of small trees and bushes; ***the hill is covered with scrub*** **(b)** area of ground covered with scrub; ***the enemy position is to the right up the scrub*** (NOTE: no plural)

scrubland ['skrʌblənd] *noun* terrain consisting mainly of scrub

Scud [skʌd] *noun* Soviet-designed surface-to-surface missile, capable of carrying a variety of warheads (for example chemical, high explosive, nuclear, etc.)

SDI [ˌes diː 'aɪ] *see* STRATEGIC DEFENCE INITIATIVE

sea [siː] *noun* mass of salt water, which covers most of the earth's surface; **sea mile** = unit of linear measurement at sea, corresponding to 2,025 yards or 1,852 metres; *see also* NAUTICAL MILE

Sea Dart ['siː ˌdɑːt] *noun* British-designed long-range naval surface-to-air missile (SAM)

seaborne ['siːbɔːn] *adjective* deployed *or* carried by ships; ***the operation will be supported by a seaborne landing near Ostend***

SEAD [siːd] = SUPPRESSION OF ENEMY AIR DEFENCE

Sea Eagle ['siː ˌiːgl] *noun* British-designed anti-ship missile

Sea Harrier ['siː ˌhærɪə] *noun* British-designed multirole fighter aircraft with a vertical take-off capability, which is designed to operate from aircraft carriers and certain other ships

Seahawk ['siːhɔːk] *noun* American-designed multi-role helicopter designed to operate from a ship

Sea King ['siː ˌkɪŋ] *noun* American-designed multirole helicopter designed for operating off ships

SEALs [siːlz] *noun (short for* SEA, AIR, LAND) American naval special forces

seaman ['siːmən] *noun* **(a)** *(formerly)* rank in the British Navy (NOTE: now called **able rating, leading rating**) **(b)** *US* junior non-commissioned rank in the navy *US*

seaman apprentice = lowest non-commissioned rank in the navy

seaplane ['si:pleɪn] *noun* aircraft designed to take off from and land on water

search [sɜ:tʃ] **1** *noun* act of searching; **search and rescue operation** = operation to look for someone and rescue them **2** *verb* to look for someone or something; **search warrant** = warrant authorizing the security forces to search a specified building or property

searchlight ['sɜ:tʃlaɪt] *noun* powerful electric light, which produces a concentrated beam for illuminating objects

Sea Skua ['si: ˌskju:ə] *noun* British-designed anti-ship missile (ASM)

Sea Sparrow ['si: ˌspærəʊ] *noun* American-designed medium-range naval surface-to-air missile (SAM)

Sea Stallion ['si: ˌstæljən] *see* CH-53

sea-strike ['si: straɪk] *noun* naval aviation role, involving the use of carrier-based aircraft to attack enemy shipping

> COMMENT: this is the maritime equivalent of **fighter ground-attack**

Sea Wolf ['si: ˌwʊlf] *noun* British-designed short-range naval surface-to-air missile (SAM)

second ['sekənd] *noun* unit of time, corresponding to a sixtieth part of one minute

second [sɪ'kɒnd] *verb* to post a serviceman to another arm *or* service *or* even to the armed forces of a foreign state, usually to provide specialist expertise *or* training; ***he was seconded to the Sultan of Oman's Armed Forces***

second in command (2IC) [ˌsekənd ɪn kə'mɑ:nd] *noun* most senior person after the commander and nominated to take command in his absence

second lieutenant (2Lt) [ˌsekənd lef'tenənt *US* lu:'tenənt] *noun* **(a)** *GB* lowest officer rank in the army or marines (usually in command of a platoon or equivalent-sized grouping) **(b)** *US* lowest officer rank in the army, marines or air force (usually in command of a platoon or equivalent-sized grouping)

secondment [sɪ'kɒndmənt] *noun* act of being seconded; ***he went on secondment to the Sultan of Oman's Armed Forces***

secret ['si:krət] **1** *adjective* **(a)** not for common knowledge; ***they got out of the camp through a secret tunnel*** **(b)** high level of security classification for documents and information; ***this information is classified as secret***; **top secret** = highest security classification for documents and information (NOTE: information is classified according to its importance: **restricted, confidential, secret, top secret, etc.**) **2** *noun* fact or information which is secret; ***he was charged with passing secrets to the enemy***

sectarian [sek'teərɪən] *adjective* relating to conflict caused by differences in religion (eg. Roman Catholic as opposed to Protestant *or* Orthodox); ***this was a sectarian attack***; ***the conflict in Bosnia was sectarian rather than ethnic***

section ['sekʃn] *noun* **(a)** sub-unit of a platoon **(b)** *GB* tactical infantry grouping of eight men (usually divided into two fireteams) **(c)** *US* tactical armoured grouping of two tanks (NOTE: in the US Army, section-sized infantry groupings are known as **squads**)

> COMMENT: British infantry sections are usually commanded by corporals

sector ['sektə] *noun* subdivision of an area of ground; ***there has been no enemy activity in this sector***

secure [sɪ'kjʊə] **1** *adjective* **(a)** *(of ground)* in your possession and prepared for defence; ***the LZ is now secure*** **(b)** *(of objects)* properly attached or fastened; ***make sure that all your kit is secure*** **(c)** *(of containers, rooms, buildings or places)* locked or otherwise protected against theft; ***put these documents in a secure place*** **(d)** *(of communications)* encoded or scrambled; ***is the line to HQ secure?*** **2** *verb* **(a)** *(of ground)* to capture or otherwise take possession of a location, and prepare it for defence; ***the objective has been secured*** **(b)** *(of objects)* to attach or fasten properly; ***he secured the field dressing to his webbing***; ***all hatches have been secured*** **(c)** to make sure that something happens;

the aim of the operation is to secure the release of the hostages

security [sɪ'kjʊərəti] *noun* **(a)** all measures taken by a unit to protect itself from surveillance or offensive action by the enemy; ***the enemy has breached our security***; **security zone** = area of ground around a grouping's positions, which is covered by its weapons systems, surveillance equipment and patrol activity **(b)** all measures taken by an organization to protect its property and personnel from attack, espionage, sabotage, theft or any other threat or danger; ***he is responsible for security on the base***; **security forces** = a state's armed forces and police force (especially in a counter-insurgency situation) **(c)** department or organization responsible for protection against theft, etc.; ***security has reported a break-in***

seek [si:k] *verb* **(a)** to look for something; **heat-seeking missile** = missile equipped with a guidance system which homes in on a source of heat (such as the jet pipes of an aircraft engine) **(b)** to try to achieve something; ***he is seeking promotion***

seeker ['si:kə] *noun* person or device which is looking for something; **radar seeker** = guidance system which uses radar to locate its targets

segregate ['segrəgeɪt] *verb* to separate a person or group from other people, and keep them apart; ***the officers were segregated from the other prisoners***

seize [si:z] *verb* **(a)** to capture; ***A Company has seized the objective*** **(b)** to take something by force; ***the security forces have seized a large quantity of explosives***

self-control [ˌself kən'trəʊl] *noun* ability to control your own emotions (eg. anger, fear, etc.)

self-inflicted wound [ˌself ɪnˌflɪktɪd 'wu:nd] *noun* wound inflicted by a person on himself (usually in order to get out of the combat zone)

self-loading rifle (SLR) [ˌself ˌləʊdɪŋ 'raɪfl] *noun* **(a)** semi-automatic rifle (that is, one that reloads itself after each shot) **(b)** British-produced 7.62mm assault weapon, based on the Belgian-designed FN-FAL

self-propelled [ˌself prə'peld] *adjective* which has a motor which makes it move; **self-propelled gun (SPG)** = artillery piece, in the form of an armoured vehicle; **self-propelled howitzer (SPH)** = howitzer in the form of an armoured fighting vehicle (AFV); **self-propelled anti-aircraft gun (SPAAG)** = armoured fighting vehicle fitted with an anti-aircraft gun (NOTE **these weapons frequently have two or more barrels)**

semi-automatic [ˌsemi ɔ:tə'mætɪk] *adjective; (of firearms)* designed to reload automatically after each shot (as opposed to bolt-action rifles, which are operated by hand); ***the government forces are equipped with semi-automatic weapons***

Semtex ['semteks] *noun* Czech-produced plastic explosive (PE)

COMMENT: Semtex is favoured by terrorists, because it has no smell and does not 'sweat'. This makes it very difficult to detect

send [send] *verb* to make something go from one place to another; ***Britain has sent two battalions to the region***; **sending nation** = nation which has sent forces to be part of a multinational force (NOTE: **sending - sent**)

senior ['si:njə] *adjective* of higher rank; ***he is senior to you***; ***the senior officers were accommodated in a hotel*** GB **the Senior Service** = the Royal Navy

seniority [ˌsi:ni'ɒrɪti] *noun* position in the rank structure; ***you do not have the seniority to do this job***

sensor ['sensə] *noun* device which is designed to detect something (such as chemical agent, movement, radiation, etc.)

sentry ['sentri] *noun* **(a)** serviceman assigned to guard a military base or installation; ***a sentry challenged us as we approached the perimeter fence*** **(b)** soldier assigned to watch for any approaching enemy, while others rest or carry out other tasks; **air sentry** = soldier assigned to watch for hostile aircraft; **chemical sentry** = soldier assigned to watch for signs of a chemical attack **(c)** **Sentry** = American-designed E-3 airborne warning and control system (AWACS) aircraft

SERE ['sɪə(r)] = *US* SURVIVAL, EVASION, RESISTANCE (TO INTERROGATION) & ESCAPE

sergeant (Sgt) ['sɑːdʒənt] *noun GB & US* non-commissioned officer (NCO) in the army, marines or air force; *see also* FLIGHT SERGEANT, MASTER SERGEANT, STAFF SERGEANT

sergeant major [ˌsɑːdʒənt 'meɪdʒə] *noun* **(a)** *GB* warrant officer in the army or marines **(b)** *US* senior non-commissioned officer (SNCO) in the army, marines or air force

Sergeant York [ˌsɑːdʒənt 'jɔːk] *noun* nickname for the American-designed M-247 self-propelled anti-aircraft gun (SPAAG)

serve [sɜːv] *verb* to be employed in the armed forces; ***men serving in the armed forces are eligible for a pension***; ***she served ten years in the RAF***

service ['sɜːvɪs] **1** *noun* **(a)** act of serving in the armed forces; ***he has ten years' service***; **service dress** = smart khaki uniform worn on formal duties and parades; **active service** = service in a war zone; **operational service** = service involving the possibility of real combat (as opposed to peacetime soldiering) (NOTE: in the British armed forces, the term **active service** is only used when the nation is officially at war; For counter-insurgency and peacekeeping operations, the term **operational service** is used); **service support** = general term for administration and logistics at small unit level **(b) the services** = the armed forces; *GB* **the Junior Service** = the Royal Air Force; **the Senior Service** = the Royal Navy **2** *verb* to repair equipment and keep it in good condition

serviceman ['sɜːvɪsmən] *noun* man serving in the armed forces; *see also* EX-SERVICEMAN

servicewoman ['sɜːvɪsˌwʊmən] *noun* woman serving in the armed forces

servicing ['sɜːvɪsɪŋ] *noun* doing work for someone, repairing or maintaining equipment; *see also* CROSS-SERVICING

set [set] **1** *adjective* ready to function; ***the ambush is now set*** **2** *noun* apparatus; ***a radio set*** **3** *verb* to put something into a certain position; ***he set the controls to manual*** (NOTE: **setting - have set**)

set off [ˌset 'ɒf] *verb* to start on a jounrey; ***they set off during cover of darkness***

setting ['setɪŋ] *noun* position in which the control knobs or switches of an apparatus are set

set up [ˌset 'ʌp] *verb* **(a)** to assemble something; ***we will set up the mortar here*** **(b)** to establish something at a location; ***the RAP has been set up in the village***

sewer ['suːə] *noun* system of underground tunnels, used to carry water from drains and toilets

SF [ˌes 'ef] *noun* machine-gun in the sustained fire role; ***the SF was dug in on the forward edge of the wood***; ***the SFs will be located on that small knoll***; *see also* SUSTAINED FIRE

Sgt = SERGEANT

shanty town ['ʃænti taʊn] *noun* area of poor dwellings built from waste materials (often found on the edges of cities in poor countries)

shake out [ˌʃeɪk 'aʊt] *verb* to deploy from march formation (eg. column *or* file) into tactical formation for advance *or* assault; ***as soon as we clear the breach, we will shake out into assault formation***

SHAPE [ʃeɪp] *noun* = SUPREME HEADQUARTERS ALLIED POWERS IN EUROPE main NATO headquarters in Europe

shaped-charge warhead [ˌʃeɪpt ˌtʃɑːdʒ 'wɔːhed] *noun* anti-tank warhead, in which the explosive is packed around an inverted metal cone (on detonation, the cone collapses inwards to form a high velocity liquid-like jet which is capable of penetrating armour); *see also* HEAT

shared ['ʃeəd] *adjective* used by several different people together; **shared use** = use of forces or supplies sent by different nations to a multinational force

shell [ʃel] **1** *noun* **(a)** artillery projectile consisting of a metal case filled with high explosive, which is designed to explode on impact with the ground or when detonated by a fuse; ***shells fell on the town during the night***; **shell-case** = metal cartridge, used to hold the propellant of an artillery shell **(b)** *US* cartridge **2** *verb* to fire artillery shells at a target; ***'hullo 2, this is 22, am being shelled, wait out!'***; ***the enemy shelled the town for several hours***

shellfire ['ʃelfaɪə] *noun* firing of shells; ***we could hear shellfire during the night***

shell-scrape ['ʃel skreɪp] *noun* shallow pit designed to offer limited protection from artillery fire to a man lying on his stomach; ***we dig shell-scrapes every time we halt***

shell shock ['ʃel ˌʃɒk] *noun* mental and physical collapse, as a result of being shelled or simply being in combat for a long period of time; ***he is suffering from shell shock***; *see also* BATTLE FATIGUE, POST-TRAUMATIC STRESS DISORDER

shemagh [ʃɪ'mɑː] *noun Arabic* traditional Arab scarf, which may be worn around the neck *or* as a headdress and is suitable for protecting the nose and mouth during sandstorms; ***during the Gulf War, it was fashionable among British troops to wear a shemagh***

Sheridan (M-551) ['ʃerɪdən] *noun* American-designed light tank

shermuly [ʃə'muːli] *noun* parachute flare, which is fired from a small hand-held disposable launcher

shield [ʃiːld] **1** *noun* **(a)** piece of metal, plastic or other material, which is designed to be held in front of your body as protection from blows or projectiles; ***the riot police were equipped with batons and shields*** **(b)** anything which is used as a shield; **human shield = (i)** group of hostages who are placed in a location, in order to deter an attack on that location; **(ii)** group of hostages, behind whom a person positions himself, in order to deter people from shooting at him **(c)** structure which is fitted to a piece of equipment or machinery, in order to protect the operator from any dangerous effects; ***he was injured because he had removed the safety shield*** **2** *verb* to protect a person by placing something between him and a source of danger; ***we were shielded from the blast by the truck***

Shilka ['ʃɪlkə] *noun* Russian nickname for a ZSU-23 anti-aircraft gun

ship [ʃɪp] *noun* large boat

shipping ['ʃɪpɪŋ] *noun* ships in general (especially commercial vessels)

shock [ʃɒk] *noun* **(a)** effect caused by the violent collision of two objects; **shock action = (i)** formerly, a charge by heavy cavalry; **(ii)** sudden or aggressive attack or counter-attack, especially by tanks; **shock troops =** elite troops who are kept in reserve in order to attack or counter-attack the enemy when he is at his most vulnerable (for example, during reorganization) **(b)** physical collapse, as a result of a serious wound or horrifying experience; ***he was suffering from shock***; *see also* SHELL SHOCK

shoot [ʃuːt] **1** *noun* field of fire; ***this position offers a good shoot into the valley*** **2** *verb* **(a)** to fire a weapon at something; ***stop or I'll shoot!*** **(b)** to kill or wound someone by firing a weapon at him; ***he was shot in the leg***; ***both commanders were shot in the fighting*** **(c)** to kill someone as a punishment; ***the deserters were taken away and shot*** (NOTE: **shooting - shot - have shot**)

shoot away [ˌʃuːt ə'weɪ] *verb* to remove part of a vehicle, aircraft, etc., by firing a weapon at it; ***the tailplane was almost shot away by cannon fire***

shooting ['ʃuːtɪŋ] *noun* **(a)** act of shooting; ***we heard shooting during the night*** **(b)** incident in which a person is shot; ***there has been a shooting***

shore [ʃɔː] *noun* land at the edge of a lake or the sea

short-range ['ʃɔːt ˌreɪndʒ] *adjective* used over short distances; ***the Exocet is a short-range missile***

short take-off and landing (STOL) [ˌʃɔːt ˌteɪk ɒf ən 'lændɪŋ] *noun* technology which enables a fixed-wing aircraft to take off and land over considerably shorter distances than those required by conventional fixed-wing aircraft

> COMMENT: vertical take-off aircraft also require a short take-off and landing capability because vertical take-off is not usually possible when the aircraft is carrying a full payload of munitions. Once these munitions have been discharged, however, a

normal vertical landing would again be possible. The acronyms STOVL (short take-off and vertical landing) and V/STOL (vertical or short take-off and landing) are used to describe these capabilities

shot [ʃɒt] *noun* **(a)** act of firing a weapon; ***we heard a shot***; ***he took two shots at the enemy tank*** **(b)** person who shoots; ***he's an excellent shot*** **(c)** small metal balls fired from a shotgun (NOTE: the word is plural in this meaning)

shotgun ['ʃɒtgʌn] *noun* gun, usually with two barrels, which fires a quantity of small metal balls (or shot)

shovel ['ʃʌvəl] *noun* simple tool consisting of a rounded metal blade attached to a long wooden handle, designed for digging holes in the ground; **shovel recce** = (slang) going to the toilet in the field; ***I'm just going on a shovel recce***; *see also* SPADE

shrapnel ['ʃræpnəl] *noun* **(a)** *(historical)* artillery shell containing ball-bearings, which become projectiles when the shell explodes **(b)** projectiles formed by fragments of an exploding artillery shell or grenade; ***he was hit in the leg by shrapnel***; ***he was blinded by a piece of shrapnel*** (NOTE: no plural)

Shrike [ʃraɪk] *noun* American-designed air-to-ground anti-radar missile (ARM)

shrine [ʃraɪn] *noun* small building *or* structure which is used for religious purposes

SIB [ˌes aɪ 'biː] *noun GB* = SPECIAL INVESTIGATION BRANCH detective branch of the military police, which investigates criminal offences committed by servicemen while they are subject to military law (eg. on MOD property *or* while on operational service); ***the SIB have been called in***

side [saɪd] *noun* **(a)** vertical surface of an object; ***he aimed at the side of the tank*** **(b)** one of two opposing states or alliances, which are involved in a war or conflict; **opposite side** *or* **other side** = enemy forces; **our side** = friendly forces; ***our side has inflicted considerable damage on the enemy positions***

Sidewinder ['saɪdwaɪndə] *noun* American-designed heat-seeking air-to-air missile (AAM)

siege [siːdʒ] *noun* act of surrounding a town or location and preventing the entry of reinforcements and supplies, in order to force the defenders to surrender or to weaken them prior to an assault; ***the siege of Leningrad lasted more than a year***; *see also* BESIEGE, INVESTMENT

COMMENT: "siege" is not normally used in modern military English; it has now been largely replaced by the noun "investment"

Sierra [si'erə] nineteenth letter of the phonetic alphabet (Ss)

sight [saɪt] **1** *noun* **(a)** ability to see; **out of sight** = no longer visible **(b)** device on a weapon, which is used by the firer to aim at a target; **iron sight** = simple sight, forming part of the basic design of the weapon and consisting of a rear-sight and a fore-sight, which are lined up with each other and the point of aim; **night sight** = sight consisting of an optical instrument (such as an image intensifier or infrared), which improves visibility at night; **optic sight** = weapon sight which gives the firer a magnified image of the target **2** *verb* to see something for the first time; ***6 Platoon have sighted the enemy***; ***the enemy destroyer was sighted on the horizon***

sighting ['saɪtɪŋ] *noun* act of seeing something (usually for the first time); ***there have been no sightings of the enemy***

SIGINT ['sɪgɪnt] *noun* = SIGNALS INTELLIGENCE intelligence obtained by listening to the enemy's radio transmissions

sign [saɪn] **1** *noun* **(a)** written words or symbols painted or printed on a board or on the surface of an object, in order to convey information (such as direction, identity of a unit, location of a minefield, etc.); ***follow the signs to Brigade HQ*** **(b)** gesture designed to convey a meaning; ***he made a sign for us to keep quiet*** **(c)** evidence of activity or the presence of something (such as blood, discarded equipment, vehicle tracks, etc.); ***there was no sign of the enemy*** **2** *verb* to write your name in a special way to show that you have approved a document; ***the report must be signed by the author***; **to sign on** = to join the armed forces for a period of time and sign a contract of employment; ***he signed on for seven years***

signal ['sɪgnəl] **1** *noun* **(a)** sign made by flags, gestures, light or any other means, in order to convey information or instructions; ***the signal to withdraw is a red flare followed by a green flare*** **(b)** message transmitted by radio; ***we have received a signal from HQ*** **(c)** electomagnetic waves transmitted by a radio transmitter; ***I am getting a very weak signal*** **(d) signals** = communications (especially radio); ***we have captured an enemy signals detachment***; **Royal Corps of Signals** = British troops who specialize in communications; **signals intelligence;** *see* SIGINT **2** *verb* **(a)** to make a sign, in order to convey information or instructions; ***he signalled to us to get down*** **(b)** to send a message by radio; ***they signalled HQ to request air support for the operation***

signaller *US* **signaler** ['sɪgnələ] *noun* **(a)** serviceman, who specializes in the use of radios and other communications equipment **(b)** *GB* title of a private in the Royal Corps of Signals; ***Signaller Jones***

signature ['sɪgnətʃə(r)] *noun* **(a)** person's name written by him/her, usually to show authorization for something; ***I need your signature on this document*** **(b)** any distinctive sign (eg. heat, light, smoke *or* radiation) which is produced *or* emitted by a weapon *or* other piece of equipment, and which reveals its location to observers *or* surveillance equipment; ***this tank produces a strong thermal signature***

signpost ['saɪnpəʊst] *noun* sign positioned at a road junction, which shows the direction (and sometimes the distance) to a town or village

silence ['saɪləns] **1** *noun* **(a)** state when there is no noise **(b)** state when no one speaks; **electronic silence** = state when all radios and other transmitting equipment (such as radar) must be switched off; **radio silence** = state when no one is allowed to transmit on the radio; **to impose radio silence** = to start radio silence; **to lift radio silence** = to end radio silence **2** *verb* **(a)** (of guards or sentries) to kill or immobilize; ***he silenced the sentry with a knife*** **(b)** *(of enemy artillery, machine-guns, fire-positions, etc.)* to destroy or immobilize; ***the battery has been silenced***

silencer ['saɪlənsə] *noun* device which is fitted to a firearm, in order to reduce the noise made when it is fired

silent ['saɪlənt] *adjective* **(a)** *(of people)* not speaking; ***they were ordered to remain silent*** **(b)** *(of things)* not making any noise; ***we made a silent approach to the objective***

silhouette [ˌsɪluː'et] **1** *noun* shape of an object when seen on the skyline or against a lighter background; ***we could see the silhouette of a tank*** **2** *verb* to appear as a silhouette; ***the tank was silhouetted on the ridge***

silo ['saɪləʊ] *noun* **(a)** underground chamber where a missile is stored and from which it can be launched **(b)** large structure (often cylindrical) used for storing grain

Silva™ compass [ˌsɪlvə 'kʌmpəs] *noun* compass which is designed to be placed onto a map in order to calculate bearings (without the need for a protractor)

simulate ['sɪmjʊleɪt] *verb* to imitate effects or conditions for training purposes

simulation [ˌsɪmjʊ'leɪʃn] *noun* act of simulating

simulator ['sɪmjʊleɪtə] *noun* apparatus designed to simulate effects or conditions for training purposes (such as the control of an aircraft, direction of artillery fire, firing of a missile, etc.); **flight simulator** = computer program which allows a user to pilot a plane, showing a realistic control panel and moving scenes, used as training programme

sink [sɪŋk] *verb* **(a)** *(of boats and ships)* to go to the bottom of the sea, river or other area of water; ***the ship sank in a storm*** **(b)** to make a boat or ship sink (especially as a result of an attack); ***the enemy have sunk HMS Sheffield*** (NOTE: **sinking - sank - have sunk**)

Sioux [suː] *noun* outdated American-designed reconnaissance helicopter

siphon *or* **syphon** ['saɪfən] *verb* to move liquid from one container to another using atmospheric pressure; one container is placed in a higher position than the other and liquid is then sucked from the higher container through a hose and directed into the lower container; the liquid will then flow freely as a result of pressure; ***we caught some men trying to siphon fuel from our truck***

siren ['saɪrən] *noun* device which makes a loud noise as a signal or warning

site [saɪt] **1** *noun* location which is selected for a particular purpose; ***this would be a good site for the RAP*** **2** *verb* to select a location for a particular purpose; ***where have you sited the machine-gun?***

SITREP ['sɪtrep] = SITUATION REPORT

situation [ˌsɪtjuː'eɪʃn] *noun* what is happening at a particular moment in time; **situation report (SITREP)** = verbal or written message describing everything of importance which is happening or has happened in a unit or sub-unit's area of responsibility

ski [skiː] **1** *noun* one of a pair of long thin pieces of wood or plastic, which a person attaches to his feet in order to move over snow; ***in winter, they patrol the border on skis*** **2** *verb* to move on skis; ***we will have to ski to the RV***

ski-jump ['skiː dʒʌmp] *noun* ramp at the end of the flight deck on an aircraft carrier, which is designed to assist take-off

skill [skɪl] *noun* ability to carry out a task or procedure (usually improved by teaching and practice); **skill at arms (SAA)** = skill in the use of weapons, especially small arms

ski-mask ['skiː mɑːsk] *noun* woollen garment which covers the head, neck and face, with holes for the eyes, which is designed to protect a person's face in extremely cold conditions, but is also sometimes used to conceal a person's identity; ***the terrorists were all wearing ski-masks***

skirmish ['skɜːmɪʃ] **1** *noun* short battle between small groups of soldiers; ***skirmishes broke out along the line*** **2** *verb* to assault or withdraw, using fire and manoeuvre; ***the section skirmished onto the forward edge of the enemy position***

Sky Flash ['skaɪ ˌflæʃ] *noun* British-designed radar-guided air-to-air missile (AAM)

Skyhawk ['skaɪhɔːk] *see* A-4

SL = START LINE

slacken ['slækən] *verb* to become less intense; ***we assaulted as soon as the enemy fire started to slacken***

SLAM [slæm] *noun* = STAND-OFF LAND ATTACK MISSILE American-designed long-range air-to-ground missile

sleeper ['sliːpə(r)] *noun* agent who lives and works in an enemy country but carries out no action until the time is right (eg. at the outbreak of war); ***we believe that a sleeper has planted a virus in our computer system***

sleeping-bag ['sliːpɪŋ ˌbæg] *noun* quilted bag used for sleeping in

sleeping sickness ['sliːpɪŋ ˌsɪknɪs] *noun* common name for the disease, African trypanosomiasis, which is spread by the tsetse fly

sling [slɪŋ] *noun* leather or webbing strap, by which a weapon can be hung from a person's shoulder so that he is free to use his hands; ***we made a rope out of rifle slings***

sling shot ['slɪŋ ˌʃɒt] *noun* *US* weapon made of a Y-shaped piece of metal with a rubber attached, used to send stones and other small projectiles over long distances (NOTE: British English for this is **catapult**)

sloop [sluːp] *noun* small armed naval ship

slope [sləʊp] **1** *noun* **(a)** area of ground, in which one part is higher than the other; ***the tank rolled down the slope*** **(b)** side of a hill; ***the enemy is dug in on the forward slope***; **forward slope** =side of a hill which is facing the enemy; ***the enemy are dug in on the forward slope***; **reverse slope** = far side of a hill, which is in dead ground to the enemy; ***the Duke of Wellington always preferred a reverse slope position*** **2** *verb* *(of ground)* to form a slope; ***the ground slopes to the south***

SLR [ˌes el 'ɑː] = SELF-LOADING RIFLE

small arms ['smɔːl ˌɑːmz] *plural noun* arms which can be carried, such as rifles, machine-guns and sub-machine-guns

smallbore ['smɔːlbɔː] *adjective* with a barrel which has a small bore; ***a smallbore shotgun***

smallpox ['smɔːlpɒks] *noun* potentially fatal infectious disease, causing fever and a severe rash on the skin, which often results in permanent scars

> COMMENT: caused by the poxvirus, smallpox is also known as *variola*. The disease has been practically eradicated from the developed world as a result of widespread vaccination programmes. However, certain nations are believed to have developed strains of the virus for use as a biological weapon

smart bomb ['smɑːt bɒm] *noun* name given by the media to precision guided munitions (PGM); *compare* DUMB BOMB

SMAW [smɔː] *noun* = SHOULDER-LAUNCHED MULTIPURPOSE ASSAULT WEAPON American-designed hand-held rocket launcher (basically an anti-tank weapon)

smoke [sməʊk] **1** *noun* **(a)** particles of carbon produced by a burning object or substance, which are suspended in the air to form a thick black or white cloud; ***clouds of smoke reduced visibility*** **(b)** any projectile or grenade, which is designed to produce smoke, in order to blind the enemy or to mask the movements of friendly forces; ***five rounds smoke, fire for effect!***; **smoke canister** = metal container containing chemicals which produce smoke **2** *verb* **(a)** to give off smoke; ***the burnt-out tank is still smoking*** **(b)** to smoke a cigarette; ***he reminded his men that they were not allowed to smoke after dark***

smoking ['sməʊkɪŋ] *noun* act of smoking a cigarette; ***smoking is not allowed after dark***

smuggle ['smʌg(ə)l] *verb* **(a)** to import illegal goods (eg. drugs, weapons) *or* import goods without paying customs duty (eg. alcohol, tobacco); ***he was prosecuted for smuggling cigarettes*** **(b)** to convey something secretly into *or* out of a location; ***they were caught trying to smuggle a girl into the barracks***

smuggler ['smʌglə(r)] *noun* person who smuggles; ***he is a well-known drug smuggler***

snake [sneɪk] *noun* creature with a very long, thin body and no legs, which often has a poisonous bite

snatch [snætʃ] *verb* to rush in and arrest or capture a person; ***our mission is to snatch an officer from the enemy position***; **snatch squad** = small group of soldiers detailed to grab and arrest someone

snipe [snaɪp] *verb* **(a)** to shoot at a person from a hidden fire-position **(b)** to shoot at selected enemy personnel such as commanders, machine-gunners, signallers, etc., as opposed to any person who happens to be within your field of fire **(c)** to shoot at enemy personnel as a form of harassment (especially when they are not actually fighting)

sniper ['snaɪpə] *noun* trained marksman, who specializes in sniping at the enemy; ***the street patrol came under sniper fire***; ***he was disabled by a sniper's bullet***

> COMMENT: snipers are usually sited away from the main force, so that they can concentrate on shooting at selected targets instead of being drawn into the general firefight. They are also less likely to be affected when the main force comes under artillery fire

snore [snɔː] *verb* to make a noise through the nose while sleeping; ***we could hear someone snoring***

snorkel ['snɔːkl] *noun* **(a)** tube used by a person to breathe through, when swimming underwater; ***he was using a snorkel*** **(b)** breathing tube, fitted to a tank for crossing rivers or landing on a beach; ***this tank can be fitted with a snorkel for river crossings***

snow [snəʊ] **1** *noun* flakes of crystallized ice, which fall from the sky; ***most of the mountain roads were blocked by snow*** **2** *verb (of snow)* to fall from the sky; ***it was snowing hard as the attack began***

snowcat ['snəʊkæt] *noun* lightweight tracked vehicle designed for use in arctic conditions

snowshoe ['snəʊʃuː] *noun* device, similar in appearance to a tennis racket, which is strapped on the foot to allow a person to walk across deep snow

SOCO ['sɒkəʊ] *noun GB* = SCENES OF CRIME OFFICER civil police officer responsible for searching for evidence at the scene of a terrorist incident; ***SOCO has arrived at the ICP***

SOCEUR ['sɒkɜː(r)] = *US* SPECIAL OPERATIONS COMMAND EUROPE

Sod's Law [ˌsɒdz 'lɔː] *noun GB* further development of Murphy's Law, which states that if something does go wrong, then it is certain to go wrong in the worst possible way; ***well, that's Sod's Law, isn't it ?***

SOF [ˌes əʊ 'ef] *noun* = SPECIAL OPERATING FORCES special forces (eg. commandos, rangers, SAS, SEALs, Spetznaz, etc)

soft-skinned ['sɒft ˌskɪnd] *adjective;* ***soft-skinned vehicle*** = vehicle which is not protected by armour (such as a jeep, lorry, truck, etc.); ***all soft-skinned vehicles were withdrawn to the rear***

soft target [ˌsɒft 'tɑːgɪt] *noun* person *or* unit *or* vehicle which is vulnerable *or* unable to defend itself properly; ***the terrorists are only interested in attacking soft targets***

soil [sɔɪl] *noun* substance, consisting of particles of rock and decayed vegetation, in which plants grow

solar still [ˌsəʊlə(r) 'stɪl] *noun* emergency method of producing water in the desert; a pit is dug in the sand and a container placed at the bottom; the pit is then covered with some plastic sheeting which is secured at the sides by heaped sand and weighted in the centre by a small stone, thereby forming an inverted cone; droplets of water form through condensation on the underside of the sheet and trickle down into the container (NOTE: condensation can be increased by placing pieces of vegetation in the pit *or* even urinating in the sand of the pit)

soldier ['səʊldʒə] *noun* ***(private) soldier*** = person serving in the army; ***a group of soldiers took command of the radio station***; ***we were trapped in the camp by enemy soldiers***; **soldier of fortune** = mercenary

soldiering ['səʊldʒərɪŋ] *noun* profession of being a soldier

solution [sə'luːʃən] *noun* **(a)** mixture of a solid substance with a liquid; ***he cleaned the wound with a solution of salt and water*** **(b)** answer to a problem; ***we've got a solution to your supply problem*** **(c)** moment when the operator of a guided weapon has the target in his sights and the guidance system is activated; ***he achieved a solution on the leading plane*** (NOTE: use with the verb **achieve**); *see also* LOCK-ON

Soman ['səʊmən] *see* GD

sonar ['səʊnɑː] *noun* system for detecting underwater objects through the transmission of sound waves, which are reflected back by the object

SOP [ˌes əʊ 'piː] = STANDARD OPERATING PROCEDURE

sortie ['sɔːti] *noun* **(a)** operational flight; ***we lost two aircraft in the last sortie*** **(b)** small offensive operation mounted by troops who are occupying a defensive position; ***we made a sortie while the enemy were reorganizing***

SOS [ˌes əʊ 'es] *noun* international distress signal, signifying an urgent request for assistance; ***they sent out an SOS***

source [sɔːs] *noun* agent *or* informer who provides intelligence; ***we've heard from a reliable source that the terrorists are planning an attack***

south [saʊθ] **1** *noun* **(a)** one of the four main points of the compass, corresponding to a bearing of 180 degrees or 3200 mils **(b)** area to the south of your location; ***the enemy are approaching from the south*** **(c) the South** = the southern part of a country **2** *adjective* relating to south; ***the South Gate***; **south wind** = wind blowing from the south **3** *adverb* towards the south; ***the enemy is moving south***

southbound ['saʊθbaʊnd] *adjective* moving or leading towards the south; ***a southbound convoy***

southerly ['sʌðəli] *adjective* **(a)** towards the south; ***the troops were heading in a southerly direction*** **(b)** *(of wind)* from the south

southern ['sʌðən] *adjective* relating to the south; ***the southern part of the country***

southward ['saʊθwəd] **1** *adjective* towards the south; ***they moved in a southward direction*** **2** *adverb US* towards the south; ***they are moving southward***

southwards ['saʊθwədz] *adverb* towards the south; ***they are moving southwards***

Soviet ['səʊviət] *adjective* of or relating to the Soviet Union; **the Soviet Union (USSR)** = the empire of communist Russia, which disintegrated in 1991

> COMMENT: the former Soviet Union is now generally known as the **Commonwealth of Independent States (CIS)**

SPAAG = SELF-PROPELLED ANTI-AIRCRAFT GUN

space [speɪs] *noun* **(a)** empty area between objects; ***there is not enough space to deploy the brigade***; **calculations in time and space** = calculations to determine how long it will take to get from one location to another **(b)** unlimited area beyond the earth's atmosphere; ***the Americans have sent another satellite into space***; **Space Command** = department of the US forces responsible for the use of satellites (eg. for surveillance, communications, GPS, missile guidance, NMD, etc)

spacecraft ['speɪskrɑːft] *noun* machine designed to travel in space

spade [speɪd] *noun* simple digging tool consisting of a metal blade attached to a long wooden handle; *see also* SHOVEL

spall [spɔːl] *noun* fragments of armour, which are broken off and blasted into the interior of an armoured vehicle, as a result of a hit by an anti-tank projectile

Spandrel ['spændrəl] *noun* Soviet-designed tube-launched, wire-guided anti-tank missile (ATGW)

spare [speə(r)] **1** *adjective* kept in order to replace something which is lost *or* damaged; ***this vehicle doesn't have a spare wheel***; **spares** = spare parts **2** *verb* **(a)** to manage without; ***we can't spare the men for this task*** **(b)** to not kill; ***only the women and children were spared***

sparkle ['spɑːk(ə)l] *verb (forward air controller jargon)* to illuminate a target with a laser target designator; ***hello G33 this is Cowboy, sparkle, sparkle, over***

Sparrow ['spærəʊ] *noun* American-designed radar-guided air-to-air missile (AAM)

sparrow-fart ['spærəʊ ˌfɑːt] *noun GB (slang)* first light; ***we'll move out at sparrow-fart***

Spartan ['spɑːtən] *noun* small British-designed armoured-personnel carrier (APC) normally used by specialist troops (eg. anti-tank, artillery, engineers)

spat [spæt] *noun* garment of fabric, which is worn over the ankle and lower leg and extends over the upper part of the shoe *or* boot, in order to keep your trousers dry and to prevent small stones and other objects going into your boots; ***the pipers were wearing kilts and white spats***

> COMMENT: spats are now usually worn as part of a ceremonial uniform, although American troops wore them in combat during World War II

spearhead ['spɪəhed] **1** *noun* leading elements of a large-scale offensive operation; ***the enemy's spearhead has reached Minden*** **2** *verb* to act as spearhead; ***3 Brigade will spearhead the attack***

special ['speʃl] *adjective* for a specific purpose; **Special Air Service (SAS)** = elite British Army special forces organization; **Special Boat Service (SBS)** = elite British special forces organization, recruited from the Royal Navy and Royal Marines; **special forces** = highly trained elite troops, who specialize in unconventional military operations (such as covert operations, intelligence gathering, raids, sabotage, etc.); **Special Investigation Branch (SIB)** = detective branch of the British military police, which investigates criminal offences committed by servicemen while they are subject to military law (eg. on MOD property *or* while on operational service); **special operations capable (SOC)** = having sufficient training and expertise to carry out specialized military tasks

spent [spent] *adjective* used; ***spent ammunition lay round the machine-gun***

Spetznaz ['spetsnæz] *noun* elite Soviet special forces organization; ***Spetznaz units are operating in this area***

SPG [ˌes piː 'dʒiː] = SELF-PROPELLED GUN

SPH [ˌes piː 'eɪtʃ] = SELF-PROPELLED HOWITZER

Spigot ['spɪgət] *noun* Soviet-designed wire-guided anti-tank missile (ATGW)

spinney ['spɪnɪ] *noun* small wood; ***we think the enemy has an OP in that spinney***

Spiral ['spaɪrəl] *noun* Soviet-designed laser-guided anti-tank missile (ATGW)

spire ['spaɪə] *noun* sharp pointed tower, usually forming part of a church

Spirit ['spɪrɪt] *see* B-2

spitlock ['spɪtlɒk] *verb* to mark the proposed layout of a trench *or* other field fortification, by digging its outline into the turf; ***the recce group had spitlocked the positions for us***

splash [splæʃ] **1** *verb (of liquids)* to be thrown in small drops onto another object *or* thing; ***he was splashed with burning petrol when the jeep exploded*** **2** *noun* impact of an explosive projectile as seen by an observer (usually the firer); ***we didn't see the splash but we heard the explosion***

splice the main brace [splaɪs ðə 'meɪnbreɪs] *verb GB* naval custom, where every man is given a measure of rum; ***the admiral told his captains to splice the main brace***

splinter ['splɪntə(r)] *noun* thin,sharp fragment; ***he was killed by a shell splinter***

spoil [spɔɪl] *noun* soil or sand which is dug out of the ground; ***the spoil from a trench is used to build the parapet and parados***

spoiling attack ['spɔɪlɪŋ əˌtæk] *noun* attack mounted on an advancing enemy force in order to disrupt its activities and prevent it carrying out its intentions; ***H-hour was delayed when the enemy mounted a spoiling attack in 3 Brigade's sector***

spook [spuːk] *noun (slang)* person involved in extremely covert operations (eg. spy, special forces, etc)

spore [spɔː(r)] *noun* reproductive body of certain bacteria; ***this warhead releases spores of anthrax into the atmosphere***

spot [spɒt] **1** *verb* **(a)** to catch sight of; ***I spotted someone moving in the garden*** **(b)** to observe and direct artillery fire (usually from an aircraft); ***he was spotting from a helicopter*** **2** *noun* location; ***this is a good spot for the mortars***

spot height ['spɒt haɪt] *noun* point marked on a map to show where a measurement of altitude has been made

spotter ['spɒtə] *noun* officer or NCO who directs artillery fire (usually from an aircraft); **spotter aircraft** *or* **plane** = aircraft used for observing and directing artillery fire

spring [sprɪŋ] *noun* **(a)** place where water comes out of the ground naturally; ***there are very few springs in these mountains*** **(b)** flexible piece of metal (often in the form of tightly coiled wire), which is used as a shock absorber *or* to to keep a catch *or* clip closed *or* to maintain tension; ***we need to replace the springs on this vehicle***

springing-mine [ˌsprɪŋɪŋ 'maɪn] *noun* anti-personnel mine, which is designed to jump into the air in order to inflict injury to a person's upper body

spur [spɜː] *noun (topographical)* ridge protruding from a hill or mountain into lower-lying ground; ***we cannot advance until the enemy have been cleared off that spur***

spy [spaɪ] **1** *noun* person who secretly tries to obtain information about the enemy, or about a foreign power; ***information about the troop movements came from our spies in the capital*** **2** *verb* to act as a spy; ***he was accused of spying for the enemy***

Sqn = SQUADRON

SQN LDR = SQUADRON LEADER

squad [skwɒd] *noun* **(a)** small grouping of servicemen, formed for a specific purpose or task (such as drill) **(b)** sub-unit of an infantry platoon **(c)** *US* tactical infantry grouping of nine men (usually divided into two fire teams) **(d)** *US* tactical armoured cavalry grouping of seven men **(e)** US Marine Corps tactical grouping of thirteen men (uually divided into three fire teams); **Squad Automatic Weapon (SAW);** *see* M-249 (NOTE: in the British

Army, a squad-sized infantry grouping is known as a **section**)

> COMMENT: American infantry squads are usually commanded by sergeants

squaddie ['skwɒdi] *noun* GB *(slang)* ordinary soldier

squadron ['skwɒdrən] *noun* **(a)** small tactical grouping of warships; ***he commanded a British squadron in the West Indies*** **(b)** air force unit consisting of two or more flights, ie between ten and eighteen aircraft; ***two squadrons of fighters were sent to intercept the bombers*** GB **squadron leader** = officer in the air force, below wing commander and above flight lieutenant (usually in command of a squadron) **(c)** GB company-sized tank grouping of three or more troops; **squadron and company group** = combined-arms grouping, based on a tank squadron (equivalent of a company team in the US Army); **company and squadron group** = combined-arms grouping based on an infantry company (equivalent of a company team in the US Army) **(d)** US battalion-sized armored cavalry grouping, consisting of three cavalry troops, one tank company and one battery

> COMMENT: The number of aircraft in a squadron will vary according to aircraft type and role. A bomber squadron may have as few as six aircraft while a fighter squadron may have as many as twenty-four. In the army, a British armoured brigade might consist of two armoured regiments and one armoured or mechanized infantry battalion or, alternatively, two infantry battalions and one armoured regiment, plus artillery and supporting arms. On operations, these units are broken down and combined into **battle groups.** As an example, an armoured battle group might consist of two squadrons of tanks and one infantry company, which are organized into two **squadron and company groups** and a **company and squadron group** under the command of the armoured regimental HQ. The exact composition will vary according to the tactical requirement at the time. In the US Army, a battle group is known as a task force, while company and squadron groups and squadron and company groups are known as company teams

squall [skwɔːl] *noun* sudden storm of wind and rain, at sea

square [skweə] **1** *adjective* having four sides of equal length; ***a square piece of wood***; **(GB) square brigade** = brigade, consisting of two armoured regiments and two battalions of armoured or mechanized infantry **2** *noun* flat area where drill is carried out; *(informal)* **square bashing** = drill practice; **grid square** = segment of a map grid formed by two eastings and two northings, normally showing an area of one square kilometre

SR-71 [ˌes ɑː sevənti 'wʌn] *noun* American-designed strategic reconnaissance aircraft (NOTE: also known as the **Blackbird**)

SS = SUBMARINE

SSB = SUBMARINE (with ballistic missiles)

SSBN = SUBMARINE (nuclear-powered, with ballistic missiles)

SSG = SUBMARINE (with guided missiles)

SSGN = SUBMARINE (nuclear-powered, with guided missiles)

S/Sgt = STAFF SERGEANT

SSM [ˌes es 'em] = SURFACE-TO-SURFACE MISSILE

SSN = SUBMARINE (nuclear-powered)

stab [stæb] *verb* to pierce someone's body with a knife or bayonet

Staballoy [steɪ'bælɔɪ] *noun* US depleted uranium (DU)

stable ['steɪb(ə)l] *noun* building used to accomodate horses *or* mules

staff [stɑːf] *noun* group of officers and other ranks, who assist the commander of a large tactical grouping (such as a brigade, division, corps, etc.), and who form his headquarters; **staff officer** = officer who serves in a staff; **chief of staff (COS)** = senior staff officer in a headquarters;

general staff = staff which has supreme control over a state's armed forces

staff college ['stɑːf ˌkɒlɪdʒ] *noun* training establishment, which prepares officers for high command

staff sergeant (S/Sgt) ['stɑːf ˌsɑːdʒənt] *noun* **(a)** GB senior non-commissioned officer (SNCO) in the army (usually employed as the quartermaster sergeant of a company or equivalent-sized grouping) **(b)** US senior non-commissioned officer (SNCO) in the army, marines or air force (NOTE: in the British Army, the infantry equivalent of staff sergeant is **colour sergeant (C/Sgt)**)

stag [stæg] *noun* GB *(slang)* period of duty as a sentry or on radio watch

stagger ['stægə(r)] *verb* **(a)** to arrange actions so that they do not happen at the same time; ***the departures of the companies were staggered at fifteen minute intervals*** **(b)** to arrange vehicles *or* aircraft *or* men so that they are not in a straight line; ***we advanced in a staggered formation***

staging area ['steɪdʒɪŋ ˌeərɪə] *noun* place along a route where troops can stop, in order to rest and reorganize before continuing their journey; ***the enemy are using that wood as a staging area***

staging camp ['steɪdʒɪŋ ˌkæmp] *noun* camp where troops are accomodated for a short period, before moving to another destination; ***you will go to a staging camp, where you will wait until we can move you forward to your battalion***

stalk [stɔːk] *verb* to creep towards a person or vehicle, in order to shoot at him or it from a close range

stand [stænd] *verb* to support yourself, using your feet and legs, in a stationary position; ***he was standing next to the tank*** (NOTE: **standing - stood**)

standard ['stændəd] **1** *adjective* **(a)** basic or normal; ***this is the standard type of respirator***; **standard issue** = for normal everyday use (as opposed to specialist tasks); ***he was wearing standard issue combat clothing*** **(b)** officially recognized as the correct way to do something; **standard operating procedure (SOP)** = set of instructions, produced by an arm, grouping or unit, which explain exactly how various duties and tasks should be carried out **2** *noun* **(a)** measure of quality, by which all similar things are judged; ***your boots are not up to standard***; ***the standard of shooting is very high*** **(b)** regimental flag (especially in cavalry regiments)

standarization [ˌstændədaɪ'zeɪʃn] *noun* making sure that all procedures, personnel and material all work in the same way; **standardization agreement** = agreement between various nations to use standard equipment, operating procedures, etc.

Standard Missile [ˌstændəd 'mɪsaɪl] *noun* American-designed long-range naval surface-to-air missile (SAM)

stand by [ˌstænd 'baɪ] *verb* to be ready to do something; ***B Company is standing by to give fire support***

stand-by ['stændbaɪ] *noun* act of standing by; **on stand-by** = ready to do something; ***the battalion is on stand-by to deploy to the Gulf***

stand down [ˌstænd 'daʊn] **1** *verb* **(a)** to stop standing to **(b)** to stop standing by; ***the Brigade was ordered to stand down*** **2** *noun* act of standing down (as part of a routine); ***stand down is thirty minutes after first light***; *compare* STAND BY, STAND TO

stand fast [ˌstænd 'fɑːst] *verb* to stop what you are doing and wait for further instructions; ***we were ordered to stand fast***

standing army [ˌstændɪŋ 'ɑːmi] *noun* the regular army of a state, as opposed to reserve forces; ***the country maintains a standing army of 100,000 men***

standing orders [ˌstændɪŋ 'ɔːdəz] *plural noun* set of rules and regulations relating to duties and discipline

standing patrol [ˌstændɪŋ pə'trəʊl] *noun* patrol sent out to occupy a covert position in no-man's-land in order to provide warning of enemy activity

stand off [ˌstænd 'ɒf] *verb* to remain at a distance from something; ***the tanks stood off in order to engage the enemy from the flank***

stand to [ˌstænd 'tuː] **1** *verb (of a unit or sub-unit)* to be awake and at battle stations, in order to receive an enemy attack; ***the battalion stood to at first light*** **2** *noun* act of standing to (as part of a routine); ***stand to is at 0545hrs***

star [stɑː] *noun* **(a)** tiny point of light, visible in the sky at night; **star shell** = illuminating round used by artillery **(b)** insignia in the shape of a star, used as a badge of rank (NOTE: in the British Army also called a **pip**)

> COMMENT: in the British Army, one star denotes a second lieutenant, two a lieutenant and three a captain. In the US Army, one star denotes a brigadier general, two a major general, three a lieutenant general and four a general

starboard ['stɑːbəd] *noun* right-hand side of an aircraft, boat or ship; ***enemy fighters approaching starboard!***; *compare* PORT

Starlifter ['stɑːlɪftə] *see* C-141

starlight scope ['stɑːlaɪt ˌskəʊp] *noun* type of image intensifier

Stars and Stripes [ˌstɑːz ənd 'straɪps] *noun* national flag of the United States of America (USA)

Starstreak ['stɑːstriːk] *noun* British-designed surface-to-air missile (SAM)

start [stɑːt] *verb* to begin an activity; ***we will start the advance at 0600hrs***; ***he started to run***; **start line (SL)** = real or imaginary line, the crossing of which marks the start of an advance, attack or other offensive operation; *see also* LINE OF DEPARTURE

Star Wars ['stɑː ˌwɔːz] *noun* media name for the American Strategic Defence Initiative (SDI); *see also* NMD

state [steɪt] **1** *noun* **(a)** condition or situation; ***he was horrified by the state of the prisoners***; **ammunition state** = quantity of ammunition held by a unit or sub-unit; **NBC state** = degree of possibility or probability that the enemy will mount a nuclear, chemical or biological attack; **vehicle state** = condition of vehicles held by a unit or sub-unit; **weapon state** = condition in which a weapon is carried (ie unloaded, made safe or made ready) **(b)** independent community of people, with its own territory, government and armed forces; *see also* COUNTRY, NATION **2** *plural noun* ***the States*** = United States of America (USA)

static ['stætɪk] *adjective* not moving, in a fixed position; **static defence** = defensive doctrine which relies on static defensive positions and the use of attrition to halt an enemy advance; *also known as* POSITIONAL DEFENCE; *compare* MOBILE DEFENCE; **static line** = method used to pull a parachute open as the parachutist jumps out of the aircraft; *compare* FREE-FALL

station ['steɪʃn] **1** *noun* **(a)** place where soldiers are based; ***he was not happy at his last station*** **(b)** *GB* base location for an air-force grouping; **station commander** = commanding officer of a RAF unit **(c)** regular stopping place on a railway line; ***the train finally arrived at the station two hours late*** **2** *verb* to send a serviceman to serve in a particular location; ***I was stationed in Germany***

stationary ['steɪʃnri] *adjective* not moving; ***he aimed at the stationary tank***

steal [stiːl] *verb* to take another person's property without his or her agreement or permission; ***someone has stolen my helmet*** (NOTE: **stealing - stole - have stolen.** The noun for this verb is **theft**)

stealth [stelθ] *adjective* referring to an aircraft which is difficult or impossible to detect by radar and other surveillance equipment, as a result of its design: for example reduced radar cross-section (RCS) and the use of materials such as radar absorbent material (RAM) and radar-absorbent structural material (RAS); ***stealth bombers were used in the operation***; *see also* B2, F117, F-22

steel [stiːl] *noun* metal, made of iron and carbon, which is used in the production of armour, weapons and vehicles; **cold steel** = use of the bayonet

steep [stiːp] *adjective; (of hills or slopes)* to slope at a high angle

Step Up ['step ˌʌp] *noun* small headquarters party, which moves forward in advance of the main party to set up a new headquarters location; once Step Up is established, the old headquarters hands over control of the battle and moves forward to join it

stern [stɜːn] *noun* rear part of a ship; *compare* BOWS

stencil ['stensɪl] *noun* **(a)** thin sheet of metal *or* plastic *or* stiff card, out of which letters *or* numbers *or* other shapes have been cut, and which is placed on the surface of an object (eg. vehicle, container, etc) and painted over to reproduce the shapes on the surface below; ***I need stencils for the letters A and G*** **(b)** stiff sheet of plastic, out of which a selection of geometrical shapes have been cut, and which is used for drawing tactical symbols on a map

steppe [step] *noun* wide area of uncultivated grassland with few trees (especially in Russia and Eurasia) (NOTE: the American equivalent is **prairie**)

stick [stɪk] *noun* **(a)** long thin piece of wood, which is broken or cut from a branch of a tree **(b)** quantity of bombs, which are released by an aircraft at the same time **(c)** group of paratroopers, who jump out of an aircraft during a single pass over the drop zone (DZ)

Stinger ['stɪŋə] *noun* American-designed hand-held surface-to-air missile (SAM)

stock [stɒk] *noun* quantity of supplies held ready for use

Stockholm Syndrome ['stɒkhəʊm ˌsɪndrəʊm] *noun* psychological reaction to fear and stress, in which hostages start to feel sympathetic towards their captors

STOL = SHORT TAKE-OFF AND LANDING

stone [stəʊn] **1** *noun* small piece of rock; ***the sentry heard stones rolling down the slope*** **2** *verb* to throw stones at a person *or* vehicle; ***the patrol was stoned by a group of youths***

stonk *or* **stonking** [stɒŋk] *noun slang* attack by artillery *or* mortars; ***we gave the enemy OP a bloody good stonking !***

stood to [ˌstʊd 'tuː] *adverb* standing to (ie awake and at battle stations); ***the battalion was stood to for most of the night***

stop [stɒp] *verb* **(a)** to finish doing something; ***he stopped working*** **(b)** to stop moving and stand still; ***stop, or I will shoot!*** **(c)** to prevent someone or something from moving; ***our orders are to stop all vehicles and check the drivers*** **(d)** to prevent the enemy from advancing or successfully completing an attack; ***the enemy have been stopped at the river***

stoppage ['stɒpɪdʒ] *noun (of automatic or semi-automatic firearms)* mechanical failure, which prevents further firing

store [stɔː] **1** *noun* **(a)** quantity of things, which are kept for future use; ***the fire destroyed our store of winter clothing*** **(b)** place used for storing things; ***he works in the clothing store*** **(c)** *US* shop; ***several stores were looted during the riot*** **(d)** **stores** = quantities of different things which are stored for a particular purpose; ***we airlifted stores to the garrison*** **2** *verb* to keep things for future use

storm [stɔːm] **1** *noun* **(a)** violent weather, consisting of high wind and rain, snow or hail; ***the sortie was cancelled because of the storm*** **(b) by storm =** using force in order to occupy an enemy position; ***the troops took the enemy positions by storm*** **2** *verb* to assault and capture a position or place; ***the town was stormed by the 7th Infantry Regiment***

STOVL = SHORT TAKE-OFF AND VERTICAL LANDING

stow [stəʊ] *verb* to pack equipment or supplies tidily into an aircraft, ship or vehicle; ***all the equipment has been stowed ready for take-off***

straddle ['strædl] *verb* **(a)** *(of troops, formations or positions)* to be positioned on either side of something; ***A Company's position straddles the main road*** **(b)** *(of artillery or mortar fire)* to land rounds on either side of a target

strafe [streɪf] *verb (of fighter aircraft)* to shoot at targets on the ground, especially along a road, or at ships at sea; ***enemy fighters strafed the advancing column***

straggle ['strægl] *verb* to be unable to keep up with your unit during a long journey or march; ***many of the soldiers were unfit and started to straggle***

straggler ['stræglə] *noun* soldier who is unable to keep up with his unit during a long journey or march; ***we captured some enemy stragglers***

strait *or* **straits** [streɪt] *noun* narrow stretch of sea connecting two larger areas of sea; ***the straits of Gibraltar*** (NOTE: often used in the plural)

strap [stræp] *noun* long thin piece of webbing or leather, which forms part of a soldier's load-bearing equipment or is used to fasten objects together

STRATCOM ['strætkɒm] *noun US* = STRATEGIC COMMAND department of the US forces responsible for intercontinental ballistic missiles (ICBM) and missile submarines

strategic [strə'tiːdʒɪk] *adjective* **(a)** relating to strategy; ***this town is of great strategic importance***; **Strategic Defence Initiative (SDI)** = American programme to develop satellites which are capable of destroying enemy missiles in space (NOTE: known in the media as **Star Wars**); *see also* NMD; **strategic mobility** = ability of forces to move over very great distances **(b)** *(of bombs and missiles)* directed at the enemy's home territory, in order to destroy both his civil and his military infrastructures, thereby reducing his ability to conduct a war; **strategic bombing** = bombing of enemy towns and cities, industrial centres or communications (such as ports and airports, railways, roads, etc.), command centres, missile sites, airfields or any other target of strategic importance; **strategic nuclear weapon** = large long-range nuclear weapon designed to destroy targets of strategic importance; *compare* TACTICAL

strategist ['strætədʒɪst] *noun* person who is concerned with strategy; ***military strategists in the high command recommended a different course of action***

strategy ['strætədʒi] *noun* art of using of large military groupings (such as armies, corps, fleets, etc.) in order to achieve long-term objectives which will affect the course of a campaign or war; ***the commander's long-term strategy was to wear the enemy down by cutting off his supply routes***; *compare* TACTICS

> COMMENT: **strategy** refers to the movement of armies in order to achieve the overall objectives of a campaign or war (for example the capture of a port, which can be used to land supplies and reinforcements for future operations), while **tactics** refers to the movement of battalions, brigades, divisions and equivalent-sized groupings, in order to achieve local objectives (for example the destruction of an enemy battalion, which is defending one of the approaches to the port)

stray round [ˌstreɪ 'raʊnd] *noun* bullet or other projectile, which misses the target at which it was aimed; ***he was killed by a stray round***

stream [striːm] *noun* small river

street [striːt] *noun* road with buildings on each side

strength [streŋθ] *noun* **(a)** state of being strong or in large numbers; ***this projectile will test the strength of the tank's armour*** **(b)** number of men, aircraft, ships or vehicles available to a grouping; **fighting strength** = number of men or vehicles available to a unit for the purposes of fighting; **at full strength** = having all the men, aircraft, ships or vehicles which one should have; **in strength** = in large numbers; ***the enemy is crossing the river in strength***; **on strength** = avaliable to a unit; ***we have 875 men on strength***

stretcher ['stretʃə] *noun* piece of fabric suspended between two poles, which is used to carry an injured person; **stretcher-bearer** = person who helps to carry a stretcher; **stretcher-case** = casualty who needs to be carried on a stretcher

strike [straɪk] **1** *noun* **(a)** *(of projectiles, especially missiles)* act of hitting a target **(b)** attack (especially by aircraft or missiles on ground targets); ***the last strike destroyed our fuel dump***; **strike aircraft** = fighter aircraft used to attack targets on the ground; *see also* ATTACK AIRCRAFT, FIGHTER-BOMBER **2** *verb* to hit someone or something; ***he was arrested for striking an officer***; ***the missile hit an enemy command post*** (NOTE: **striking - struck**)

string [strɪŋ] *noun* thin line of twisted fibres, normally used for binding objects together; **string of mines** = several mines which are connected in such a way that the detonation of one will cause all the others to detonate too

strip [strɪp] *verb* **(a)** *(of people)* to take off all your clothing **(b) to strip down =** to take a weapon to pieces (for cleaning)

stripe [straɪp] *noun (slang)* chevron

stripwood ['strɪpwʊd] *noun* long thin wood

strobe [strəʊb] *noun* lamp which produces intermittent flashes of very bright light and is used by someone on the ground to attract the attention of aircraft; ***we switched on the strobe when we heard the helicopter***

strong point ['strɒŋ ˌpɔɪnt] *noun* key point in a defensive position, which is usually heavily fortified and well-armed

Styx [stɪks] *noun* NATO name for Soviet-designed P-15 long-range anti-ship missile (ASM)

SU-24 [ˌes juː twenti 'fɔː] *noun* Soviet-designed fighter-bomber (NOTE: known to NATO as **Fencer**)

SU-25 [ˌes juː twenti 'faɪv] *noun* Soviet-designed ground-attack aircraft (NOTE: known to NATO as **Frogfoot**)

SU-27 [ˌes juː twenti 'sevən] *noun* Soviet-designed fighter aircraft (NOTE: known to NATO as **Flanker**)

sub [sʌb] = SUBMARINE

subaltern ['sʌbəltən] *noun* *GB* lieutenant or second lieutenant

sub-lieutenant (Sub-Lt) [ˌsʌblef'tenənt] *noun* *GB* junior officer in the navy

Sub-Lt = SUB-LIEUTENANT

sub-machine-gun [ˌsʌbmə'ʃiːngʌn] *noun* small hand-held machine-gun, which is carried as a personal weapon

submarine [ˌsʌbmə'riːn] *noun* warship designed to move and operate under water, armed with torpedoes or nuclear weapons; ***submarines attacked and sank three of our ships***; ***their ship was torpedoed by an enemy submarine***; **hunter-killer submarine =** submarine which is designed to locate and destroy enemy submarines

submariner [sʌb'mærɪnə] *noun* sailor who serves on a submarine

submerge [səb'mɜːdʒ] *verb* to go or position something under water; ***the submarine has submerged***

submunitions [ˌsʌbmjuː'nɪʃənz] *noun* small projectiles, which are often used in clusters; *see also* TERMINALLY GUIDED SUBMUNITIONS

subordinate [səb'ɔːdɪnət] **1** *adjective* **(a)** of a lower rank (than another person); ***all subordinate commanders will attend the O Group*** **(b)** working under another person's command or supervision; ***you are subordinate to Captain Jones for this operation*** **2** *noun* person who works under another person's command or supervision; ***he is always rude to his subordinates***

subsonic [sʌb'sɒnɪk] *adjective* travelling at less than the speed of sound; ***this is a subsonic projectile***

substantive [səb'stæntɪv] *adjective; (of rank)* permanent (as opposed to acting or temporary); ***he has the substantive rank of colonel***

sub-unit ['sʌb ˌjuːnɪt] *noun* grouping, which forms part of a larger grouping

> COMMENT: a section is a sub-unit of a platoon; a platoon is a sub-unit of a company; a company is a sub-unit of a battalion

subway ['sʌbweɪ] *noun* **(a)** tunnel under a road **(b)** *US* underground railway

suffer ['sʌfə] *verb* to experience discomfort, pain or unhappiness; ***the civilian population suffered many casualties***; ***we have suffered heavy losses***; ***the enemy is suffering from low morale***; ***after three months on the front line he suffered a breakdown***

suffering ['sʌfrɪŋ] *noun* experience of discomfort, pain or unhappiness

suicide bomb ['suːɪsaɪd ˌbɒm] *noun* terrorist bombing tactic, where a terrorist carries an explosive device *or* drives a vehicle containing an explosive device up to a target (eg. security force base) and initiates it, deliberately killing himself in the process.

Sukhoi ['sʊkɔɪ] *noun* Soviet-designed fighter aircraft

Sultan ['sʌltən] *noun* British-designed armoured vehicle, which is designed to be used as a mobile command post

summit ['sʌmɪt] *noun* highest point of a hill or mountain

Sunray ['sʌnreɪ] *noun (radio terminology)* commander of a unit or sub-unit; ***Sunray will be at your location in ten minutes***

sunrise ['sʌnraɪz] *noun* time at which the sun appears over the horizon in the morning; *see also* DAWN, FIRST LIGHT

sunset ['sʌnset] *noun* time at which the sun disappears below the horizon in the evening; *see also* DUSK, LAST LIGHT

Super Etendard [ˌsuːpər 'etəndɑːd] *noun* French-designed multirole fighter aircraft, designed to operate from an aircraft carrier

superior [suː'pɪəriə] **1** *adjective* **(a)** of higher rank than another person; ***he is always rude to superior officers*** **(b)** bigger or stronger than something else; ***we were attacked by a superior force*** **(c)** of better quality than something else; ***our night-viewing equipment is superior to the enemy's*** **2** *noun* person who holds a higher rank than another person; ***you must obey your superiors***; *compare* INFERIOR

superiority [suːˌpɪərɪ'ɒrɪtɪ] *noun* state of being superior; **air superiority** = having sufficient fighter aircraft to prevent the enemy from using his air assets effectively; **local superiority** = having more troops than the enemy on one part of the battlefield, even though the enemy force as a whole may be equal in strength *or* even superior to your own

supernumerary [ˌsuːpə'njuːmərərɪ] **1** *adjective* additional to the establishment of a grouping; ***all supernumerary personnel will return to their own units*** **2** *noun* an extra *or* unwanted person *or* thing; ***all supernumeraries were ordered to move to the rear***

superpower ['suːpəpaʊə] *noun* extremely powerful country with great economic strength and large armed forces; ***the USA is the world's single superpower***

supersonic [ˌsuːpə'sɒnɪk] *adjective* capable of travelling faster than the speed of sound

Super Stallion ['suːpə ˌstæljən] *see* CH-53

supervise ['suːpəvaɪz] *verb* to control or guide the actions or work of other people; ***he is supervising the digging of the latrines***

supervisor ['suːpəvaɪzə] *noun* person who supervises other people

supervisory [ˌsuːpə'vaɪzəri] *adjective* controlling or guiding the actions or work of other people; ***he has been given a supervisory job***

supply [sə'plaɪ] **1** *noun* **(a)** act of supplying something; ***he is responsible for the supply of food*** **(b)** quantity of equipment, materiel, etc., which is available for use; ***we have a large supply of fuel*** **(c) supplies** = items which an army needs in order to carry out its tasks (such as ammunition, food, fuel, etc.); ***the enemy is short of supplies***; **supply depot** = military establishment, where supplies are stored; **supply dump** = temporary store of supplies in the field **2** *verb* to provide a person or group with the things they need; ***we haven't been supplied with NBC suits***

support [sə'pɔːt] **1** *noun* **(a)** assistance or help; ***B Company are calling for support***; **support company** = company of an infantry battalion, consisting of specialist platoons (for example anti-tank, mortar, reconnaissance, etc.); **support weapons** = specialist weapons held by an infantry unit (such as anti-tank weapons, machine-guns, mortars, etc.); **air support** = **(i)** attack by aircraft in support of ground troops; **(ii)** any assistance given by aircraft to ground troops; **close air support** = attack by aircraft on a target which is close to friendly ground forces; **fire support** = additional fire provided by another unit or arm; **in support** = providing or ready to provide support; **naval gunfire support (NGS)** = indirect fire provided by warships in support of ground forces **(b)** units or sub-units which provide support; ***Brigade can't send us any support*** **(c)** fire support; ***we are providing support to C Company during phase 3***; **mutual support** = ability of two *or* more defensive positions *or* groupings *or* vehicles to give fire support to each other **2** *verb* **(a)** to assist or help another person or group **(b)** to provide fire support to another grouping; ***Company B will support us***

supporting arms [sə'pɔːtɪŋ ˌɑːmz] *noun* arms which support the teeth arms (for example, engineers, signals, transport); *compare* TEETH ARMS

suppress [sə'pres] *verb* to fire at an enemy, in order to prevent him using his weapons; *see also* NEUTRALIZE

> COMMENT: when suppressing an enemy, it is not necessary to kill him. The object is simply to make him keep his head down

suppression [sə'preʃn] *noun* act of suppressing

supreme [su'priːm] *adjective* most senior; ***the Supreme Commander, NATO forces in Europe***; **Supreme Headquarters, Allied Powers in Europe (SHAPE)** = the main NATO headquarters in Europe; *see also* SACEUR, SACLANT

surface ['sɜːfɪs] **1** *noun* **(a)** outside of an object **(b)** top part of the earth (ie the land or sea); **air-to-surface missile (ASM)** = missile designed to be fired from an aircraft at a target on the ground or on the surface of the sea; **surface-to-air missile (SAM)** = anti-aircraft missile designed to be fired from the ground or from a ship; **surface-to-surface missile (SSM)** = missile designed to be fired from a launcher on the ground or on a ship at a target on the ground; **surface vessel** = boat or ship which travels on the surface of water (as opposed to a submarine) **2** *verb (of submarines)* to return to the surface of the sea after being under water; ***we dropped depth charges in order to try to force the submarine to surface***

surgeon ['sɜːdʒən] *noun* doctor who specializes in surgery; **surgeon-captain** = naval medical officer with the rank of captain

surgery ['sɜːdʒəri] *noun* process of treating illness or injury by cutting into a person's body in order to repair or remove damaged tissue or organs; ***he will need surgery***

surgical ['sɜːdʒɪkl] *adjective* relating to surgery; ***a surgical team is on stand-by***

surprise [sə'praɪz] **1** *noun* **(a)** unexpected action or event; ***the raid was a complete surprise to the enemy*** **(b)** act of surprising someone; ***surprise will be vital to the success of this operation***; **surprise attack** = attack which is mounted on the enemy when he is not expecting it **2** *verb* **(a)** to do something unexpected to another person; ***we surprised him as he was stealing petrol*** **(b)** to mount a surprise attack; ***we surprised the enemy as they were crossing the river***

surrender [sə'rendə] **1** *noun* act of surrendering; ***we don't know what happened to him after the surrender***; ***at the surrender, the defeated enemy general gave up his sword***; **unconditional surrender** = surrender where the side which is surrendering is not permitted to dictate any of its own terms or conditions; ***they demanded the unconditional surrender of the whole battalion*** **2** *verb* to stop fighting and hand oneself over to the enemy; ***7 Brigade has surrendered***; ***two thousand soldiers surrendered to our unit***

surround [sə'raʊnd] *verb* **(a)** to be on all sides of something; ***the village is surrounded by woods*** **(b)** to position your forces on all sides of an enemy, so that he is unable to escape or be reinforced; ***6 Brigade is surrounded***

surveillance [sɜː'veɪləns] *noun* **(a)** any method which can be used to locate the enemy or observe his activities and movements or listen to his radio transmissions; ***the general places great importance on good surveillance*** **(b)** people or equipment involved in surveillance; ***this manoeuvre is designed to confuse the enemy surveillance***

surveyor [sə'veɪə(r)] *noun* assistant to a forward observation officer (FOO); *also known as an* OP/Ack

survival [sə'vaɪvəl] *noun* act or process of surviving; **survival area** = concealed location, to which a unit deploys when war is imminent, because the unit's peacetime location is probably registered as a target and may be attacked as soon as hostilities begin; **survival course** = series of lessons, lectures and practical exercises on how to survive in a particular situation; ***I am going on a survival course***

survive [sə'vaɪv] *verb* to remain alive, in spite of a dangerous situation or life-threatening injury; ***he survived the crash***

survivor [sə'vaɪvə] *noun* person who survives; ***there were no survivors from the massacre***

sustainability [səˌsteɪnə'bɪləti] *noun* ability of a force to remain equipped and ready for action during the whole of an operation

sustained fire (SF) [səˌsteɪnd 'faɪə] *noun* fire from a machine-gun, which has been mounted on a tripod and fitted with a special sight, so that it can engage registered targets at long ranges, even when visibility is poor; ***a machine-gun in the sustained fire role***

swamp [swɒmp] *noun* thick woodland growing on wet ground, much of which is permanently under water

sweep [swi:p] **1** *noun* search of an area of ground or sea **2** *verb* to search an area of ground or sea (especially for mines); ***the channel has been swept for mines*** (NOTE: **sweeping - swept**)

Swingfire ['swɪŋfaɪə] *noun* British-designed wire-guided anti-tank missile (ATGW), usually fired from a variant of the AFV-432

sword [sɔ:d] *noun* weapon with a long blade, formerly used in warfare, but now ceremonial; ***at the surrender, the defeated enemy general gave up his sword***; **sword of honour** = sword presented to the best student in a class at a military college

synagogue ['sɪnəgɒg] *noun* building used for religious worship by Jews

synchronize ['sɪŋkrənaɪz] *verb* **(a)** to make (actions, etc.) happen at the same time; ***the attacks were not synchronized properly*** **(b)** to adjust clocks or watches, so that they are all showing the same time; ***the commanders forgot to synchronize their watches at the O Group***

syphon *see* SIPHON

syrette [sɪ'ret] *noun* device similar to a syringe, containing an individual dose of a drug *or* vaccine, which is designed to be carried by a soldier so that he can inject himself in an emergency; ***each man was issued with three syrettes of atropine***

syringe [sɪ'rɪndʒ] *noun* device, consisting of a tube, plunger and needle, which is used to inject liquid into a person's body *or* to extract blood *or* other fluids; ***we found a syringe in his locker***

TANGO - Tt

T-54 [ˌtiː fɪfti ˈfɔː] *noun* 1950s-era Soviet-designed main battle tank (MBT) (NOTE: plural is **T-54s** [ˌtiː fɪfti ˈfɔːz])

T-62 [ˌtiː sɪksti ˈtuː] *noun* 1960s-era Soviet-designed main battle tank (MBT) (NOTE: plural is **T-62s** [ˌtiː sɪksti ˈtuːz])

T-64 [ˌtiː sɪksti fɔː] *noun* 1970s-era Soviet-designed main battle tank (MBT) (NOTE: plural is **T-64s** [ˌtiː sɪksti ˈfɔːz])

T-72 [ˌtiː sevənti ˈtuː] *noun* simpler version of the Soviet-designed T-64 main battle tank, produced for export to Warsaw Pact countries and other allies of the Soviet Union (NOTE: plural is **T-72s** [ˌtiː sevənti ˈtuːz])

T-80 [ˌtiː ˈeɪti] *noun* 1980s-era Soviet-designed main battle tank (MBT) (NOTE: plural is **T-80s** [ˌtiː ˈeɪtiz])

TA [ˌtiː ˈeɪ] = TERRITORIAL ARMY

TAA [ˌtiː eɪ ˈeɪ] = TACTICAL ASSEMBLY AREA

tab [tæb] **1** *noun* **(a)** small coloured patch worn on each side of the collar; ***the general is the one with the red tabs*** **(b)** *(infantry slang)* march; ***we had a long tab to our pick-up point*** **2** *verb (infantry slang)* to go on foot; ***we were tabbing for most of the night***

Tabun [təˈbʊn] *see* GA

TAC *or* **TAC HQ** [tæk *or* ˌtæk eɪtʃ ˈkjuː] = TACTICAL HEADQUARTERS

TACC [ˌtiː eɪ siː ˈsiː] *noun US* = TACTICAL AIR COMMAND CENTRE principal command centre for air operations in an operational theatre

tac-sign [ˈtæk saɪn] *noun* = TACTICAL SIGN **(a)** small unobtrusive signpost (often marked with symbols instead of words); ***just follow the tac-signs*** **(b)** identification symbol painted on a vehicle; ***our tac-sign is a black triangle***

TACP or Tac-P [tæk ˈpiː] *noun US* = TACTICAL AIR CONTROL PARTY small group, trained to direct close air support; ***we've got a TACP attached to us for Phase 1***; *see also* FORWARD AIR CONTROLLER (FAC)

TACSAT [ˈtæksæt] *noun* = TACTICAL SATELLITE RADIO secure radio system, in which the signal is transmitted to a satellite before being redirected to another radio with the correct receiving equipment

tactic [ˈtæktɪk] *noun* **(a)** combination of firepower, formation and manoeuvre, which is used to achieve a military objective **(b) tactics** = art of employing military forces on the battlefield; *compare* STRATEGY

> COMMENT: **tactics** refers to the movement of battalions, brigades, divisions and equivalent-sized groupings, in order to achieve local objectives, while **strategy** refers to the movement of armies in order to achieve the overall objectives of a campaign or war

tactical [ˈtæktɪkl] *adjective* **(a)** relating to tactics; **tactical bound** = distance which ensures that one group is close enough to support another group without the risk of both coming under effective fire from the same enemy; ***Platoon HQ was moving a tactical bound behind the point section***; **tactical withdrawal** = withdrawal from the enemy as part of a planned manoeuvre **(b)** relating to the battlefield; **tactical area of responsibility (TAOR)** = area of ground defended and patrolled by a unit or other tactical grouping; **tactical bombing** = bombing carried out in direct support of ground forces; **tactical exercise without troops** = *see* TEWT; *US* **tactical fighter**

wing (TFW) = tactical air-force grouping of three fighter squadrons plus supporting arms; **tactical headquarters (TAC** *or* **TAC HQ)** = small mobile headquarters, used by a commander when he is moving around the battlefield; **tactical mobility** = ability to move forces to respond to an enemy attack; **tactical nuclear weapon** = small nuclear weapon designed to destroy enemy forces on the battlefield; **tactical reserves** = reserve forces kept for use in the battlefield; **tactical situation** = positions, strengths and known or probable intentions of both friendly forces and enemy forces; *compare* STRATEGIC

tactician [tæk'tɪʃn] *noun* person who is an expert at tactics

tail [teɪl] *noun* **(a)** rear elements of a large military force *or* grouping; ***the enemy's tail is now extremely vulnerable to air attack*** **(b)** *(informal)* rear end of an aircraft; ***Look out ! There's a bogey on your tail !***

tailplane ['teɪlpleɪn] *noun* small wing-like structure at the rear of an aircraft; ***the tailplane was almost shot away by cannon fire***

take [teɪk] *verb* **(a)** to acquire; ***he took a cigarette from the packet*** **(b)** to capture; ***the enemy has taken the bridge***; ***the patrol took two prisoners*** **(c)** to remove; ***someone has taken my rifle*** **(d)** to carry with you; ***the patrol is taking a night viewing device*** **(e)** to be accompanied by; ***we took a local farmer as a guide*** (NOTE: **taking - took - have taken**)

take action [ˌteɪk 'ækʃn] *verb* to do something

take cover [ˌteɪk 'kʌvə] *verb* to hide, or to seek protection from enemy fire

take off [ˌteɪk 'ɒf] *verb (of aircraft)* to leave the ground; ***the fighters took off at first light***

take-off ['teɪk ɒf] *noun* action of an aircraft taking off from the ground; ***take-off at 0630hrs***; ***the plane crashed on take-off***; ***take-off was delayed by fog***; **short take-off and landing (STOL)** = technology which enables a fixed-wing aircraft to take off and land over considerably shorter distances than those required by conventional fixed-wing aircraft; **vertical take-off and landing (VTOL)** = technology which enables a fixed-wing aircraft to take off and land from a stationary position (ie without the need for a runway)

> COMMENT: vertical take-off is not usually possible when the aircraft is carrying a full payload of munitions. In such cases, the aircraft would need to take off from a runway like any conventional fixed-wing aircraft. Most vertical take-off aircraft, however, require a considerably shorter distance to take off than do conventional aircraft, and would therefore be able to use stretches of road or grass areas as runways. Once the aircraft has discharged its munitions it would be able to carry out a normal vertical landing. The acronyms STOVL (short take-off and vertical landing) and V/STOL (vertical or short take-off and landing) are used to describe these capabilities

take out [ˌteɪk 'aʊt] *verb* to kill or destroy; ***B troop took out six APCs***

talc [tælk] *noun* clear plastic sheeting, which is used to cover maps and which may be written upon or marked

tally *or* **tally ho** ['tælɪ həʊ] *adverb (air force terminology)* I have seen an enemy aircraft *or* other target

> COMMENT: this expression is taken from the sport of foxhunting

tandem warhead [ˌtændəm 'wɔːhed] *noun* anti-tank warhead, consisting of two shaped charges positioned one behind the other, which is designed to defeat explosive reactive armour (ERA); the first charge activates the ERA and the second charge then penetrates the main armour underneath

Tango ['tæŋgəʊ] twentieth letter of the phonetic alphabet (Tt)

tank [tæŋk] *noun* **(a)** armoured fighting vehicle fitted with tracks and a gun; *(of artillery)* **tank action** = using a gun as a direct-fire weapon against tanks; **tank transporter** = large wheeled vehicle, designed to carry a tank or other armoured vehicle over long distances by road; **main battle tank (MBT)** = heavily armoured tank, fitted with a large-calibre gun, which is primarily designed to destroy enemy tanks; **light tank** = another name for a tracked armoured reconnaissance vehicle (CVRT) **(b)** large container or structure designed to hold liquid or gas; ***a shell hit a fuel storage tank*** **(c)** part of an aircraft *or* ship *or* vehicle which contains its fuel; ***the plane exploded when a round hit its fuel tank***; **drop tank** = additional fuel tank for an aircraft, which can be jettisoned when empty

tanker ['tæŋkə] *noun* **(a)** aircraft, ship or vehicle fitted with a tank or tanks designed to carry liquid or gas (especially fuel) **(b)** soldier in an armored unit

tannoy ['tænɔɪ] *noun* system of loudspeakers in a ship *or* building, which are used to make announcements

TAOR [ˌtiː eɪ əʊ 'ɑː] = TACTICAL AREA OF RESPONSIBILTY

tape [teɪp] **1** *noun* **(a)** strip of fabric or plastic, used to bind objects together or as a marker or for other purposes; **magnetic tape** = tape used for recording sound, images or computer data; **mine tape** = white or florescent tape, designed for marking lanes through a minefield or for marking a boundary **(b)** magnetic tape, used for recording sound; ***he was listening to a tape of military music***; ***they played back the tape of the conversation*** **2** *verb* **(a)** to record on magnetic tape; ***we have taped some of the enemy radio transmissions*** **(b) to tape off** = to use mine tape as a barrier or boundary; ***we have taped off the area of the explosion***

Taps [tæps] *noun US* nickname for the bugle-call "Last Post"

target ['tɑːgɪt] **1** *noun* any object or area which is shot at, fired upon or bombed; ***they dropped six bombs on the target***; ***two of our shells missed the target***; **soft-target** = person *or* unit *or* vehicle which is vulnerable *or* unable to defend itself properly; ***the terrorists are only interested in attacking soft targets*** **2** *verb* to select as a target; ***the enemy have denied that they were targeting civilians***; **hard-target** = to move across ground in such a way as not to present an easy target to the enemy; ***we had to hard-target across the square***

tarmac ['tɑːmæk] road surface made of a mixture of tar and gravel (NOTE: American English is **asphalt**)

tarp [tɑːp] = TARPAULIN

tarpaulin [tɑː'pɔːlɪn] *noun* waterproof sheet used to protect an object from dust or rain

tartan ['tɑːtən] *noun* traditional Scottish fabric pattern of coloured lines and checks; ***some Scottish regiments wear tartan flashes on their bonnets***

task [tɑːsk] **1** *noun* something which must be done; ***he failed to complete his task***; ***we have been given the task of collecting information on underground organizations*** **2** *verb* to allocate a task; ***B Company has been tasked for this mission***

task force ['tɑːsk ˌfɔːs] *noun* **(a)** *US* combined arms grouping based on an infantry or tank battalion (NOTE: British equivalent is battle group. US Marine Corps equivalent is battalion landing team (BLT)); **balanced task force** = two tank companies and two companies of mechanized infantry; **mech-heavy task force** = three infantry companies and one tank company; **tank-heavy task force** = three tank companies and one company of infantry **(b)** large combined arms grouping formed for a specific operation or campaign; ***the government is sending a task force to the area***; *see also* JOINT TASK FORCE **(c)** outdated British Army term for an armoured brigade

> COMMENT: an American task force often organizes its companies into combined arms groupings known as company teams , containing a mix of tank platoons and mechanized infantry platoons. The exact composition will nd on the tactical requirement at the time. The British equivalents of company teams are known as company and squadron groups and squadron and company groups

team [tiːm] *noun* group of people who work together

tear gas ['tɪə ˌgæs] *noun* chemical agent which irritates the eyes and makes people choke; *see also* CS GAS

technical ['teknɪkl] *adjective* relating to instruments, machinery, radios, weapons, etc.; **technical arrangements** *or* **technical agreements** = agreements reached between commanders of different NATO forces regarding the implementation of higher level agreements on the ground (NOTE: also called **detailed support arrangements**) *GB* **technical quartermaster (TQM)** = officer (with a quartermaster commission) responsible for all technical equipment and machinery held by a battalion or equivalent-sized grouping; *GB* **technical quartermaster sergeant (TQMS)** = warrant officer who assists the technical quartermaster of a battalion or equivalent-sized grouping; *US* **technical sergeant** = senior non-commissioned officer in the air force

technician [tek'nɪʃn] *noun* person trained in the repair and maintenance of technical equipment; **junior technician** = non-commissioned rank in the air force (equivalent to an experienced or well-qualified private soldier in the army)

teeth arms ['tiːθ ˌɑːmz] *noun* branches of the armed forces which actually fight (such as armour, artillery, infantry); *compare* SUPPORTING ARMS

TEL [ˌtiː iː 'el] *noun* = TRANSPORTER-ERECTOR-LAUNCHER vehicle designed to carry and launch a surface-to-surface missile (SSM); *see also* MEL

telescope ['telɪskəʊp] *noun* optical instrument formed of a single long tube with lenses at both ends, designed for looking at distant objects; ***he examined the bridge through a powerful telescope***; *compare* FIELD GLASSES

telescopic [ˌtelɪ'skɒpɪk] *adjective* **(a)** relating to a telescope; **telescopic sight** = small telescope used as a sight for a rifle **(b)** made in sections which slide together, for ease of carriage or storage; ***telescopic antenna***

template ['templeɪt] *noun* **(a)** simple mathematical instrument for working out the danger area for a field-firing range, consisting of a thin piece of plastic cut to specific angles and measurements; ***on this course, officers are shown how to construct and apply range templates*** **(b)** range danger area, marked on a map by means of a template; ***that road is inside the template***

tenable ['tenəbl] *adjective* possible to defend; ***our position is no longer tenable***

tent [tent] *noun* portable shelter made of waterproof fabric, which is supported by poles; ***the unit will be housed in tents for the period of the exercises***

TEREC ['terek] *noun* = TACTICAL ELECTRONIC RECONNAISSANCE SYSTEM airborne radar receiving system used for the location of enemy radar sites

terminally guided ['tɜːmɪnəli ˌgaɪdɪd] *adjective* which guides itself automatically towards its own target; **terminally guided submunitions** = *see* TGSM; **terminally guided warhead (TGW)** = anti-tank missile which uses a radar seeker to search for suitable targets

terminate ['tɜːmɪneɪt] *verb* **(a)** to finish something; ***we will have to terminate the conference now*** **(b)** *US* to kill; ***he has been terminated***

terrain [tə'reɪn] *noun* **(a)** ground; **key terrain** = ground which you must occupy or control in order to achieve your mission **(b)** type of land (such as desert, farmland, mountains, woodland, etc.); ***the attack will be launched through wooded terrain***; **all-terrain vehicle** = vehicle which is capable of operating in all types of terrain

terrace ['terəs] *noun* one of a series of level areas constructed on the side of a hill, in order to cultivate crops

terraced housing [ˌterəst 'haʊzɪŋ] *noun* line of houses along a street *or* road, which are all joined to each other

terrier ['terɪə(r)] *noun* *GB informal* member of the Territorial Army (TA); ***we're being relieved by a battalion of terriers***

territorial [ˌterɪ'tɔːriəl] **1** *adjective* relating to the territory of a state; *GB* **Territorial Army (TA)** = volunteer force of part-time soldiers, designed to reinforce the regular army in the event of war; **territorial waters** = area of sea coming under the jurisdiction of a state; ***the ship was attacked in French territorial waters*** **2** *noun* member of the Territorial Army; ***200 territorials were sent to the area***

territory ['terɪtri] *noun* area or region coming under the control or jurisdiction of a state or military force; ***the squad wandered into enemy-occupied territory***

terrorism ['terərɪzm] *noun* use of physical violence to intimidate a government or the general public, in order to achieve political objectives

terrorist ['terərɪst] *noun* person involved in terrorism

> COMMENT: the use of this word is very much a question of perception. A **terrorist** in one person's view may very well be seen as a **freedom fighter** by another person holding opposing views

tetanus ['tetənəs] *noun* sometimes fatal bacterial disease, an infection affecting the nervous system caused by *Clostridium tetani* in the soil; it affects the spinal cord and causes spasms in the muscles which occur first in the jaw

TEWT ['tjuːt] *noun* = TACTICAL EXERCISE WITHOUT TROOPS exercise involving only the command elements of a tactical grouping, who examine an area of ground in order to plan and then discuss a hypothetical military operation; ***we are going on a TEWT tomorrow***

TEZ [ˌtiː iː ˌzed *US* ˌtiː iː 'ziː] = TACTICAL EXCLUSION ZONE

TF [tiː 'ef] = TASK FORCE

TFW = TACTICAL FIGHTER WING

TGSM [ˌtiː dʒiː es 'em] *noun* = TERMINALLY GUIDED SUBMUNITIONS small anti-armour projectiles, which are released by a missile over a target area and which then independently seek out and attack their own targets

TGW [ˌtiː dʒiː 'dʌb(ə)ljuː] = TERMINALLY GUIDED WARHEAD

theatre *US* **theater** ['θiːətə] *noun* area in which operations are being carried out; ***strategic mobility implies the ability of forces to move rapidly between theatres***

theft [θeft] *noun* act of stealing

thermal ['θɜːməl] *adjective* relating to heat; **thermal image (TI)** = image produced by equipment which can identify the varying levels of heat given off by different objects; **thermal imager (TI)** = optical instrument which produces a thermal image; **thermal imaging sight (TIS)** = weapon sight designed around a thermal imager

thermos ['θɜːmɒs] *noun* container designed to keep drinks hot for several hours; ***I took a thermos of tea out to the gun position***

threat [θret] *noun* **(a)** something which is dangerous or hostile; ***the partisans are posing a major threat to our supply routes*** **(b)** enemy forces; ***the main threat is from the east*** **(c)** statement declaring a person's intention to do harm; ***he was making threats to other people in the bar***

threaten ['θretən] *verb* **(a)** to manoeuvre against; ***the enemy is threatening our left flank*** **(b)** to say that you intend to do harm to someone; ***he threatened me; he threatened to shoot me***

throat-mike ['θrəʊt maɪk] *noun* radio microphone which is strapped to the user's throat and utilizes the vibrations from his vocal cords

thrust [θrʌst] **1** *noun* advance; ***G2 reports a strong enemy thrust in the direction of Prague*** **2** *verb* to move forward with force; ***the invaders thrust on towards the capital***

thumbs-up [θʌmz 'ʌp] *noun* gesture, consisting of a clenched fist with the thumb pointing upwards, which indicates that everything is alright, *or* that the next phase of an activity may proceed; ***once the minefield was breached, he gave a thumbs-up to the company commander***; **to give something the thumbs-up** = to approve a course of action; ***the operation has been given the thumbs-up***

Thunderbolt ['θʌndəbəʊlt] *see* A-10

thunderbox ['θʌndəbɒks] *noun (slang)* latrine, usually consisting of a box-seat positioned over a pit; ***as a punishment, you can clean out the thunderbox***

thunderflash ['θʌndəflæʃ] *noun* pyrotechnic device, producing a small explosion but no shrapnel or other dangerous fragments, which is designed to simulate artillery or grenade explosions on training exercises

TI [ˌtiː 'eɪ] = THERMAL IMAGE; THERMAL IMAGER

tick [tɪk] *noun* small insect which attaches itself to an animal's skin in order to suck its blood

tidal ['taɪdəl] *adjective* affected by tides; ***the river is tidal as far as Carrick-on-Suir***

tide [taɪd] *noun* rise and fall of the sea which takes place twice a day; **high tide** = point at which the tide has risen to its highest level; ***the landing will take place at high tide***; **low tide** = point at which the tide has fallen to its lowest level; ***the low tide left the landing craft stranded on the rocks***

tilt-switch ['tɪlt swɪtʃ] *noun* device for initiating an explosive device (especially booby traps), consisting of a small glass *or* plastic container, fitted with a positive and a negative electric wire and half-filled with mercury; when the container is moved, the mercury flows over the exposed ends of the two wires, completing the electrical circuit and thus initiating the explosion

> COMMENT: terrorist bombs which have been designed to be initiated by timer *or* remote control, are often fitted with a tilt-switch as well, in case anyone tries to remove *or* defuse the device

timer ['taɪmə(r)] *noun* device for arming *or* initiating an improvised explosive device (IED) at a pre-set time; ***they used a small alarm clock as a timer***

timing ['taɪmɪŋ] *noun* time at which an event is scheduled to occur; ***all the timings were changed at the last moment***; ***he sent a list of timings to HQ***

> COMMENT: military timings are always given using the **twenty-four hour clock,** usually followed by the word **hours** which is abbreviated to **hrs.** Thus, 8.15am is 0815hrs, 1pm is 1300hrs, 6.30pm is 1830hrs, etc. NATO forces normally use **Greenwich Mean Time (GMT)** for their timings. This is known as **Zulu time** (for example: **H-Hour at 0645Z**). The time of the country in which one is operating is known as **local time**

tin [tɪn] *noun GB* metal container in which food or drink is hermetically sealed for storage over long periods; ***we found some tins of meat left in the camp*** (NOTE: in American English, **can** is more usual)

tire [taɪə] *US* = TYRE

TIS [ˌtiː eɪ 'es] = THERMAL IMAGING SIGHT

T-junction [ˌtiː 'dʒʌŋkʃn] *noun* place where two roads meet at right angles to each other

TNT [ˌtiː en 'tiː] *noun* = TRINITROTOLUENE type of high explosive

TOGS [tɒgz] = THERMAL OBSERVATION GUNNERY SIGHT

Tomahawk ['tɒməhɔːk] *noun* American-designed cruise missile

Tomcat ['tɒmkæt] *see* F-14

ton [tʌn] *noun* **(a)** *GB (long ton)* unit of weight corresponding to 1,016.05 kilograms **(b)** *US (short ton)* unit of weight corresponding to 907.19 kilograms; **metric ton (tonne)** = unit of weight corresponding to 1,000 kilograms

tonne [tʌn] *noun (metric ton)* unit of weight corresponding to 1,000 kilograms

top secret ['tɒp ˌsiːkrət] *adjective* highest security classification for documents and information; ***that information is classified top secret***; ***he left some top-secret documents on the back seat of a taxi***

torch [tɔːtʃ] **1** *noun* hand-held battery-powered device for producing light; ***he used a torch to attract attention***; *see also* FLASHLIGHT **2** *verb (informal)* to set fire to something; ***they torched the village***

Tornado [tɔː'neɪdəʊ] *noun* British/German/Italian-designed fighter aircraft; **Tornado F-3** = long-range interceptor; **Tornado GR.1** = multirole fighter optimized for ground-attack

torpedo [tɔː'piːdəʊ] **1** *noun* underwater missile, designed to explode when it hits a ship; **torpedo-boat** = small fast-moving warship, designed to fire torpedos at other ships; **aerial torpedo** = torpedo designed to be dropped by aircraft **2** *verb* to hit (and sink) a ship using a torpedo; ***their ship was torpedoed by an enemy submarine***

torture ['tɔːtʃə] **1** *noun* deliberate act of inflicting pain on a person (usually in order to extract information); ***the enemy is known to use torture***; ***the prisoner died***

under torture **2** *verb* to deliberately inflict pain on another person; ***the rebels have been torturing civilians***

toss-bombing ['tɒs ˌbɒmɪŋ] attack where bombs are released as the aircraft is making a shallow climb at high speed; the bombs' trajectories then carry them forward a considerable distance before they hit the ground, making it unnecessary for the aircraft to pass directly over its target

TOT [ˌtiː əʊ 'tiː] = TIME OVER TARGET

touch [tʌtʃ] **1** *verb* to come into physical contact with another thing; ***he couldn't touch the dead man*** **2** *noun* physical contact; ***I can't stand the touch of a dead body***; **in touch =;** *(radio terminology)* radio contact with another call-sign; ***are you in touch with 33 ?***; **to get in touch =** to establish radio contact with another call sign; ***get in touch with 22B and ask them for a SITREP***

touch down [ˌtʌtʃ 'daʊn] *verb (of aircraft)* to land; ***the aircraft touched down at 1500 hrs***

tour [tʊːə] *noun* period of operational duty; ***the regiment has just completed its second tour of duty in the region***

tourniquet ['tɜːnɪkeɪ] *noun* act of twisting a stick through a bandage which is bound around a limb, in order to constrict the artery and thus reduce the bleeding from a serious wound; ***he applied a tourniquet***

> COMMENT: a *tourniquet* can do more harm than good if it is incorrectly applied

TOW [təʊ] *noun* = TUBE-LAUNCHED, OPTICALLY-TRACKED, WIRE-GUIDED MISSILE American-designed anti-tank missile

tow [təʊ] *verb* to move a vehicle, aircraft or ship by pulling it; ***we had to tow the tank off the battlefield***

tower ['taʊə] *noun* tall structure, usually built for observation or defence

town [taʊn] *noun* large settlement

toxic ['tɒksɪk] *adjective* poisonous; ***clouds of toxic gas rose from the burning supply dump***

TP = TROOP

TPFDL [ˌtiː piː ef diː 'el] *noun US* = TIME PHASED FORCE DEPLOYMENT LIST document showing the order in which units and groupings deploy to an area of operations

TPr = TROOPER

TQM [ˌtiː kjuː 'em] *GB* = TECHNICAL QUARTERMASTER

TQMS [ˌtiː kjuː em 'es] *GB* = TECHNICAL QUARTERMASTER SERGEANT

TR-1 [ˌtiː ɑː 'wʌn] *noun* American-designed high-altitude reconnaissance aircraft

trace [treɪs] *noun* piece of transparent paper or plastic, marked with boundaries, positions, routes, and other information relating to an operation, which is designed to be placed over a map as a means of briefing the participants (NOTE: also called an **overlay**)

tracer ['treɪsə] *noun* bullet which is designed to ignite after firing and burn in flight, so that the fall of shot can be observed (NOTE: also called an **incendiary bullet**)

track [træk] **1** *noun* **(a)** marks on the ground, made by the movement of a person or vehicle; ***we followed the tracks of the convoy*** **(b)** rough path or road; ***someone is moving along the track***; ***there are several tracks through the wood*** **(c)** railway line; ***the track has been blown up in several places*** **(d)** moving band of metal links fitted around the wheels of a tank or other armoured vehicle, enabling it to move over soft or uneven ground; ***the tank came off the road when it lost a track***; *see also* HALF-TRACK **2** *verb* **(a)** to follow the track of a person or vehicle; ***the deserters were tracked to the local railway station*** **(b)** to follow the movement of an aircraft, vehicle or ship using surveillance equipment or a missile guidance system; ***they were unable to track the aircraft***; ***we are being tracked***

tracked [trækd] *adjective; (of armoured vehicles)* fitted with tracks

tracker dog ['trækə dɒg] *noun* dog trained to follow the smell of a person

tracking ['trækɪŋ] *noun* the act of following the movement of an aircraft, vehicle or ship using surveillance equipment or a missile guidance system

trade [treɪd] *noun* **(a)** general term for the business of buying and selling goods (especially between different countries); ***the war has severely disrupted all trade in the region*** **(b)** *(air-force jargon)* targets (especially enemy aircraft); ***Hello Fruitbat, this is Merlin. I have some trade for you, north-west of Minden.***

traffic ['træfɪk] *noun* **(a)** vehicles moving on a road; ***the convoy was delayed by heavy traffic*** **(b) air traffic** = aircraft moving in the air; ***there will be an accident if air traffic is not reduced*** **(c)** mass of messages on radio; ***there is too much unnecessary traffic on this net***

trail [treɪl] **1** *adjective* relating to the subsequent waves of an advancing force, which are in a position to reinforce the leading elements *or* take over the lead when required; ***the enemy trail units were broken up by our airstrikes***; *see also* FOLLOW-ON FORCES **2** *noun* **(a)** rough path or track; ***the patrol made its way up the trail*** **(b)** marks on the ground, made by the movement of a person or vehicle; ***we followed the trail left by enemy soldiers*** **(c)** structure at the rear of an artillery piece, which enables it to be towed by a vehicle; ***he fell over the trail of the gun*** **(d)** the way in which a bomb falls behind an aircraft after it has been dropped, because the aircraft's forward speed is greater than that of the bomb

trailer ['treɪlə] *noun* vehicle with no engine, designed to be towed by another vehicle

train [treɪn] **1** *noun* **(a)** several railway carriages towed by a railway engine; ***the battalion will move by train*** **(b)** column of vehicles carrying supplies, which accompanies a military force; ***the enemy has captured our train*** **2** *verb* **(a)** to teach or instruct; ***he has been trained in the use of explosives*** **(b)** *(of artillery)* to point a gun; ***they trained their guns on the town***

trainer ['treɪnə] *noun* aircraft used for training

training ['treɪnɪŋ] *noun* teaching and practice of military skills; ***we were sent to signals school for training***; ***the unit spent two weeks in Norway undergoing Arctic training***; **physical training (PT)** = activities and exercises designed to improve or maintain physical fitness

traitor ['treɪtə] *noun* person who assists an enemy power against the interests of his own state

trajectory [trə'dʒektəri] *noun* curved flight of a projectile from the weapon to the point of impact; ***mortars fire projectiles with a very high trajectory***

transceiver [træn'siːvər] *noun* combined radio receiver and transmitter

transfer ['trænsfə] *noun* action of moving someone or something to a different position; **transfer of authority** = the action of passing authority over forces from one commander to another, or from a national command to a NATO command

transit ['trænzɪt] *noun* movement from one location to another; **transit camp** = camp providing temporary accommodation for people who are moving from one location to another; **in transit** = moving from one location to another; ***the equipment was damaged in transit***

transmission [trænz'mɪʃn] *noun* act of sending a radio signal

transmit [trænz'mɪt] *verb* **(a)** to send a radio signal; ***we were unable to transmit the signal*** **(b)** to infect with a disease; ***the disease is transmitted by a parasite***

transmitter [trænz'mɪtə] *noun* apparatus used to send a radio signal; ***we found a transmitter hidden in the attic of the farmhouse***

transport 1 ['trænspɔːt] *noun* **(a)** act of moving people or things by aircraft, ship or vehicle; ***the transport of the brigade will be carried out by aircraft*** **(b)** aircraft, ship or vehicle used to transport people or things; ***they stayed in the camp for ten days, waiting for transport*** **2** [træns'pɔːt] *verb* to move people or things by aircraft, ship or vehicle; ***the tanks were transported by train***

transportation [ˌtrænzpə'teɪʃn] *noun* = TRANSPORT

trap [træp] **1** *noun* deception or trick which encourages a person to place himself in a dangerous situation from which there is no escape; ***B Company have walked straight into a trap*** **2** *verb* to place a person in a dangerous situation from which there is no escape; ***3 Brigade has been trapped by the enemy encirclement***; ***the pilot was trapped in his cockpit***

TRAP [træp] = TACTICAL RECOVERY OF AIRCRAFT, EQUIPMENT AND PERSONNEL

TRAP [træp] *noun* = TACTICAL RECOVERY OF AIRCRAFT AND PERSONNEL mission to recover an aircraft and its crew, after being shot down or crashing in enemy territory

trapdoor ['træpdɔː] *noun* small door or hatch in a ceiling, floor or roof

traveller ['trævələ(r)] *noun* **(a)** person who travels from one place to another **(b)** another name for gipsy

traverse [trə'vɜːs] **1** *noun* a pair of right-angled bends in a trench, which is designed to prevent anyone firing up the entire length (in the event of the enemy capturing part of the trench) **2** *verb* **(a)** to move across an area of ground; ***we had to traverse 200m of open field*** **(b)** *(of guns)* to move the barrel sideways when aiming or firing; ***traverse left!***

treachery ['tretʃəri] *noun* act of betraying your country or comrades

treason ['triːzn] *noun* act which threatens the interests or security of your own state

treeline ['triːlaɪn] *noun* **(a)** line of trees; ***there is an enemy OP in that treeline*** **(b)** edge of a forest or wood; ***we stopped at the treeline*** **(c)** altitude above which trees cannot survive; ***the patrol moved back down to the treeline***

trembler ['tremblə(r)] *noun* device designed to initiate an explosive device at the slightest movement; *see also* TILT-SWITCH

trench [trentʃ] *noun* narrow hole or channel dug into the ground, in order to provide protection from enemy fire; **communication trench** = trench used for movement from one fire trench to another; **fire trench** = trench used by infantrymen as a fire position; **slit trench** = another term for fire trench; **trench foot** = severe fungal infection of the feet, caused by wearing wet boots over a long period

> COMMENT: the length of a trench can vary from a few metres to several kilometres, depending upon the tactical requirement at the time. During the First World War (1914-1918), both the Allies and the Germans occupied trench systems which extended, without a break, from the North Sea to the Alps

trews [truːz] *noun* trousers of tartan cloth, worn by some Scottish regiments

triage ['triːɑːdʒ] *noun* process of assessing a casualty's priority for medical treatment according to the nature of his injuries; ***we'll set up triage over therep***

triangulate [traɪ'æŋgjʊleɪt] *verb* ***a*** to locate a radio using direction-finding equipment, by taking bearings on its emissions from three different locations and then seeing where the bearings intersect on a map **b** to calculate your position by working out the back-bearings from three known *or* probable reference points and then seeing where the back-bearings intersect on a map

triangulation point [traɪˌæŋgjʊ'leɪʃn ˌpɔɪnt] *noun* small concrete or stone pillar, designed to serve as a firm base for cartographers' surveying instruments (NOTE: also called a **trig point**)

> COMMENT: triangulation points are permanent structures, and are represented on maps by a triangle with a dot in the centre

tribal ['traɪbəl] *adjective* relating to tribes; ***the civil war is essentially a tribal conflict***

tribe [traɪb] *noun* group of families *or* communities sharing a common language *or* dialect, distinct ethnic *or* religious links, and a strong sense of group identity and loyalty to their own leaders; ***in Nigeria, the two main tribes are the Ibo and the Yoruba***

> COMMENT: the word *tribe* has rather a primitive connotation, and is really only applicable to communities in certain developing countries, especially Africa

tributary ['trɪbjʊtəri] *noun* river or stream which flows into a larger river

tricolour US tricolor ['traɪkʌlə(r) *or* 'trɪkələ(r)] *noun* flag consisting of three different blocks of colour (usually side by side); ***the French national flag is a tricolour of blue, white and red***

trigger ['trɪgə] *noun* moving lever which releases the firing mechanism of a gun; **trigger-happy** = lacking in judgement when using firearms, willing to shoot at random

trig point ['trɪg ˌpɔɪnt] *see* TRIANGULATION POINT

trinitrotoluene [traɪˌnaɪtrəʊ'tɒljʊiːn] *noun see* TNT

trip [trɪp] *verb* to stumble or fall as a result of catching your legs in something; ***he tripped over an ammunition box***; **trip-flare** = flare which is activated by a trip-wire; **trip-wire** = wire which is stretched horizontally close to the ground, in order to activate an explosive device, trip-flare or other device when someone trips over it

triplicate ['trɪplɪkət] *noun* a third copy of a document; **in triplicate** = in three copies

tripod ['traɪpɒd] *noun* three-legged stand designed to support a weapon or other piece of equipment

Triple-A [ˌtrɪpəl 'eɪ] = ANTI-AIRCRAFT ARTILLERY

troop (Tp) [truːp] *noun* **(a)** *GB* platoon-sized armoured grouping of three or more tanks **(b)** *GB* platoon-sized artillery grouping of two or more guns **(c)** *GB* platoon-sized grouping in certain supporting arms, such as engineers **(d)** *US* company-sized armored cavalry grouping of three or more platoons

trooper (Tpr) ['truːpə] *noun GB* private soldier in an armoured regiment (NOTE: also used as a title: **Trooper Williams**)

troops [truːps] *noun* soldiers in general; ***troops are being deployed in the region***; ***the enemy fell back, their troops were tired and demoralized***; ***British troops entered the capital on Friday morning***

troopship ['truːpʃɪp] *noun* ship designed or adapted to transport troops

tropical ['trɒpɪkl] *adjective* **(a)** relating to the Tropics; ***we had to get used to tropical conditions*** **(b)** designed for use in hot climates; ***tropical clothing will be issued for the operation***

Tropics ['trɒpɪks] *noun* ***the Tropics*** = the region between latitudes 23° 28N (Tropic of Cancer) and 23° 28S (Tropic of Capricorn)

truce [truːs] *noun* agreement by both sides to stop fighting; ***both sides agreed to sign a truce***; **flag of truce** = white flag displayed by soldiers wishing to surrender, or by a messenger indicating to the enemy that they should stop shooting; *see also* ARMISTICE, CEASEFIRE

truck [trʌk] *noun* large wheeled vehicle designed to transport men, equipment or supplies (NOTE: British English also uses the word **lorry**)

tsetse fly ['tetsi flaɪ] *noun* African insect, whose bite can cause African trypanosomiasis *or* sleeping sickness; ***that region is infested with tseste flies***

TU-16 [ˌtiː juː sɪk'stiːn] *noun* Soviet-designed medium bomber aircraft (NOTE: known to NATO as **Badger**)

TU-22 [ˌtiː juː twenti 'tuː] *noun* Soviet-designed medium bomber aircraft (NOTE: known to NATO as **Blinder**; a strategic variant of this aircraft is known as **Backfire**)

TU-95 [ˌtiː juː naɪnti 'faɪv] *noun* Soviet-designed strategic bomber aircraft (NOTE: known to NATO as the **Bear**)

TU-160 [ˌtiː juː wʌn 'sɪksti] *noun* Soviet-designed strategic bomber aircraft (NOTE: known to NATO as the **Blackjack**)

tube [tjuːb] *noun* cylindrical container; ***a tube of cam-cream***; **torpedo tube** = barrel through which a torpedo is fired from a submarine; **tube-launched, optically-tracked, wire-guided missile;** *see* TOW

tumulus ['tjuːmjʊləs] *noun* small man-made mound, usually marking the site of an ancient grave (NOTE: plural is tumuli)

tunic ['tjuːnɪk] *noun* close-fitting jacket, worn as part of a ceremonial uniform; ***the soldiers wore red tunics***

tunnel ['tʌnl] **1** *noun* man-made passage dug under the ground or through a hill; ***terrorists have blown up the railway tunnel***; ***we found a system of tunnels under the enemy position*** **2** *verb* to dig a tunnel; ***they managed to escape by tunnelling under the prison wall***

turf [tɜːf] *noun* layer of grass and the soil surrounding its roots, which can be removed from the ground intact *or* in sections; ***the turf is used to camouflage the parapet and parados***

turning movement ['tɜːnɪŋ ˌmuːvmənt] *noun* manoeuvre designed to force an enemy to change his positions to meet a new threat, usually achieved by advancing on him from an unexpected direction (eg. from a flank)

turret ['tʌrɪt] *noun* revolving gun compartment on an aircraft or armoured fighting vehicle (AFV) or warship

twenty-four hour clock [ˌtwenti fɔː aʊə 'klɒk] *see* TIMING

twin [twɪn] *adjective* fitted as a pair; **twin-mounted machine-guns** = two machine-guns mounted coaxially

2IC [ˌtuː aɪ 'siː] = SECOND-IN-COMMAND

two-up [tuː 'wʌp] *adverb* tactical formation in which two sub-units are leading abreast of each other, and the third sub-unit is following; ***we'll be assaulting two-up***; *compare* ONE-UP

COMMENT: this formation is suitable for an assault

typhoid fever ['taɪfɔɪd ˌfiːvə(r)] *noun* infection of the intestine, caused by Salmonella typhi in food and water

typhus ['taɪfəs] *noun* infectious fever caused by the Rikettsia bacterium, which is transmitted by lice

COMMENT: epidemics of typhus are very common in wartime due to a breakdown in hygiene and sanitation

tyre *US* **tire** [taɪə] *noun* circular rubber cover containing an air-filled inner tube, which is fitted to a vehicle wheel

UNIFORM - Uu

UAV [ˌjuː eɪ 'viː] *noun* = UNMANNED AERIAL VEHICLE another name for a drone (NOTE: also known as **remotely piloted vehicle (RPV)**)

UFO [ˌjuː ef 'əʊ *or* 'juːfəʊ] *noun* = UNIDENTIFIED FLYING OBJECT any unexplained object which is seen flying through the air or detected on a radar screen (NOTE: this term is usually applied to suspected alien spacecraft)

UGS [ˌjuː dʒiː 'es] = UNATTENDED GROUND SENSOR

UH-1 [ˌjuː eɪtʃ 'wʌn] *see* HUEY

UH-60 [ˌjuː eɪtʃ 'sɪksti] *see* BLACKHAWK

UHF [ˌjuː eɪtʃ 'ef] = ULTRA HIGH FREQUENCY

UK [ˌjuː 'keɪ] = UNITED KINGDOM

UKLF [ˌjuː keɪ el 'ef] = UNITED KINGDOM LAND FORCES

UKLO [ˌjuː keɪ el 'əʊ] = UNITED KINGDOM LIAISON OFFICER

ULC [ˌjuː el 'siː] *noun* large metal container pre-packed with artillery rounds, designed to be transported onto the battlefield

ultra high frequency (UHF) [ˌʌltrə ˌhaɪ 'friːkwənsi] *noun* range of radio frequencies from 300 - 3,000 megahertz (Mhz)

umpire ['ʌmpaɪə] **1** *noun* person assigned to observe a military training exercise and to assess the performance of those taking part; ***he was acting as umpire*** **2** *verb* to act as an umpire; ***he is umpiring the exercise***

UN [ˌjuː 'en] = UNITED NATIONS

unarmed [ʌn'ɑːmd] *adjective* without weapons; **unarmed combat** = fighting using the hands, arms and feet, but not guns; ***marines receive special training in unarmed combat***

unarmoured [ʌn'ɑːməd] *adjective* (vehicle) which is not protected by armour (such as a jeep, lorry, truck, etc.) (NOTE: also called **soft-skinned**)

unattached [ʌnə'tætʃt] *adjective* not attached; **unattached personnel** = people who are not members of or attached to a specific unit

unauthorized [ʌn'ɔːθəraɪzd] *noun* not authorized; ***unauthorized entry is prohibited***

unclassified [ʌn'klæsɪfaɪd] *adjective; (of documents or information)* without a security classification (such as restricted, secret, etc.); ***this information is unclassified***

> COMMENT: unclassified information is information which may be passed to the media and the general public

unconditional surrender [ˌʌnkənˌdɪʃənl sə'rendə] *noun* surrender where the side which is surrendering is not permitted to dictate any of its own terms or conditions; ***they demanded the unconditional surrender of the whole battalion***

unconscious [ʌn'kɒnʃəs] *adjective* not awake and unaware of your surroundings as a result of illness or injury; ***one of the casualties is unconscious***

undercarriage ['ʌndəkærɪdʒ] *noun* structure to which the wheels of an aircraft are attached; ***the pilot was unable to lower the undercarriage***

underground ['ʌndəgraʊnd] **1** *adjective* **(a)** constructed or designed to operate beneath the surface of the ground; ***an underground railway*** **(b)** relating to a group or movement which is working secretly against the established authority or an occupying power; ***I have been given the task of collecting information on underground organizations*** **2** *noun* **(a)** underground railway **(b)** group or movement which is working secretly against the established authority or an occupying power; ***the commandos were working with the local underground***

undergrowth ['ʌndəgrəʊθ] *noun* bushes and plants growing beneath the trees of a wood or forest; ***someone is moving through the undergrowth on our left***

underpass ['ʌndəpɑːs] *noun* road which passes beneath another road (by means of a tunnel *or* bridge)

underslung load [ˌʌndəslʌŋ 'ləʊd] *noun* load of equipment or supplies which is carried suspended from a helicopter

UNHCR [ˌjuː en eɪtʃ siː 'ɑː(r)] = UNITED NATIONS HIGH COMMISSION FOR REFUGEES

Uniform ['juːnɪfɔːm] twenty-first letter of the phonetic alphabet (Uu)

uniform ['juːnɪfɔːm] **1** *adjective; (of pattern, shape, size, weight, etc.)* exactly the same, identical; ***the armour is of uniform thickness all over the vehicle*** **2** *noun* standard military clothing worn by members of the same arm or grouping

uninhabited [ˌʌnɪn'hæbɪtɪd] *adjective* not lived in; ***the village is uninhabited***

Union Jack *or* **Union flag** [ˌjuːniən 'dʒæk *or* 'flæg] *noun* the national flag of Great Britain (NOTE: the term **Union flag** is more correct, but **Union Jack** is more usual)

unit ['juːnɪt] *noun* **(a)** military grouping with its own organization and command structure **(b)** standard quantity; ***a kilometre is a unit of linear measure***

> COMMENT: in the army, a unit normally refers to a battalion or equivalent-sized grouping

United Kingdom (UK) [juːˌnaɪtɪd 'kɪŋdəm] *noun* country formed of Great Britain and Northern Ireland

United Nations (UN) [juːˌnaɪtɪd 'neɪʃnz] *noun* international organization dedicated to the promotion of world peace, and able to call upon its member states to contribute military forces for international peacekeeping operations

United States *or* **United States of America (US** *or* **USA)** [juːˌnaɪtɪd 'steɪts] *noun* large country in North America, the world's single superpower; ***the United States were not involved in the peace talks***

unload [ʌn'ləʊd] *verb* **(a)** to remove ammunition from a weapon; ***they were ordered to unload*** **(b)** to remove a load from an aircraft, ship or vehicle; ***the enemy attacked while we were unloading the ship***

unmanned [ʌn'mænd] *adjective* **(a)** *(of an aircraft)* designed to fly without a pilot (that is, by remote control) **(b)** *(of an installation)* not needing people to man it; ***the rebroadcasting station is unmanned***

unmetalled *US* **unmetaled** ['ʌnmetəld] *adjective; (of roads and tracks)* without a surface of asphalt *or* tarmac *or* other strengthening materials (eg. gravel *or* small stones); *see also* ***DIRT ROAD*** *or* ***TRACK***

UNMO ['ʌnməʊ] = UNITED NATIONS MILITARY OBSERVER

unobtainable [ˌʌnəb'teɪnəbl] *adjective; (radio terminology)* not in radio contact; ***B Company is unobtainable at the moment***

UNPA ['ʌnpʌ(r)] = UNITED NATIONS PROTECTED AREA

UNPF [ˌjuː en piː 'ef] = UNITED NATIONS PEACE FORCES

UNPROFOR [ʌn'prəʊˌfɔː(r)] = UNITED NATIONS PROTECTION FORCE

UNSC [ˌjuː en es 'siː] = UNITED NATIONS SECURITY COUNCIL

UNSCR [ˌjuː en es siː 'ɑː(r)] = UNITED NATIONS SECURITY COUNCIL RESOLUTION

unserviceable (u/s) [ʌn'sɜːvɪsəbl] *adjective; (of equipment)* damaged or defective (so that it does not work properly); ***the radio is unserviceable***

untenable [ʌn'tenəbl] *adjective* impossible to defend; ***our position is untenable***

update **1** ['ʌpdeɪt] *noun* fresh information; ***here is the latest intelligence***

update **2** [ʌp'deɪt] *verb* to give someone fresh information; ***I need to update you on the latest intelligence***

updraught ['ʌpdrɑːft] *noun* strong upward current of air

upgrade [ʌp'greɪd] *verb* to improve the design or capability of something

upper case [ˌʌpə 'keɪs] *noun* capital letters written as A, B, C, etc. (NOTE: the opposite, i.e. small letters written as a, b, c, etc., is **lower case**)

upstream [ʌp'striːm] *adverb* in the opposite direction to that in which a river or stream is flowing; ***we moved upstream***; ***the enemy are crossing upstream of the town***; *compare* DOWNSTREAM

upwind [ʌp'wɪnd] *adverb* in a position where the wind is blowing from your own location towards another location; ***fortunately, our position was upwind of the chemical attack***; *compare* DOWNWIND

urban ['ɜːbən] *adjective* relating to towns and cities; *compare* RURAL

urgent ['ɜːdʒənt] *adjective* requiring immediate action or attention; ***we have received an urgent message from HQ***

US [ˌjuː 'es] *adjective* referring to the United States of America; ***US troops landed last night***

u/s [ˌjuː 'es] = UNSERVICEABLE

USA [ˌjuː es 'eɪ] = UNITED STATES OF AMERICA; UNITED STATES ARMY

USAF [ˌjuː es eɪ 'ef] = UNITED STATES AIR FORCE

USAFE = UNITED STATES AIR FORCE IN EUROPE

USAREUR = UNITED STATES ARMY IN EUROPE

USEUCOM = US EUROPEAN COMMAND

USMC [ˌjuː es em 'siː] = UNITED STATES MARINE CORPS

USN [ˌjuː es 'en] = UNITED STATES NAVY

USS ['juː es es] *abbreviation* = UNITED STATES SHIP prefix given to all ships of the United States Navy; ***I served on board the USS Saratoga***

USSR [ˌjuː es es 'ɑː] *noun* = UNION OF SOVIET SOCIALIST REPUBLICS full official title of the former Soviet Union

utility [juː'tɪləti] *adjective* designed for general use; *US* **utility helicopter** = helicopter designed to transport men, equipment or supplies

U-turn [juː 'tɜːn] *noun* act of turning a vehicle sharply around, so that it is facing in the direction from which it has just come; ***the tank did a U-turn and disappeared behind the church***

Uzi ['uːzi] *noun* Israeli-designed 9mm sub-machine-gun

VICTOR - Vv

vaccinate ['væksɪneɪt] *verb* to give someone a vaccine which prevents him or her from contracting a disease; ***we were vaccinated against anthrax***; *see also* INOCULATE

vaccination [ˌvæksɪ'neɪʃn] *noun* act of vaccinating someone; ***the troops were given anthrax vaccinations***; *see also* INOCULATION

vaccine ['væksiːn] *noun* substance, containing the germs of a disease, which provides a person with immunity to that disease

V-agent ['viː ˌeɪdʒənt] *noun* persistent nerve agent

valley ['væli] *noun* area of low ground flanked by hills, usually with a river running through it

van [væn] *noun* **(a)** light motor vehicle designed for carrying goods; ***the bomb was hidden in a small white van*** **(b)** vanguard; ***the general was moving in the van of the advancing force***

vanguard ['væŋgɑːd] *noun* leading elements of the main body of an advancing force

> COMMENT: the **vanguard** should not be confused with the **advance guard** which moves ahead of the main body

vantage point ['vɑːntɪdʒ ˌpɔɪnt] *noun* place from which one can observe a thing or area; ***that hill is an excellent vantage point***

vapor *see* VAPOUR

vapour *US* **vapor** ['veɪpə] *noun* particles of liquid or other substance suspended in air; ***this chemical agent is used in the form of a vapour***

variable-time fuse (VT) [ˌveəriəbl 'taɪm ˌfjuːz] *noun* fuse fitted to an artillery shell, which causes it to explode at a specified height above the ground

variant ['veəriənt] *noun* model which is different from the original design; ***the Russians are testing a new variant of the T-80***

VCP [ˌviː siː 'piː] = VEHICLE CHECK-POINT

VD [ˌviː 'diː] = VENEREAL DISEASE

VDU [ˌviː diː 'juː] = VISUAL DISPLAY UNIT

vector ['vektə(r)] **1** *noun* course taken by an aircraft; ***vector two-three-nine for CAP*** **2** *verb (of air traffic controllers, fighter controllers, etc) to direct a pilot;* ***he vectored the aircraft to its CAP position***

vegetation [ˌvegɪ'teɪʃn] *noun* plants in general; ***there is very little vegetation on the island***

vehicle ['viːɪkl] *noun* machine which moves on land; **vehicle state =** condition of vehicles held by a unit or sub-unit; **all-terrain vehicle =** vehicle which is capable of operating in all types of terrain; **armoured vehicle** *or* **armoured fighting vehicle (AFV) =** vehicle which is protected by armour; **soft-skinned vehicle =** vehicle which is not protected by armour (such as a jeep, lorry, truck, etc.); **re-entry vehicle =** warhead of a surface-to-surface missile which is designed to travel through space on its way to its target

vehicle check-point (VCP) [ˌviːɪkl 'tʃek pɔɪnt] *noun* **(a)** place on a road where soldiers or policemen stop vehicles in order to search them or to check the identity of the occupants; ***we set up a VCP at the crossroads*** **(b)** persons manning a vehicle check-point; ***the VCP was attacked by partisans***

veiled speech [ˌveɪld 'spiːtʃ] *noun* act of speaking on a telephone or radio, in such a way as to conceal the true meaning of the conversation, without actually using a code

veld *or* **veldt** [velt] *noun (South Africa)* uncultivated grassland

velocity [vəˈlɒsəti] *noun* speed at which an object travels; **high velocity** = travelling faster than the speed of sound; **low velocity** = travelling slower than the speed of sound; **muzzle velocity** = speed of a projectile, at the moment that it leaves the muzzle of a weapon

venereal disease (VD) [vəˈnɪəriəl dɪˌziːz] *noun* disease which is passed by sexual contact (for example gonorrhoea, syphilis, etc.)

venomous ['venəməs] *adjective; (of snakes, insects and some other creatures)* having a poisonous bite *or* sting; ***I don't think this snake is venomous***

verbal ['vɜːbl] *adjective* spoken (as opposed to written or other forms of communication); ***all verbal requests should be confirmed in writing***

verification [ˌverɪfɪ'keɪʃn] *noun* process of establishing if something is accurate or true; ***we need verification of the report***

verify ['verɪfaɪ] *verb* **(a)** to establish if something is accurate or true; ***we need to verify the report*** **(b)** to confirm that something is accurate or true; ***he verified the allegation***

vertical take-off and landing [ˌ'vɜːtɪkl ˌteɪk ɒf ənd 'lændɪŋ] *noun* technology which enables a fixed-wing aircraft to take off and land from a stationary position (ie without the need for a runway); ***the Harrier has a vertical take-off and landing capability***

> COMMENT: vertical take-off is not usually possible when the aircraft is carrying a full payload of munitions. In such cases, the aircraft would need to take off from a runway like any conventional fixed-wing aircraft. Most vertical take-off aircraft, however, require a considerably shorter distance to take off than do conventional aircraft, and would therefore be able to use stretches of road or grass areas as runways. Once the aircraft has discharged its munitions it would be able to carry out a normal vertical landing. The acronyms STOVL (short take-off and landing) and V/STOL (vertical or short take-off and landing) are used to describe these capabilities

very high frequency (VHF) [ˌveri ˌhaɪ 'friːkwənsi] *noun* range of radio frequencies from 30 - 300 megahertz (Mhz)

Very light ['vɪəri ˌlaɪt] *noun* illuminating flare which is fired from a Very pistol

Very pistol ['vɪəri ˌpɪstl] *noun* pistol designed to fire an illuminating flare into the air

vessel ['vesl] *noun* boat or ship

vet [vet] **1** *noun* **(a)** person who is qualified to give medical treatment and surgery to animals; ***the guard dog was taken to the vet*** **(b)** *US (informal)* veteran; ***a party of vets is visiting the base*** **2** *verb* to check a person's history and family and social connections, in order to establish whether they are suitable for a job or to have access to classified information; ***he will have to be vetted*** (NOTE: **vetting - vetted**)

veteran ['vetrən] *noun* **(a)** person with considerable combat experience; ***they replaced the battalion of recruits with veterans of the last campaign*** **(b)** *US* ex-serviceman

> COMMENT: in recent years, the media have started to apply this term to anyone who has taken part in a military operation, however short the duration (for example Falklands veterans, Gulf veterans, etc.)

vetting ['vetɪŋ] *noun* act of vetting a person; ***he was given a thorough vetting***

VHF = VERY HIGH FREQUENCY

vice-admiral [ˌvaɪs 'ædmərəl] *noun GB* senior officer in the navy, above a rear-admiral

vice admiral [ˌvaɪs 'ædmərəl] *noun US* senior officer in the navy

vicinity [vɪ'sɪnəti] *noun* area which surrounds a place; ***enemy special forces are operating in the vicinity of Linz***

victim ['vɪktɪm] *noun* person who is killed or injured as the result of an action or occurrence

Victor ['vɪktə] twenty-second letter of the phonetic alphabet (Vv)

victor ['vɪktə] *noun* person who is victorious

victorious [vɪk'tɔːriəs] *adjective* relating to a military force or state which has defeated an enemy; ***the victorious troops looted the town***; ***the victorious army conquered one state after another***

victory ['vɪktəri] *noun* defeat of an enemy in battle or war; ***it was a decisive victory which changed the outcome of the war***; ***Marlbrough won a series of victories in Northern Europe***; **pyrrhic victory =** victory in which the losses suffered by the winning side are so high, that they outweigh the advantages gained by winning the battle

view [vjuː] *noun* area which is visible from a particular location

Viggen ['vɪgən] *see* SAAB-37

vigilance ['vɪdʒɪləns] *noun* act of guarding against a possible danger or threat; ***we need to show extra vigilance tonight***

vigilant ['vɪdʒɪlənt] *adjective* alert to a possible danger or threat

vigor *see* VIGOUR

vigorous ['vɪgərəs] *adjective* showing or demanding strong physical effort; ***paratroops undergo a vigorous training course***

vigour *US* **vigor** ['vɪgə] *noun* strong physical effort; ***the attack was not pressed home with sufficient vigour***

Viking ['vaɪkɪŋ] *see* S-3

village ['vɪlɪdʒ] *noun* small rural settlement

virus ['vaɪrəs] *noun* germ cell which infects the cells of living organisms, thus causing disease; **computer virus =** secret code which is fed into an existing programme, in order to sabotage a computer system by destroying *or* disrupting data stored on it

visible ['vɪzəbl] *adjective* able to be seen; ***the tanks were clearly visible***

visibility [ˌvɪzə'bɪləti] *noun* amount of what is visible; ***visibility was poor because of the fog***; ***visibility is down to two hundred metres***

vision ['vɪʒn] *noun* ability to see; ***he suffered a temporary loss of vision***

visor ['vaɪzə] *noun* movable shield attached to a helmet, designed to protect the face while allowing the wearer to see

visual ['vɪʒuəl] *adjective* relating to sight; **visual display unit (VDU) =** apparatus similar to a television, attached to a computer, which shows data on a screen; **to have a person on visual =** to be able to see a person

vital ['vaɪtl] *adjective* **(a)** of the greatest importance; ***it is vital that you capture that position*** **(b)** essential to the outcome of a matter; **vital ground =** area of ground which, if captured by the enemy, will make it impossible for a unit or sub-unit to fulfill its mission

> COMMENT: the **vital ground** of a sub-unit (such as a platoon) will often constitute the **ground of tactical importance** of its higher formation (that is, a company). If a unit's vital ground is captured, then that unit has effectively lost its part of the battle

voice procedure ['vɔɪs prəˌsiːdʒə] *noun* standard words and expressions which are used when talking on a radio; ***'hello 22, this is 2, use correct voice procedure, out!'***

volley ['vɒli] **1** *noun* act of firing several weapons at the same time, in order to produce a concentration of fire; ***they fired several volleys into the crowd*** **2** *verb* to fire several guns together

volume ['vɒljuːm] *noun* quantity of sound given out by a radio or other apparatus

volunteer [ˌvɒlən'tɪə] **1** *noun* **(a)** person who offers to do a task; ***I need a volunteer to take a message back to headquarters*** **(b)** person who joins the armed forces because he wishes to, rather than because he is conscripted; ***most of the men in the battalion are volunteers*** **2** *verb* to offer to carry out a task (usually one which is dangerous or unpleasant); ***he volunteered to take the message back to HQ***

vomit ['vɒmɪt] *verb* to bring up food from one's stomach; ***the gas made him vomit***

voyage ['vɔɪɪdʒ] *noun* journey made by a ship

VR55 [ˌviː ɑː fɪfti 'faɪv] *noun* NATO name for Soviet-produced nerve agent

VSI [ˌviː es 'aɪ] = VERY SERIOUSLY INJURED

V/STOL = VERTICAL OR SHORT TAKE-OFF AND LANDING

VT [ˌviː 'tiː] = VARIABLE-TIME FUSE

VTOL = VERTICAL TAKE-OFF AND LANDING

Vulcan ['vʌlkən] *noun* **(a)** nickname for the American-designed M-61A1 20mm anti-aircraft cannon **(b)** obselete British-designed strategic bomber aircraft

vulnerable ['vʌlnərəbl] *adjective* **(a)** *(of people)* easy to injure or kill; ***we are extremely vulnerable in this position*** **(b)** *(of things)* easy to damage or destroy; ***this vehicle is vulnerable to small-arms fire*** **(c)** *(of groupings)* easy to outmanoeuvre or overrun; ***our left flank is now extremely vulnerable***

VX [ˌviː 'eks] *noun* American-produced type of nerve agent

> COMMENT: the chemical composition of VX is still secret

WHISKY - Ww

WAC [wæk] = WEAPONS-AIMING COMPUTER

wade [weɪd] *verb* to walk through water; ***the company had to wade the river***

wadi ['wɒdi] *noun (in Arabic countries)* dry river-bed or gully (in desert regions); ***the mortar line was sited in a wadi***

wage ['weɪdʒ] *verb* ***to wage war on someone*** = to fight a war against someone

wait out ['weɪt 'aʊt] *phrase; (radio terminology)* I am too busy to give you further information at the moment, but I will call you as soon as I am able to; ***'hello 2, this is 22, contact, grid 021944, wait out!'***; ***'hello 3, this is 33d, am being shelled, wait out!'***

war [wɔː] *noun* **(a)** armed conflict between nations; ***war broke out in the Middle East***; **to declare war on someone** = to state officially that you are in a state of war with someone; **war correspondent** = journalist or reporter who is attached to a military force, in order to report on a war; **war crime** = act which violates international rules of war; **war games** = military training exercise; **war reserves** = stocks of equipment and supplies kept to be available immediately in case of war; **civil war** = war fought between citizens of the same country; **to be on a war footing** = to be at full strength and fully equipped and prepared to fight a war; ***the battalion is now on a war footing*** **(b)** *(used in names of particular wars)* ***the Crimean War***; ***the First World War***; ***War is nothing more than the continuation of policy by other means - Clausewitz***

wardroom ['wɔːdrʊm] *noun* officers' mess on a warship

warehouse ['weəhaʊs] *noun* large building used for storing goods

warfare ['wɔːfeə] *noun* war (in general); ***arctic warfare***; ***nuclear warfare***; **anti-air warfare (AAW)** = naval term for air defence; **chemical warfare** = warfare involving the use of chemical weapons; **chemical-warfare unit** = specialist unit trained to detect the presence of chemical weapons and to decontaminate persons, equipment and vehicles which have been affected *US* **chemical and biological warfare (CBW)** = warfare using both chemical and biological weapons; **electronic warfare (ELW** *or* **EW)** = location and suppression of an enemy's electronic equipment; *see also* ELECTRONIC COUNTERMEASURES, ELECTRONIC COUNTER-COUNTERMEASURES

wargame ['wɔːgeɪm] *verb* to test the viability of an operational plan, by playing it out on a map and calculating likely enemy responses; ***we wargamed several different scenarios***

War Graves Commission ['wɔːr greɪvz kəˌmɪʃən] *noun* official British organization responsible for setting up and maintaining cemeteries for servicemen who die *or* are killed in wartime

warhead ['wɔːhed] *noun* explosive head of a missile or other projectile; **chemical warhead** = explosive part of a missile used as a means of delivering a chemical agent; *see also* SHAPED-CHARGE WARHEAD, TANDEM WARHEAD

warn [wɔːn] *noun* **(a)** to inform another person of a danger or threat; ***we warned him of the increased chemical threat*** **(b)** to inform another person that his actions or conduct are unacceptable and that he will be punished if it happens again; ***he was warned about his behaviour***

warning ['wɔːnɪŋ] *noun* **(a)** act of warning someone; ***we have received a warning of a probable nuclear strike***; **warning order** = message which warns a unit or sub-unit of a future operation or task, and provides sufficient information for the unit to start making its preparations; **warning signal** = signal such as a red light, which warns that something has gone wrong **(b)** official record that a person has been warned about his actions or conduct; ***he was given a warning***

> COMMENT: apart from the task itself, the most important piece of information in a warning order is the timing "no move before ..."

warrant ['wɒrənt] *noun* document which authorizes a person to do something; **arrest warrant** = warrant authorizing the security forces to arrest a specified person *GB* **rail warrant** = official document which entitles a serviceman to a free railway ticket; **Royal Warrant** = authority by which a warrant officer holds a rank in the armed forces; **search warrant** = warrant authorizing the security forces to search a specified building or property

warrant officer (WO) ['wɒrənt ˌɒfɪsə] *noun* **(a)** *GB* senior non-commissioned officer in the army or air force who holds his or her rank by Royal Warrant GB **warrant officer first class (WO1)** = regimental sergeant major or someone of equivalent seniority GB **warrant officer second class (WO2)** = company sergeant major or regimental quartermaster sergeant or someone of equivalent seniority **(b)** *US* senior non-commissioned officer who holds a special rank because his or her job requires a greater level of responsibility than that which is normally expected of senior enlisted personnel

warring ['wɔːɪŋ] *adjective* actively involved in armed conflict; ***negotiations between the warring factions have collapsed***

Warrior ['wɒriə] *noun* British-designed 1980s-era infantry fighting vehicle (IFV)

Warsaw Pact [ˌwɔːsɔː 'pækt] *noun* military alliance, consisting of the Soviet Union and other communist countries of Eastern Europe (for example Czechoslovakia, Hungary, Poland, etc.), which disintegrated following the collapse of communism at the end of the 1980s

warship ['wɔːʃɪp] *noun* armoured ship, equipped with guns or missiles, which is designed for fighting at sea; *see also* AIRCRAFT CARRIER, BATTLESHIP, CRUISER, DESTROYER, SUBMARINE, etc.

Warthog ['wɔːthɒg] *noun* unofficial nickname for the American-designed A-10 ground-attack aircraft

wartime ['wɔːtaɪm] *noun* period during which a war is fought

wash [wɒʃ] **1** *noun* disturbance on the surface of water, left by a passing boat or ship; ***we followed the wash of the destroyer*** **2** *verb* to clean oneself or an object with water; ***he is washing his clothes***; ***he washed the blood off his hands***

wastage ['weɪstɪdʒ] *noun* **(a)** act of wasting a resource **(b)** amount that has been wasted

waste [weɪst] **1** *noun* act of using more of a resource (such as ammunition, fuel, manpower, water, etc.) than is necessary; ***that was a waste of ammunition*** **2** *verb* **(a)** to use a resource unnecessarily; ***cease fire! You are wasting ammo*** **(b)** to use more of a resource than is necessary; ***you are wasting fuel by driving in such a low gear*** **(c)** *(slang)* to kill someone; ***he got wasted***

watch [wɒtʃ] **1** *noun* **(a)** small clock which is normally attached to a person's wrist **(b)** period of daily duty on a ship; **forenoon watch** = period of duty from 0800-1200hrs; **afternoon watch** = period of duty from 1200-1600hrs; **first dogwatch** = period of duty from 1600-1800hrs; **second dogwatch** = period of duty from 1800-2000hrs; **first watch** = period of duty from 2000-2359hrs; **middle watch** = period of duty from 0001-0400hrs; **morning watch** = period of duty from 0400-0800hrs; **radio watch** = period of duty, which is spent listening to a radio; **officer of the watch** = officer on duty **(c)** period of guard duty; **to keep watch** = to watch for the approach of danger, while your comrades sleep or carry out other tasks **(d)** detachment of men assigned to guard a location **2** *verb* **(a)** to look at something which is happening; ***we watched the enemy as they were crossing the river*** **(b)** to look at an area of ground, in order to see any activity which might occur there; ***the OP was ordered to watch the main road*** **(c)** to look at a person, in order to see if he does something; ***we were told to watch the crowd of rioters*** **(d)** to guard a person or thing; ***he was ordered to watch the prisoners***; *see also* OBSERVE

watchkeeper ['wɒtʃˌkiːpə] *noun* **(a)** (navy) duty officer on a warship who, in the event of an unforseen incident, is qualified to make command decisions until a more senior officer takes over; ***he's just been awarded his watchkeeper's ticket*** **(b)** (army) operational appointment, in which an officer or non-commissioned officer has limited control over a headquarters department while the normal staff officer is resting or engaged in other tasks; ***I acted as a G4 watchkeeper in Bosnia***

watch-tower ['wɒtʃ taʊə] *noun* tower from which one can watch an area of ground

watercourse ['wɔːtəkɔːs] *noun* canal, river, stream or dry river-bed

waterfall ['wɔːtəfɔːl] *noun* place where a river or stream flows over a cliff or rocks

waterproof ['wɔːtəpruːf] **1** *adjective; (of clothing, footwear, etc.)* designed to prevent the passage of water; ***I've got a waterproof sleeping-bag cover*** **2** *verb* to make something waterproof; ***he is waterproofing his boots***

watertight ['wɔːtətaɪt] *adjective; (of compartments, doors, joints, etc.)* designed to prevent the passage of water; ***all the compartments in the boat are watertight***

waterway ['wɔːtəweɪ] *noun* canal or navigable river

wave [weɪv] **1** *noun* **(a)** moving ridge of water; ***a huge wave broke over the ship*** **(b)** one of several tactical groupings which are advancing or attacking, one behind the other; ***waves of bombers attacked the town*** **2** *verb* **(a)** to raise your hand and move it about as a greeting; ***the girls waved at the soldiers as they marched past*** **(b)** to raise your arm and move it as a signal; ***he waved the men away*** **(c)** to display something by raising it and moving it about; ***the enemy were waving white flags***

way [weɪ] *noun* **(a)** road, path, track or any other natural or man-made feature which allows movement; ***we could not find a way through the marsh*** **(b)** method; ***that's not the way to do it*** **(c)** direction; ***B Company HQ is that way*** **(d)** route; ***do you know the way to the dressing station?***

waypoint ['weɪpɔɪnt] *noun* place or feature on the ground which is used as a navigational reference point (especially with satellite navigation systems); ***our next waypoint is the church at grid 637921***

W/Cdr = WING COMMANDER

WCP [ˌdʌb(ə)ljuː siː 'piː] *noun* = WEAPON COLLECTION POINT location set up by a peacekeeping force to collect weapons from soldiers who have been involved in an armed conflict

weak [wiːk] *adjective* **(a)** *(of people)* not strong; ***he was very weak through loss of blood*** **(b)** *(of groupings)* not at full strength; ***we have a weak brigade in front of us***

weapon ['wepən] *noun* **(a)** any object which is designed to kill or injure (such as a bayonet, grenade, rifle, etc.); ***he has lost his weapon***; **area weapon =** weapon which can deliver a quantity of projectiles over a wide area and thus effectively engage several targets simultaneously (eg. machine-gun, artillery, mortar, cluster bomb); **weapons-grade uranium =** another name for depleted uranium (DU) *or* Staballoy; **weapon state =** condition in which a weapon is carried (ie unloaded, made safe or made ready) **(b)** any object which is used to kill or injure (such as a broken bottle, knife, piece of wood, etc.); ***a wide variety of weapons were taken from the rioters*** **(c)** any object which is designed to cause damage (such as a bomb, missile, rocket, etc.); ***the enemy are threatening to use nuclear weapons***; **weapon system =** weapon which utilizes sophisticated technology (such as a guided missile); **weapon systems officer (WSO) =** crewman in a fighter aircraft, who navigates the aircraft and operates its weapons systems

weaponry ['wepənri] *noun* weapons in general

weather ['weðə] *noun* daily changes in the condition of the earth's atmosphere (such as rain, sunshine, wind, etc.)

webbing ['webɪŋ] *noun* **(a)** strong fabric used to make belts, equipment pouches, rifle slings, etc. **(b)** set of equipment pouches attached to a belt or harness; ***he has lost his webbing***

wedge [wedʒ] *noun* tactical formation in the shape of a triangle (eg. one sub-unit leading as point, with the other two sub-units following abreast of each other)

weight [weɪt] *noun* the heaviness of an object

well [wel] *noun* man-made hole in the ground from which water is obtained

wellington boot ['welɪŋtən buːt] *noun* **(a)** waterproof rubber boot which reaches up to the knee **(b)** elegant leather boot, which reaches up to the knee but is worn covered by the trouser leg, as part of a ceremonial uniform *or* mess kit

Wessex ['wesɪks] *noun* British-made utility helicopter

west [west] **1** *noun* **(a)** one of the four main points of the compass, corresponding to a bearing of 270 degrees or 4800 mils **(b)** area to the west of your location; ***the enemy are approaching from the west*** **(c)** **the West** = Europe and North America **(d)** the western part of a country **2** *adjective* relating to the west; ***the West Gate***; **west wind** = wind blowing from the west **3** *adverb* towards the west; ***the enemy is moving west***

westbound ['westbaʊnd] *adjective* moving or leading towards the west; ***a westbound convoy***

westerly ['westəli] *adjective* **(a)** towards the west; ***they pushed forward in a westerly direction*** **(b)** *(of wind)* from the west

western ['westən] *adjective* relating to the west; ***the western part of the country***; **Western European Union (WEU)** = a group of European countries linked together for mutual protection; the Union is now seen as the European Union's future defence arm, and it now includes several Eastern European countries as associate members

West Point ['west ˌpɔɪnt] *noun* US Army officer training establishment; ***he is a graduate of West Point***

westward ['westwəd] **1** *adjective* towards the west; ***a westward direction*** **2** *adverb US* towards the west; ***they are moving westward***

westwards ['westwədz] *adverb* towards the west; ***they are moving westwards***

WEU = WESTERN EUROPEAN UNION

WFP [ˌdʌb(ə)ljuː ef 'piː] *noun* = WORLD FOOD PROGRAMME United Nations organization responsible for the distribution of food and other humanitarian aid in disaster areas and war zones

WG CDR = WING COMMANDER

wheel [wiːl] **1** *noun* round piece which turns round an axle, and on which a vehicle runs; ***the mine damaged the front wheels of the truck*** **2** *verb* to swing round in line; ***the brigade wheeled left-handed and advanced towards Essingen***

wheelbarrow ['wiːlˌbærəʊ] *noun* **a** small cart with one wheel, which is designed to be pushed by a person on foot (normally used by gardeners, builders and farmers); ***we brought up the ammunition in an old wheelbarrow*** **b** small unmanned tracked vehicle, which is operated by remote control and which can be fitted with a CCTV camera and other instruments *or* tools for examining suspected improvised explosive devices (IED); ***the wheelbarrow was destroyed in the explosion***

wheeled [wiːld] *adjective (of vehicles)* fitted with wheels; ***this route is not suitable for wheeled vehicles***

Whisky *US* **Whiskey** ['wɪski] twenty-third letter of the phonetic alphabet (Ww)

whisper ['wɪspə] *verb* to speak very quietly

whistle ['wɪsl] **1** *noun* **(a)** instrument which is blown through to produce a clear shrill noise; ***he blew his whistle as a signal to advance*** **(b)** noise produced by a whistle or by blowing air through your lips; ***we heard a whistle, and then the noise of small-arms fire*** **2** *verb* to produce the sound of a whistle; ***he whistled to show that the coast was clear***

white phosphorus (WP) [ˌwaɪt 'fɒsfərəs] *noun* **(a)** chemical substance which burns on contact with oxygen, producing dense clouds of white smoke **(b)** smoke-producing projectile, or grenade containing white phosphorus

COMMENT: projectiles and grenades containing white phosphorus are usually painted light green, with red lettering and markings

WHO [ˌdʌb(ə)ljuː eɪtʃ 'əʊ] *noun* = WORLD HEALTH ORGANIZATION United Nations organization dealing with health matters

WIA [ˌdʌb(ə)ljuː aɪ 'eɪ] = WOUNDED IN ACTION

wilco ['wɪlkəʊ] *adverb (radio terminology)* = WILL COMPLY I will carry out your instructions; ***'hello22, this is 2, move now, over' - '22, wilco, out'***

Wildcat ['waɪldkæt] *noun* German-designed wheeled self-propelled anti-aircraft gun (SPAAG)

wilderness ['wɪldənəs] *noun* uninhabited and uncultivated area or region

Wild Weasel [ˌwaɪld 'wiːzl] *noun* *US* air-force role, involving the use of radar-detecting equipment and anti-radar missiles (ARM) to suppress enemy surface-to-air missile sites

winch [wɪntʃ] *verb* to lift or drop from a helicopter using a rope; ***the injured man was winched to safety***; ***two crewmembers were winched down to the forward position***

wind [wɪnd] *noun* strong movement of air; ***the high winds brought down two aerials***; **head wind** = wind blowing in the opposite direction to that in which an aircraft or ship is travelling; **tail wind** = wind blowing in the same direction as that in which an aircraft or ship is travelling

windage ['wɪndɪdʒ] *noun* **(a)** effect of wind on a projectile in flight **(b)** allowance made for wind when aiming a weapon

wind-chill ['wɪnd tʃɪl] *noun* effect of cold wind on a person when the air temperature is low, making him even colder; **wind-chill factor** = method of calculating the risk of hypothermia by adding the speed of the wind to the number of degrees of temperature below zero; ***the wind-chill factor is dangerously high at the moment***

wing [wɪŋ] *noun* **(a)** thin horizontal structure extending from either side of an aircraft, in order to support it in flight; **delta wing** = triangular wing; **fixed-wing aircraft** = aircraft with wings; **rotary-wing aircraft** = helicopter **(b)** air-force grouping of several squadrons *US* **tactical fighter wing (TFW)** = tactical air-force grouping of three fighter squadrons plus supporting arms *GB* **wing commander (W/Cdr)** = senior officer in the air force, above a squadron leader (usually in command of a wing)

wingman ['wɪŋmæn] *noun* pilot of the other aircraft, when you are flying as a pair; ***my wingman was hit by a surface-to-air missile***

wipe out [ˌwaɪp 'aʊt] *phrasal verb* to kill all the members of a grouping; ***B Company has been almost wiped out*** (NOTE: this verb is normally used in the passive)

wire ['waɪə] *noun* cord-like material made of metal; **wire-guided missile** = missile, which remains connected to its firing post by a length of wire, through which signals are transmitted in order to control its flight onto the target; **barbed wire** = wire with sharp spikes attached to it, used as an obstacle; **barbed-wire entanglement** = obstacle to infantry made out of coils of barbed wire; **razor wire** = wire with a sharp cutting edge, similar in use to barbed wire; *see also* CONCERTINA WIRE

wire-cutters ['waɪə ˌkʌtəz] *noun* tool for cutting wire; ***he dropped his wire-cutters***; ***remember to bring a pair of wire-cutters*** (NOTE: wire-cutters, like scissors, are always plural and come in **pairs** or **sets**)

wireless ['waɪələs] *noun* obsolete term for radio

wiring party ['waɪərɪŋ ˌpɑːti] *noun* detachment of soldiers sent out to construct or repair a barbed-wire obstacle

withdraw [wɪð'drɔː] *verb* **(a)** to move away from the enemy; ***B Company is withdrawing*** **(b)** to move back towards your own forces or territory; ***the enemy withdrew across the border***; *see also*

RETIRE, RETREAT (NOTE: **withdrawing - withdrew - have withdrawn**)

> COMMENT: the word **retreat** is normally used when one is forced to move back (for example, because one has been defeated or your position has become untenable), whereas **retire** *or* **withdraw** imply rearward movement as part of a planned manoeuvre or in order to occupy a better position. Consequently, **retire** or **withdraw** are sometimes used instead of **retreat** because they sound more positive

withdrawal [wɪð'drɔːl] *noun* act of withdrawing; **tactical withdrawal** = withdrawal from the enemy as part of a planned manoeuvre; *see also* RETREAT

WO1 [ˌdʌb(ə)ljuː əʊ 'wʌn] = WARRANT OFFICER FIRST CLASS

WO2 [ˌdʌb(ə)ljuː əʊ 'tuː] = WARRANT OFFICER SECOND CLASS

wood [wʊd] *noun* **(a)** area of ground covered by trees; ***we spent the night in a wood*** **(b)** material obtained from trees; ***the handguard is made of wood***

wooden ['wʊdən] *adjective* made of wood

woodland ['wʊdlənd] *noun* terrain consisting mainly of woods or forest

working-parts [ˌwɜːkɪŋ 'pɑːts] *plural noun* internal mechanism (usually consisting of several different parts) of an automatic *or* semi-automatic weapon, which moves backwards and forwards to cock the weapon, feed a round into the breech, fire the round and extract the empty cartridge case; ***on the command "Unload !", remove the magazine, pull the working-parts to the rear several times and then look inside***

wound [wuːnd] **1** *noun* serious injury, usually involving a cut or other penetration of the skin and flesh; ***he has a shrapnel wound to his leg***; ***he died of his wounds***; **self-inflicted wound** = wound inflicted by a person on himself (usually in order to get out of the combat zone) **2** *verb* to inflict a wound; ***he was wounded in the leg***

wounded ['wuːndɪd] **1** *adjective* suffering from a wound; ***wounded soldiers were removed to the field hospital*** **2** *noun* ***the wounded*** = people who have had a wound; ***the dead and wounded were removed from the battlefield***

WP [ˌdʌb(ə)ljuː 'piː] = WHITE PHOSPHORUS

wreck [rek] **1** *noun* **(a)** accidental destruction of a ship (usually by running onto rocks); ***here is the report on the wreck of HMS Ardent*** **(b)** remains of a ship which has been wrecked; ***most of the cargo was removed from the wreck*** **(c)** remains of a destroyed or badly damaged aircraft or vehicle; ***we took cover behind a tank wreck*** **2** *verb* to destroy or badly damage an aircraft, ship or vehicle; ***the ship has been wrecked***; ***we took cover behind a wrecked tank***

wreckage ['rekɪdʒ] *noun* pieces of an aircraft, ship or vehicle which has been wrecked

Wren [ren] *noun* GB *(informal)* female member of the Royal Navy

WSO = WEAPON SYSTEMS OFFICER

WVR [ˌdʌb(ə)ljuː viː 'ɑː] *adverb* = WITHIN VISUAL RANGE close enough to an enemy aircraft to see it with the naked eye; ***we'll need WVR missiles***; *compare* BVR

X-RAY - Xx, YANKEE - Yy, ZULU - Zz

XO [ˌeks ˈəʊ] *US* = EXECUTIVE OFFICER

X-ray [ˈeks ˌreɪ] twenty-fourth letter of the phonetic alphabet (Xx)

X-ray [ˈeks ˌreɪ] **1** *noun* **(a)** electromagnetic radiation of a very short wavelength, which is able to pass through the human body and can thus produce photographs of internal injuries **(b)** X-ray photograph; ***you will need a chest X-ray*** **2** *verb* to produce an X-ray photograph; ***we will have to X-ray your leg***

YAK-38 [ˌjæk θɜːti ˈeɪt] *noun* Soviet-designed multirole fighter aircraft with a vertical take-off capability, designed to operate from aircraft carriers (NOTE: known to NATO as **Forger**)

YAK-41 [ˌjæk fɔːtɪ ˈwʌn] *noun* Soviet-designed multirole fighter aircraft with a vertical take-off capability, designed to operate from aircraft carriers (NOTE: known to NATO as **Freestyle**)

Yank [jæŋk] *noun (informal)* American soldier

Yankee [ˈjæŋki] twenty-fifth letter of the phonetic alphabet (Yy)

yard [jɑːd] *noun* **(a)** unit of linear measure corresponding to 3 feet or 0.9144 metres; ***there are 1,760 yards in a mile*** **(b)** area of enclosed ground attached to a building; ***the vehicles were parked in the yard next to HQ***

yd = YARD (NOTE: plural is **yds**)

yeoman [ˈjəʊm(ə)n] *noun* **(a)** *GB* petty officer in charge of signals **(b)** *US* petty officer responsible for clerical duties

yeomanry [ˈjəʊmənri] *noun* *GB* *(historical)* unit of volunteer cavalry

> COMMENT: certain armoured regiments in the Territorial Army (TA) retain their historical title of Yeomanry

yield [jiːld] **1** *noun* amount of explosive power produced by a nuclear weapon **2** *verb* **to yield ground** = to withdraw, to go back from; ***the brigade was forced to yield the high ground in front of Skalice***

> COMMENT: the explosive yield of nuclear weapons is measured in **kilotons** *or* **megatons**

Y-junction [ˈwaɪ ˌdʒʌŋkʃn] *noun* place where a single road divides into two; *see also* FORK

yob *or* **yobbo** [jɒb *or* ˈjɒbəʊ] *noun* *GB* *(slang)* unsophisticated youth *or* young man, who behaves in a rude, inconsiderate and often aggressive manner; ***the patrol was attacked by a crowd of yobs; we've been getting a lot of aggro from the local yobbos***

youth [juːθ] *noun* person of an approximate age between 13 and 18 years; ***the patrol was attacked by a crowd of youths*** (NOTE: plural is **youths** [juːðz])

zap [zæp] *verb* *US (slang)* to shoot dead; ***he zapped three of the enemy***

zariba [zəˈriːbə(r)] *noun* *Arabic* defensive enclosure made from pieces of thorn bush; ***the guerilla base was protected by a thick zariba***

zero [ˈziːrəʊ] **1** *noun* figure 0 (nought or nil) **2** *verb* to ensure that a weapon is accurate, by firing a few rounds at a target and then adjusting the sights as required; ***3 Section are zeroing their weapons***

Zeus-23 [ˌzjuːs twenti ˈθriː] *noun* *US* informal nickname for the Soviet-designed ZSU-23-4 self-propelled anti-aircraft gun

zone [zəʊn] *noun* area or region which has some specific importance or purpose; **demilitarized zone (DMZ)** = area or region in which the presence of military forces is forbidden under the terms of a treaty or other international agreement; **exclusion zone** = area or region, defined by a state or by international agreement, which the armed forces or shipping of another state are not allowed to enter; **free fire zone** = area of ground in which any person or vehicle should be considered hostile and may therefore be shot at; **no-fly zone** = airspace defined by a state or by international agreement, which the aircraft of another state are not allowed to enter; **war zone** = region where a war is in progress; *see also* LANDING ZONE

ZSU [ˌzed es 'juː *US* ˌziː es 'juː] *noun* Soviet-designed series of self-propelled anti-aircraft guns; **ZSU-23-4** = 1960s-era weapon, fitted with four radar-controlled 23mm cannon (NOTE: the Russian nickname for this is the **Shilka**)

Zulu ['zuːluː] twenty-sixth letter of the phonetic alphabet (Zz)

zulu muster ['zuːluː ˌmʌstə] *noun* location in the field where vehicles are kept when not in use

Zulu time ['zuːluː ˌtaɪm] *noun* Greenwich Mean Time (GMT); that is, the local time on the meridian at Greenwich, London, which is used to calculate international time; ***H-Hour at 0600Z***

COMMENT: Greenwich Mean Time or Zulu time is used by NATO forces on operations

Zuni ['zuːni] *noun* American-designed unguided rocket, designed to be fired by an aircraft at a ground target

zilch [zɪltʃ] *noun* *US (slang)* nothing; ***we observed the bridge for six hours but saw zilch***

SUPPLEMENT

The Phonetic Alphabet

Certain letters of the alphabet sound very similar, especially when a person is talking on a telephone or radio. The phonetic alphabet is designed to prevent confusion, by using a distinctive word to represent each letter.

Aa	Alpha*	['ælfə]
Bb	Bravo	['brɑːvəʊ]
Cc	Charlie	['tʃɑːlɪ]
Dd	Delta	['deltə]
Ee	Echo	['ekəʊ]
Ff	Foxtrot	['fɒkstrɒt]
Gg	Golf	[gɒlf]
Hh	Hotel	[həʊ'tel]
Ii	India	['ɪndɪə]
Jj	Juliet	['dʒuːlɪət]
Kk	Kilo	['kiːləʊ]
Ll	Lima	['liːmə]
Mm	Mike	[maɪk]
Nn	November	[nə'vembə(r)]
Oo	Oscar	['ɒskə(r)]
Pp	Papa	['pɑːpə]
Qq	Quebec	[kwɪ'bek]
Rr	Romeo	['rəʊmɪəʊ]
Ss	Sierra	[sɪ'erə]
Tt	Tango	['tæŋgəʊ]
Uu	Uniform	['juːnɪfɔːm]
Vv	Victor	['vɪktə(r)]
Ww	Whisky**	['wɪskɪ]
Xx	X-ray	['eksreɪ]
Yy	Yankee	['jæŋkɪ]
Zz	Zulu	['zuːluː]

* *Alfa in US English*
** *Whiskey in US English*

Numbers

0	zero	['zɪərəʊ]	10	ten	[ten]
1	one	[wʌn]	11	eleven	[ɪ'levən]
2	two	[tuː]	12	twelve	[twelv]
3	three	[θriː]	13	thirteen	[θɜː'tiːn]
4	four	[fɔː(r)]	14	fourteen	[fɔː'tiːn]
5	five	[faɪv]	15	fifteen	[fif'tiːn]
6	six	[sɪks]	16	sixteen	[sɪk'stiːn]
7	seven	['sev(ə)n]	17	seventeen	[ˌsevən'tiːn]
8	eight	[eɪt]	18	eighteen	[eɪ'tiːn]
9	nine	[naɪn]	19	nineteen	[naɪn'tiːn]

Note: when speaking on the radio, 9 is often pronounced ['naɪnə(r)]

20	twenty	['twentɪ]	30	thirty	['θɜːtɪ]
21	twenty-one		31	thirty-one	
22	twenty-two		32	thirty-two	
23	twenty-three		40	forty	
24	twenty-four		50	fifty	
25	twenty-five		60	sixty	
26	twenty-six		70	seventy	
27	twenty-seven		80	eighty	
28	twenty-eight		90	ninety	
29	twenty-nine		100	one hundred	

101	one hundred and one	400	four hundred
110	one hundred and ten	500	five hundred
115	one hundred and fifteen	600	six hundred
120	one hundred and twenty	700	seven hundred
125	one hundred and twenty-five	800	eight hundred
200	two hundred	900	nine hundred
300	three hundred		

1,000 one thousand [wʌn 'θaʊzənd]

1,001	one thousand and one	2,000	two thousand
1,010	one thousand and ten	3,000	three thousand
1,015	one thousand and fifteen	10,000	ten thousand
1,025	one thousand and twenty-five	10,200	ten thousand, two hundred
1,100	one thousand, one hundred or eleven hundred	10,250	ten thousand, two hundred and fifty
1,105	one thousand, one hundred and five	15, 000	fifteen thousand
1,115	one thousand, one hundred and fifteen	20,000	twenty thousand
1,150	one thousand, one hundred and fifty	25,000	twenty-five thousand
1,155	one thousand, one hundred and fifty-five	100,000	one hundred thousand
1,200	one thousand, two hundred or twelve hundred	250,000	two hundred and fifty thousand
1,500	one thousand, five hundred or fifteen hundred	255,000	two hundred and fifty-five thousand

1,000,000 one million [wʌn 'mɪlɪən]

Timings

1. To avoid confusion, the twenty-four-hour clock is always used in military timings:

1am	= 0100	2pm	= 1400
8.15am	= 0815	8.45pm	= 2045

2. Verbal timings are given as follows:

1400	= fourteen hundred	1515	= fifteen fifteen
1435	= fourteen thirty-five	1528	= fifteen twenty-eight
1500	= fifteen hundred		

3. A single 0 in the timing is normally pronounced as "**zero**":

0800 = zero eight hundred
0805 = zero eight zero five

Note: This does not apply to a single 0 at the end:

1110 = eleven ten
1620 = sixteen twenty

Note: A single 0 at the beginning of a timing is sometimes pronounced like the letter **O**:

0500	= O five hundred	0830	= O eight thirty
0605	= O six zero five	0955	= O nine fifty-five

*Note:*In order to avoid confusion, it is always better to say "**zero**":

0500	= zero five hundred	0830	= zero eight thirty
0605	= zero six zero five	0955	= zero nine fifty-five

4. Midnight is usually avoided as a timing for obvious reasons. When it is used, it may be given in several different ways:

2400 = twenty-four hundred
2359 = twenty-three fifty-nine
0001 = zero zero zero one

Note: Timings between midnight and 0100 are given as follows:

0005 = zero zero zero five
0015 = zero zero fifteen
0035 = zero zero thirty-five

5. In order to show that it is a timing, the word "**hours**" is usually added to the end. In written timings, this is abbreviated to **hrs**:

0300hrs = zero three hundred hours
1210hrs = twelve ten hours

6. On operations, NATO forces normally use Greenwich Mean Time (GMT), which is also known as **Zulu time,** regardless of the time of the country in which they are operating:

1010Z = ten ten hours Zulu time

Note: Other time zones around the world are identified by different letters of the alphabet. The time of the country in which one is operating is also known as "**local time**", for example: *The general will be arriving at 1430 hours local time.*

British Military Rank Structure

Commissioned Ranks:

Army	**Marines**	**Navy**	**Air Force**
General	General	Admiral	Air Chief Marshal
Lieutenant-General	Lieutenant-General	Vice-Admiral	Air Marshal
Major-General	Major-General	Rear-Admiral	Air Vice Marshal
Brigadier	Brigadier	Commodore	Air Commodore
	Colonel	Captain (senior)	
Colonel	Lieutenant-Colonel	Captain (junior)	Group Captain
Lieutenant-Colonel	Major	Commander	Wing Commander
Major	Captain	Lieutenant-Commander	Squadron Leader
Captain	Lieutenant	Lieutenant	Flight Lieutenant
Lieutenant	Second-Lieutenant	Sub-Lieutenant	Flying Officer
Second-Lieutenant		Midshipman	Pilot Officer

Non-commissioned Ranks:

Army	**Marines**	**Navy**	**Air Force**
Warrant Officer (First Class)	Warrant Officer (First Class)	Warrant Officer	Warrant Officer
Warrant Officer (Second Class)	Warrant Officer (Second Class)		
Colour Sergeant *or* Staff Sergeant	Colour Sergeant	Chief Petty Officer	Flight Sergeant
Sergeant	Sergeant	Petty Officer	Sergeant
Corporal	Corporal	Leading Rating	Corporal
Lance-Corporal	Lance-Corporal		
	Marine	Able Rating	Junior Technician
			Leading Aircraftman
Private		Ordinary Rating	Aircraftman

American Military Rank Structure

Commissioned Ranks:

Army	Marines	Navy	Air Force
General	General	Admiral	General
Lieutenant General	Lieutenant General	Vice Admiral	Lieutenant General
Major General	Major General	Rear Admiral (Upper Half)	Major General
Brigadier General	Brigadier General	Rear Admiral (Lower Half)	Brigadier General
Colonel	Colonel	Captain	Colonel
Lieutenant Colonel	Lieutenant Colonel	Commander	Lieutenant Colonel
Major	Major	Lieutenant Commander	Major
Captain	Captain	Lieutenant	Captain
First Lieutenant	First Lieutenant	Lieutenant Junior Grade	First Lieutenant
Second Lieutenant	Second Lieutenant	Ensign	Second Lieutenant

Non-commissioned Ranks:

Army	Marines	Navy	Air Force
Warrant Officer	Warrant Officer	Warrant Officer	Warrant Officer
Command Sergeant Major	Sergeant Major	Master Chief Petty Officer	Chief Master Sergeant
Sergeant Major	Master Gunnery Sergeant		
First Sergeant	First Sergeant	Senior Chief Petty Officer	Senior Master Sergeant
Master Sergeant	Master Sergeant		
Sergeant First Class	Gunnery Sergeant	Chief Petty Officer	Master Sergeant
Staff Sergeant	Staff Sergeant	Petty Officer First Class	Technical Sergeant
Sergeant	Sergeant	Petty Officer Second Class	Staff Sergeant
Corporal	Corporal	Petty Officer Third Class	Senior Airman
Private First Class	Lance Corporal	Seaman	Airman First Class
Private	Private First Class	Seaman Apprentice	Airman

Formal Orders

The British Army and the US Army both use the same standard format for issuing orders. This can be applied to any type of operation or task and is designed to ensure that no important points are omitted. The following sequence is always used:

1. GROUND:

A detailed description of the terrain over which the operation or task will be carried out.

2. SITUATION:

a. Enemy Forces: i.e. locations, strengths, organization, current activity and future intentions
b. Friendly Forces: i.e. overall plan of the higher formation and locations and tasks of neighbouring groupings
c. Attachments and Detachments: i.e. any sub-units which are attached to the grouping for this operation. and any of the grouping's own sub-units which have been detached for other tasks.

3. MISSION:

A simple and concise statement, which explains exactly what the grouping is trying to achieve (e.g. "our mission is to capture the bridge at grid 324599")

4. EXECUTION:

a. Concept of Operations: i.e. a general outine of how the operation is intended to proceed
b. Detailed Tasks: i.e. specific tasks allocated to each sub-unit of the grouping
c. Coordinating Instructions: e.g. timings, orders for opening fire, indirect-fire support, actions to be carried out in the event of something going wrong, etc.

5. ADMINISTRATION AND LOGISTICS:

General administrative details such as: ammunition, equipment, food and water, medical facilities, etc.

6. COMMAND AND SIGNAL:

a. Command: i.e. command structure of the grouping and nomination of alternative commanders in the event of casualties
b. Signal: e.g. radio frequencies, codes and code-words, report lines, passwords, etc.

see next page for example orders

Example Formal Orders

6 Platoon, which is part of B Company, is about to take part in a battalion attack. The platoon commander has prepared the following orders:

1. GROUND:

The feature we are attacking is Ladna Hill, which runs from north to south along the 44 easting, from the 07 northing to the 04 northing. The company objective is the northern end of the feature. The northern slope is quite steep and consists of grass, with scattered gorse bushes. 500 metres to the north of Ladna Hill is a road, which will form our line of departure. To the north of the road is an area of dead ground, which will be used as our FUP.

2. SITUATION:

a. Enemy Forces:

Ladna Hill is occupied by a company of the 7th Infantry Regiment. They are well dug-in and are expected to stand and fight. There is a platoon position on the southern end of the feature, facing south-west, and a second platoon on the centre of the hill facing west. The third platoon is on the northern end facing north and their position is our company objective. The platoon has two sections forward and one in reserve. The forward right section (as we look at it) is in the area of grid 433064. The reserve section is in the area of grid 437063. The forward left section (as we look at it) is in the area of grid 437067. There are four trenches in this position, two of which are on the forward edge of a patch of gorse, while the other two are on a grass slope approximately 50 metres to the rear. The forward left section is our platoon objective.

b. Friendly Forces:

The battalion has been ordered to capture Ladna Hill by 1200hrs tomorrow. The attack will be in two phases. In phase 1, B Company will capture the platoon position on the northern end of the hill. This must be accomplished by first light. In phase 2, A and C Companies will assault the other two enemy platoons in the centre and south, while B Company provides fire support from the northern end. B Company's attack will be carried out in darkness. 5 Platoon will assault the forward right section, while we assault the forward left. Once both section positions have been captured, 7 platoon will move through us, in order to assault the rear section. There are no friendly units to our left. When the entire position is secure, 5 and 7 Platoons will prepare to provide fire support for phase 2 of the battalion attack, while we remain on the northern slope in order to cover the rear.

c. Attachments and Detachments:

41 Field Battery will be on call to B Company during phase 1 and an FOO will be attached to Company HQ. We will have L/Cpl. Smedhurst from the Mortar Platoon to act as MFC.

3. MISSION:

6 Platoon's mission is to capture the enemy section position at grid 437067.

continued.../

4. EXECUTION:

a. General Outline:

The company will leave this location at 2000hrs and move to the FUP at grid 433074. Once there, the platoon will deploy into assault formation, with 1 Section forward left, 3 Section forward right and 2 Section in reserve. Platoon HQ will be between 1 and 3 Sections. 5 Platoon will be on our right and 7 Platoon to the rear.

At 2130hrs, the company will move forward to the road which forms our line of departure. H-Hour is at 2200hrs. We will then advance directly towards our objective. Once we come under effective enemy fire, the platoon will skirmish by sections up to the forward edge of the enemy position.

Then the sections will break down into fireteams in order to assault the individual trenches. As soon as the rear trenches have been taken, the platoon will go firm. Once both our objective and 5 Platoon's objective are secure, 7 Platoon will move forward to assault the depth enemy section. When the entire company objective is secure, the platoon will reorganize and occupy the enemy trenches in order to cover the northern approaches to the hill. Meanwhile, 5 and 7 Platoons will move forward to take up their fire positions for phase 2.

b. Detailed Tasks:

1 Section will be forward left in the assault. Your objectives are the two left-hand trenches. On reorganization, you will occupy all the trenches on our objective.

3 Section will be forward right in the assault. Your objectives are the two right-hand trenches. You are also to keep the platoon commander informed of 5 Platoon's progress. On reorganization, you will occupy the trenches on 5 Platoon's objective.

2 Section will be in reserve. You will move to the rear of 1 Section and will also cover our left flank. Remember, there are no friendly units to our left. You must be prepared to support either of the two forward sections as required. On reorganization, you will occupy the trenches on 7 Platoon's objective. L/Cpl. Smedhurst will move with platoon headquarters.

c. Coordinating Instructions:

(1) Timings:
1700 - meal
1800 - last light
1930 - ready to move
2000 - move to FUP
2130 - move to line of departure
2200 - H-Hour
0615 - first light (company objective must be taken by then)

(2) Indirect Fire Support: sections may request mortar fire on the platoon net. We have also been allocated twenty illuminating rounds.

continued/...

Example Formal Orders - *continued*

5. ADMINISTRATION AND LOGISTICS:

a. Ammunition: in addition to his magazines, each man will carry 2 bandoliers, 4 anti-personnel grenades and 2 WP grenades.
b. Equipment: bergens are to be left at this location.
c. Rations: a hot meal will be provided at 1700hrs. Each man will carry two ration packs. Water will be supplied at 1700hrs.
d. Medical: the RAP will be at grid 401079

6. COMMAND AND SIGNAL:

a. Command: in the event of the platoon commander and platoon sergeant becoming casualties, the order of seniority is Cpl. Smith, Cpl. Hobbs, Cpl. Rigby.
b. Frequencies: as shown in the company signals instructions.
c. Call-signs: L/Cpl. Smedhurst's call-sign is 42D
d. Code-words: Peter Ross - 5 Platoon's objective secure
Jock Scott - 6 Platoon's objective secure
Willy Gunn - 7 platoon's objective secure
e. Password: Brick - Church

Military Grouping Symbols

1. Arms

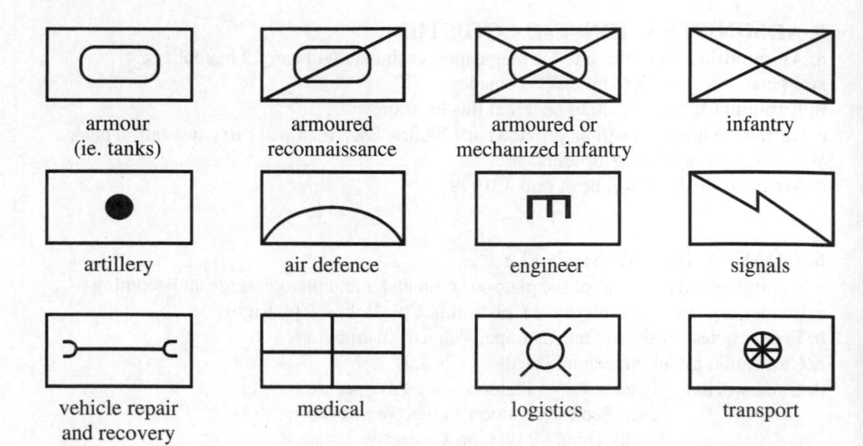

2. Grouping Size

Section/Squad	●	Brigade	X
Platoon	●●●	Division	XX
Company	I	Corps	XXX
Battalion	II	Army	XXXX
Regiment	III		

3. Examples

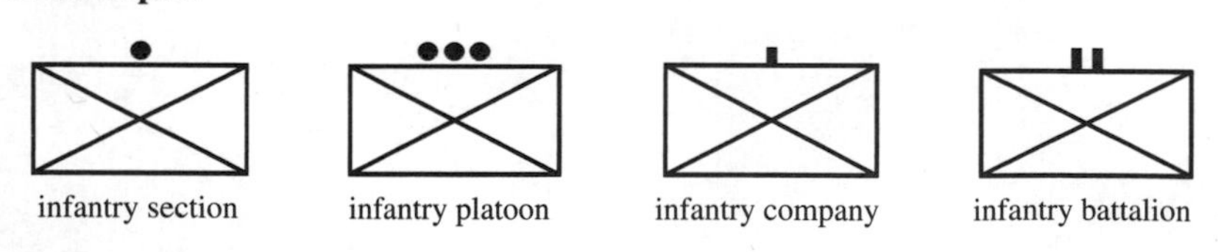

Notes:

1. In the British Army:

a. A platoon-sized grouping of tanks and certain supporting arms is known as a troop
b. A company-sized grouping of tanks and certain supporting arms is known as a squadron
c. A battalion-sized grouping of tanks and certain supporting arms is known as a regiment

2. In the US Army:

a. A company-sized armored cavalry grouping is known as a troop
b. A battalion-sized armored cavalry grouping is known as a squadron

3. In most armies, a company-sized grouping of artillery is known as a battery